T0296880

Unmanned Aerial Vehicles and Multidisciplinary Applications Using AI Techniques

Bella Mary I. Thusnavis
Karunya Institute of Technology and Sciences, India

K. Martin Sagayam
Karunya Institute of Technology and Sciences, India

Ahmed A. Elngar
Faculty of Computers and Artificial Intelligence, Beni-Suef University, Egypt

A volume in the Advances in Computational Intelligence and Robotics (ACIR) Book Series

Published in the United States of America by
IGI Global
Engineering Science Reference (an imprint of IGI Global)
701 E. Chocolate Avenue
Hershey PA, USA 17033
Tel: 717-533-8845
Fax: 717-533-8661
E-mail: cust@igi-global.com
Web site: http://www.igi-global.com

Library of Congress Cataloging-in-Publication Data

Names: Thusnavis, Mary Bella I., 1985- editor. | Sagayam, K. Martin, 1987- editor. | Elngar, Ahmed A., editor.
Title: Unmanned aerial vehicles and multidisciplinary applications using AI techniques / Bella Mary I. Thusnavis, K. Martin Sagayam, and Ahmed A. Elngar, editors.
Description: Hershey, PA : Engineering Science Reference, an imprint of IGI Global, [2022] | Includes bibliographical references and index. | Summary: "This comprehensive edited book covers Artificial techniques, pattern recognition, machine and deep learning--based methods and techniques applied to different real time applications of unmanned aerial vehicles (UAV) with the goal to synthesize the scope and importance of machine learning and deep learning models in enhancing UAV capabilities, solutions to problems and numerous application areas"-- Provided by publisher.
Identifiers: LCCN 2021035397 (print) | LCCN 2021035398 (ebook) | ISBN 9781799887638 (h/c) | ISBN 9781799887645 (s/c) | ISBN 9781799887652 (ebook)
Subjects: LCSH: Drone aircraft--Industrial applications. | Aerial surveillance.
Classification: LCC TL685.35 .U555 2022 (print) | LCC TL685.35 (ebook) | DDC 629.133/39--dc23
LC record available at https://lccn.loc.gov/2021035397
LC ebook record available at https://lccn.loc.gov/2021035398

This book is published in the IGI Global book series Advances in Computational Intelligence and Robotics (ACIR) (ISSN: 2327-0411; eISSN: 2327-042X)

British Cataloguing in Publication Data
A Cataloguing in Publication record for this book is available from the British Library.

All work contributed to this book is new, previously-unpublished material.
The views expressed in this book are those of the authors, but not necessarily of the publisher.

For electronic access to this publication, please contact: eresources@igi-global.com.

Advances in Computational Intelligence and Robotics (ACIR) Book Series

ISSN:2327-0411
EISSN:2327-042X

MISSION

While intelligence is traditionally a term applied to humans and human cognition, technology has progressed in such a way to allow for the development of intelligent systems able to simulate many human traits. With this new era of simulated and artificial intelligence, much research is needed in order to continue to advance the field and also to evaluate the ethical and societal concerns of the existence of artificial life and machine learning.

The **Advances in Computational Intelligence and Robotics (ACIR) Book Series** encourages scholarly discourse on all topics pertaining to evolutionary computing, artificial life, computational intelligence, machine learning, and robotics. ACIR presents the latest research being conducted on diverse topics in intelligence technologies with the goal of advancing knowledge and applications in this rapidly evolving field.

COVERAGE

- Brain Simulation
- Cyborgs
- Computational Logic
- Artificial Intelligence
- Computer Vision
- Cognitive Informatics
- Intelligent Control
- Robotics
- Computational Intelligence
- Neural Networks

The Advances in Computational Intelligence and Robotics (ACIR) Book Series (ISSN 2327-0411) is published by IGI Global, 701 E. Chocolate Avenue, Hershey, PA 17033-1240, USA, www.igi-global.com. This series is composed of titles available for purchase individually; each title is edited to be contextually exclusive from any other title within the series. For pricing and ordering information please visit http://www.igi-global.com/book-series/advances-computational-intelligence-robotics/73674. Postmaster: Send all address changes to above address.

Titles in this Series

For a list of additional titles in this series, please visit:
http://www.igi-global.com/book-series/advances-computational-intelligence-robotics/73674

Computer Vision and Image Processing in the Deep Learning Era
A. Srinivasan (SASTRA Deemed To Be University, India)
Engineering Science Reference • © 2022 • 325pp • H/C (ISBN: 9781799888925) • US $270.00

Artificial Intelligence for Societal Development and Global Well-Being
Abhay Saxena (Dev Sanskriti Vishwavidyalaya, India) Ashutosh Kumar Bhatt (Uttarakhand Open University, India) and Rajeev Kumar (Teerthanker Mahaveer University, India)
Engineering Science Reference • © 2022 • 292pp • H/C (ISBN: 9781668424438) • US $270.00

Demystifying Federated Learning for Blockchain and Industrial Internet of Things
Sandeep Kautish (Lord Buddha Education Foundation, Nepal) and Gaurav Dhiman (Government Bikram College of Commerce, India)
Engineering Science Reference • © 2022 • 240pp • H/C (ISBN: 9781668437339) • US $270.00

Applied AI and Multimedia Technologies for Smart Manufacturing and CPS Applications
Emmanuel Oyekanlu (Drexel University, USA)
Engineering Science Reference • © 2022 • 300pp • H/C (ISBN: 9781799878520) • US $270.00

Artificial Intelligence of Things for Weather Forecasting and Climatic Behavioral Analysis
Rajeev Kumar Gupta (Pandit Deendayal Energy University, India) Arti Jain (Jaypee Institute of Information Technology, India) John Wang (Montclair State University, USA) Ved Prakash Singh (India Meteorological Department, Ministry of Earth Sciences, Government of India, India) and Santosh Bharti (Pandit Deendayal Energy University, India)
Engineering Science Reference • © 2022 • 277pp • H/C (ISBN: 9781668439814) • US $270.00

For an entire list of titles in this series, please visit:
http://www.igi-global.com/book-series/advances-computational-intelligence-robotics/73674

701 East Chocolate Avenue, Hershey, PA 17033, USA
Tel: 717-533-8845 x100 • Fax: 717-533-8661
E-Mail: cust@igi-global.com • www.igi-global.com

Table of Contents

Detailed Table of Contents

Unmanned vehicles/systems technology is an emerging technology in recent years. Unmanned vehicles play a more significant role in many civil and military applications, such as remote sensing, surveillance, precision agriculture, and rescue operations rather than manned systems. The UAVs can gather photographs faster and more accurately than satellite imagery, allowing for more accurate assessment. This study provides a comprehensive overview of UAV civil applications, including classification and requirements. Also encompassed are research trends, critical civil challenges, and future insights on UAVs.

In order to commercialize in the industry, various sensors and electrical gadgets are used to keep prices low in a few fields. Unmanned aerial vehicles (UAVs) can be utilized for surveillance, pesticide and insecticide application, and bioprocessing mistake detection to save money and improve the abilities of agricultural experts. Both single-mode and multi-mode UAV systems will perform admirably in this application. This chapter examines the constraints of the internet of things and UAV connectivity in remote areas, as well as smart agriculture application scenarios. In addition, the benefits and uses of employing the internet of things (IoT) and UAVs in agriculture were discussed. On the basis of several elements such as geographical, technological, and business, a system model has been presented. For various IoT applications, the architecture includes enabling technologies, scalability, intelligence, and supportability. Finally, interoperability issues are examined in depth in order to uncover the complications that arise during coordination between UAV and IoT components.

Agriculture is considered to be the driving force of the Indian economy. Production of crops is considered to be one of the complex phenomena as they are influenced by the agro-climatic parameters. From novice to experienced farmers, at times, fail to figure out the suitable crop for their lands, leading to financial loss. This is because of the dynamic change in soil nutrient levels and climatic conditions. Hence, it is important to predict crops according to the presence of the nutrients in a land. Recommending the crops to a farm after considering the nutrients levels of the soil and predicting the yield will largely help the landowner in taking necessary steps for marketing and storage in the future. These results will further assist the industries to plan the logistics of their business who are working in partnership with these landowners. In this work, pH and other soil nutrients are estimated from the input ortho images to recommend crops that can grow well under the given circumstances.

The objective of this chapter is to propose a model of an automated city crime-health management that can be implemented in future smart cities of developing countries. The chapter discusses how a suitable amalgamation of existing technologies such as IoT, artificial intelligence, and machine learning can output an efficient system of unmanned city management systems, thereby facilitating indirect engendering of innovative scopes for technology workers and researchers and alleviating the living standards within the city fabrics, catalyzing infrastructure development. In this chapter, the authors have structured an ideal UAV-matrix layout for city fabric surveillance built over the scopes of artificial intelligence. Succinctly, this chapter provides a platform that would galvanize the possibilities and that could be reimagined to structure a more resourceful working model of new emerging smart cities and enlighten the settings of existing ones.

In the real world, smart city traffic management is a difficult phenomenon. Introducing the internet of things into traffic management systems in smart cities is a huge challenge. Smart city definitions differ from city to city and country to country, depending on the city's level of growth, willingness to change and reform, finances, and ambitions. Unmanned aerial vehicles (UAVs) have been used in a variety of applications for civil and defense infrastructure management. These uses include crowd surveillance, transportation, emergency management, and building design inspection. In smart cities, a variety of transport options exist with respect to public transport and private transport connectivity. The mathematical modelling-based vehicular network enables automobile manufacturers to incorporate smart features into vehicles at a low cost, boosting their market competitiveness. This proposal addresses the challenges concerning the surveillance system for smart city traffic management systems (TMSs).

Over the last few years, the way people trade information and communicate with one another has changed tremendously. In business communication, social media channels such as Facebook, Twitter, and YouTube are becoming increasingly significant. Nevertheless, the study into online brand fan page is primarily focused on using website platforms rather than social media platforms. As a result, more research is needed to analyze UAV businesses' fan page engagement behavior in order to grow their fan base and further induce a fan's buying behavior using the honeycomb model's views. Consumers who have participated in an online brand fan page are the study's target group. A web-based survey was used to collect data. Identity, conversation, presence, sharing, reputation, relationships, and groups all had a significant beneficial effect on brand equity, according to the findings. This study confirms the impact of perceived value in improving various fan page behaviors, which aids in the identification and implementation of an online engagement plan for purchase.

In this chapter, the author bases his research projects on his authentic mixed multidisciplinary applied mathematical model for transformation projects. His mathematical model, named the applied holistic mathematical model for project (AHMM4P), is supported by a tree-based heuristics structure. The AHMM4P is similar to the human empirical decision-making process and applicable to any type of project, aimed to support the evolution of organisational, national, or enterprise transformation initiatives. The AHMM4P can be used for the development of the enterprise information systems and their decision-making systems, based on artificial intelligence, data sciences, enterprise architecture, big data, and machine learning. The author tries to prove that an AHMM4P-based action research approach can unify the currently frequently used siloed machine learning trends.

In this chapter, the design, modeling, and control of a UAV was presented. The conceptual design stages of the UAV were analyzed in detail. UAVs as observers in the sky will remain important for the indefinite future. Agriculture, water quality monitoring, disease detection, crop monitoring, yield predictions, and drought monitoring are just a few of the applications. Healthcare, microbiological and laboratory samples, drugs, vaccines, emergency medical supplies, and patient transportation can all be delivered using drones.

Preface

This comprehensively edited book will cover the development of Unmanned Aerial Vehicle (UAV) for multidisciplinary applications using Artificial Intelligence (AI) Techniques. UAV broadens its applications to which AI techniques are applied. Due to its rapid and cost-effective deployment, UAV is used in various applications. Unmanned Aerial Vehicle (UAV) have extended the freedom to operate and monitor the activities from remote locations. It has advantages of flying at low altitude, small size, high resolution, lightweight, and portability. UAV and artificial intelligence have started gaining attentions of academic and industrial research. UAV along with machine learning has immense scope in scientific research and has resulted in fast and reliable outputs. Deep learning-based UAV has helped in real time monitoring, data collection and processing, and prediction in the computer/wireless networks, smart cities, military, agriculture, and mining. This comprehensive edited book covers Artificial techniques, pattern recognition, machine, and deep learning-based methods and techniques applied to different real time applications of UAV. The main aim is to synthesize the scope and importance of machine learning and deep learning models in enhancing UAV capabilities, solutions to problems and numerous application areas. This book aims to provide the state of the art of UAV as well as involved AI techniques and give an insight of the major comprehensive study in the use of deep learning, machine learning in UAV.

Owing to the scope and diversity of topics covered, the book will be of interest not only to researchers and theorists, but also to professionals, technology specialists and methodologists dealing with various applications of UAV. The book also aims to provide a roadmap to recent research areas in the field of UAV. Also, the book is designed to be the first reference choice at research and development centers, academic institutions, university libraries and any institutions interested in analyzing the UAV and the AI techniques therein. Academicians and research scholars are other projected audience who identify applications, tools and methodologies through qualitative/quantitative results, literature reviews, and reference citations.

The following describes the chapters included in this book.

Chapter 1: UAVs for Multidisciplinary Application – Introduction

Unmanned vehicles/system is an emerging technology in recent years. Unmanned vehicles play a more significant role in many civil and military applications, such as remote sensing, surveillance, precision agriculture, and rescue operations rather than manned systems. The UAVs can gather photographs faster and more accurately than satellite imagery, allowing for more accurate assessment. This study provides a comprehensive overview of UAV civil applications, including classification and requirements. Also encompassed research trends, critical civil challenges, and future insights on UAVs. This chapter covers the usage of UAVs in multidisciplinary applications.

Chapter 2: A Review on Applications of Unmanned Aerial Vehicles and Internet of Things Towards Smart Farming

In order to commercialize in the industry, various sensors and electrical gadgets are used to keep prices low in a few fields. Unmanned Aerial Vehicles (UAVs) can be utilized for surveillance, pesticide and insecticide application, and bioprocessing mistake detection to save money and improve the abilities of agricultural experts. Both single-mode and multi-mode UAV systems will perform admirably in this application. This chapter examines the constraints of the Internet of Things and UAV connectivity in remote areas, as well as smart agriculture application scenarios. In addition, the benefits and uses of employing the Internet of Things (IoT) and UAVs in agriculture were discussed. On the basis of several elements such as geographical, technological, and business, a system model has been presented. For various IoT applications, the architecture includes enabling technologies, scalability, intelligence, and supportability. Finally, interoperability issues are examined in depth in order to uncover the complications that arise during coordination between UAV and IoT components.

Chapter 3: Recommendation of Crops and Its Yield Prediction by Assessing Soil Health From Ortho-Photos

Agriculture is considered to be the driving force of the Indian economy. Production of crops is considered to be one of the complex phenomena as they are influenced by the agro-climatic parameters. From novice to experienced farmers, at times, fail to figure out the suitable crop for their lands, leading to financial loss. This is because of the dynamic change in the soil nutrient levels and climatic conditions. Hence it is important to predict crops according to the presence of the nutrients in a land. By recommending the crops to a farm after considering the nutrients levels of the soil and predicting the yield will largely help the land owner in taking necessary steps for marketing and storage in the future. These results will further assist the industries to plan the logistics of their business who are working in partnership with these land owners. In this work, pH and other soil nutrients are estimated from the input ortho images to recommend crops that can grow well under the given circumstances.

Chapter 4: Design and Implementation of Amphibious Unmanned Aerial Vehicle Systems for Agriculture Applications

Today, drone systems have become emerging technology for agriculture applications as an unmanned aerial vehicle (UAV). It helps the farmers in crop monitoring and production. Drones are used to reduce human resources and to control pollution in the agriculture field. In real-time, drones are suitable for working in the agriculture field during strong winds and even in various climate conditions. This chapter proposes an amphibious unmanned aerial vehicle (UAV) system design and implementation for agriculture applications. Drones are useful to avoid deforestation in Indian countries. The system software is used to construct the drone's structure and application to design the drone architecture. The estimated simulation results are used to calculate the drones' efficiency using their weight, flying time, and power consumption. The flying capacity is most important to maintain the stability of the UAV system for all applications. In this chapter, three different UAV system phases have been discussed, i.e., design of drones, the building of payload, and evaluation of drones using the software. This chapter helps the beginners to understand the necessary calculations of the drone design along with thrust values, select the propellers sizes, calculates the drone's flying time, stability, and power consumption for different sizes of the drone discussed.

Chapter 5: Scope of UAVs for Smart Cities – An Outlook

New technologies are always remarkable for the sustainability of human beings and for their enhancement. Now a days Unmanned Aerial Vehicles (UAV's) are highly used on a wide range for commercial as well as defense purpose. UAVs are also called 'Drones'. With the phenomenon scope of application, drones have reached to a very high level in each and every field and smart cities being no exception. UAV provide many services for the development of smart cities like traffic control, natural disaster management, monitoring, transportation, infrastructure, mapping, air quality and many other parameters are also there. UAV with high resolution cameras and advanced techniques has many properties i.e., less time consuming, highly efficient, data to collect and analyze etc. Collection of data is very fast and accurate, even analyzation of any task given is very authentic. They are also considered as an aid to surveillance for security purposes.

Chapter 6: Urban Intelligence and IoT-UAV Applications in Smart Cities – Unmanned Aerial Vehicle-Based City Management, Human Activity Recognition, and Monitoring for Health

The objective of this chapter is to propose a model of an automated city crime-health management that can be implemented in future smart cities of developing countries. The chapter discusses how a suitable amalgamation of existing technologies such as IoT, Artificial Intelligence and Machine Learning can output an efficient system of unmanned city management systems, thereby facilitating indirect engendering of innovative scopes for technology workers and researchers, and alleviating the living standards within the city fabrics, catalyzing infrastructure development. In this chapter, the authors have structured an ideal UAV-matrix layout for city fabric surveillance built over the scopes of Artificial Intelligence. Succinctly, this chapter provides a platform, which would galvanize the possibilities, that could be reimagined to structure a more resourceful working model of new emerging smart cities, and enlightening the settings of existing ones. Scopes like UAV based healthcare and crime monitoring using motion tracking and human activity recognition will be examined and discussed in this chapter.

Chapter 7: Mathematical Modeling of Unmanned Aerial Vehicle for Smart City Vehicular Surveillance Systems

In the real world, smart city traffic management is a difficult phenomenon. Introducing the Internet of Things idea into traffic management systems in smart cities is a huge challenge. Smart city definitions differ from city to city and country to country, depending on the city's level of growth, willingness to change and reform, finances, and ambitions. Unmanned aerial vehicles (UAV) have been used in a variety of applications for civil and defence infrastructure management. These uses include crowd surveillance, transportation, emergency management, and building design inspection. In Smart cities a variety of transport options with respect to, public transport and private transport connectivity. The mathematical modelling based vehicular net-work enables automobile manufacturers to incorporate smart features into vehicles at a low cost, boosting their market competitive-ness. This proposal addresses the challenges concerning the surveillance system for smart city Traffic Management System (TMS).

Chapter 8: Unmanned Aerial Vehicle Brand Fan Page Engagement Behavior Analytics

Over the last few years, the way people trade information and communicate with one another has changed tremendously. In business communication, social media channels such as Facebook, Twitter, and YouTube are becoming increasingly significant. Nevertheless, the study into online brand fan page is primarily focused on using website platforms rather than social media platforms. As a result, more research is needed to analyze UAV businesses' fan page engagement behavior in order to grow their fan base and further induce a fan's buy behavior using the honeycomb model's views. Consumers who have ever participated in an online brand fan page are the study's target group. A web-based survey was used to collect data. Identity, Conversation, presence, sharing, reputation, relationships and groups all had a significant beneficial effect on brand equity, according to the findings. This study confirms the impact of perceived value in improving various fan page behaviors, which aids in the identification and implementation of an online engagement plan for purchase.

Chapter 9: Business Transformation and Enterprise Architecture Projects – Machine Learning Integration for Projects (MLI4P)

In this chapter, the author bases his research projects on his authentic mixed multidisciplinary applied mathematical model for transformation projects. His mathematical model, named the Applied Holistic Mathematical Model for Project (AHMM4P), which is supported by a tree-based heuristics structure. The AHMM4P is similar to the human empirical decision-making process and applicable to any type of project, aimed to support the evolution of organisational, national or enterprise transformation initiatives. The AHMM4P can be used for the development of the enterprise information systems and their decision-making systems, based on Artificial Intelligence, Data Sciences, Enterprise Architecture, Big Data and Machine Learning. The author tries to prove that an AHMM4P-based Action Research approach can unify the currently frequently used siloed machine learning trends.

Chapter 10: Future Trends and Challenges of UAVs – Conclusion

In this chapter, the design, modeling and control of a UAV is presented. The conceptual design stages of the UAV are analyzed in detail. UAV as observers in the sky will remain important for the indefinite future. UAV being an efficient and successful device across the field, it places a vital role in Military; Drone has been used for carrying IEDs to Destroy enemy areas. Agriculture, Drones, remote sensing applications from tree species, water quality monitoring, disease detection, crop monitoring, yield predictions, and drought monitoring are just a few of the data sources. Health Care, Microbiological and laboratory samples, drugs, vaccines, emergency medical supplies, and patient transportation can all be delivered using drones.

The readers of this book will be benefited about the evolution, usage, challenges of UAV for multidisciplinary applications. The book aims to enable the readers to realize the importance of UAV, their limitations and future scope. The proposed book focuses to publish original research outcomes towards agriculture, smart cities, crime monitoring, and healthcare using various technological advancements. Hence, the readers will gain insights to taxonomy of challenges, issues and future research directions in this regard. This book helps the beginners to understand the drone design, interoperability issues and showcase the challenges and future research directions for existing practitioner. The readers will gain exposure to a novel paradigm where UAV, IoT and AI are merged together to solve real-time problems in smart cities, agriculture and in many other applications. Further, the book aims to bring together state-of-the-art innovations, research activities (both in academia and industry), and the corresponding standardization impacts of machine learning and deep learning so as to make the readers aware of the requirements and promising technical options to enrich and boost research activities in this area.

Bella Mary I. Thusnavis
Karunya Institute of Technology and Sciences, India

K. Martin Sagayam
Karunya Institute of Technology and Sciences, India

Ahmed A. Elngar
Faculty of Computers and Artificial Intelligence, Beni-Suef University, Egypt

Chapter 1
UAVs for Multidisciplinary Applications:
Introduction

Govarthan R.
Karunya Institute of Technology and Sciences, India

Hariharan S.
Karunya Institute of Technology and Sciences, India

John Paul
https://orcid.org/0000-0002-3371-1277
Karunya Institute of Technology and Sciences, India

Thusnavis Bella Mary
Karunya Institute of Technology and Sciences, India

K. Martin Sagayam
https://orcid.org/0000-0003-2080-0497
Karunya Institute of Technology and Sciences, Coimbatore, India

Ahmed A. Elngar
https://orcid.org/0000-0001-6124-7152
Beni-Suef University, Egypt

ABSTRACT

Unmanned vehicles/systems technology is an emerging technology in recent years. Unmanned vehicles play a more significant role in many civil and military applications, such as remote sensing, surveillance, precision agriculture, and rescue operations rather than manned systems. The UAVs can gather photographs faster and more accurately than satellite imagery, allowing for more accurate assessment. This study provides a comprehensive overview of UAV civil applications, including classification and requirements. Also encompassed are research trends, critical civil challenges, and future insights on UAVs.

DOI: 10.4018/978-1-7998-8763-8.ch001

INTRODUCTION

UAV commonly known as drone is refer to as emerging miniature aircraft, which is highly used to transporting goods, (Sivakumar and TYJ, 2021). Surveillance, research, communications and photography. It is a powered aerial system that does not seek a human operator on board. It is a modern military unmanned vehicle that can fly autonomously or RF (Radio Frequency) remote controller from a particular distance. The use of drones will drastically grow by 2020 because of its significant role. It has the ability to fly on the ground, underwater and in the air. Every Drone has its own range and speed depending on its purpose. Its miniature has more advantages and is adopted in various domains. Every drone is manufactured aerodynamically for better performance. Some Drones are Pre-programmed to operate without the assistance of humans for particular works. Various equipment is attached with it, like cameras, Sensors, Electronic Devices, Global Position System depending on its purpose. It is a mini prototype of Giant aircraft which is more capable and affordable.

MOTIVATION

Humans are always inspired by nature, which makes humans creative. The inventor was inspired by a small bird that is more capable of flying around swiftly. This inspired the invention of a small aircraft for military purposes. Later it emerged and advanced in both civil and military applications. Abraham Karem is widely known as the "Father of UAV" for his contribution to the development of drones. Initially adopted for military surveillance, with human interference. Currently, universities, government agencies, defense companies and private organizations are conducting immense research which has been widespread across various domains, including agriculture, disaster zone mapping, E-commerce, and hurricane forecasting and still more.

Features

1. UAVs are highly enhancing devices in most because of their significant properties.
2. It is cost-efficient and also user-friendly.
3. It has an ability to fly in any climatic condition, so it is highly used during Calamities for surveillance and rescue.
4. A simple drone is built using various electronics components like motors, propellers, batteries, transmitters and receivers, sensors and wings. Some

more components are included depending on this purpose. The UAV follows the law of Thermodynamics and the laws of physics.

a. UAV has a wide range of applications across the domains.
b. Highly Capable Device with-friendly approach.

CLASSIFICATION

UAVs can be classified on the basis of their performance characteristics. It is categorized on different parameters such as mass, durability, coverage area, velocity and wing loading. The price, wing span and height are also a few categories which can be considered to compare. Engine type and maximum power developed will be other categories which vary depending on its applications. These classes are narrowed by the survey. Furthermore, more classification can be done if we extend the usage in various applications. Classification by performance characteristics is useful for designers, manufacturers and customers because it enables these groups to attain their needs with the performance aspects of UAVs (Arjomandi et al., 2006).

Important Performance Characteristics

1. Mass
2. Durability and coverage area
3. Height
4. Wing Loading

Classification by Mass

The altitude of a UAV differs based on purpose and applications. Every domain has its own expectations. People classify them by size, range, and endurance, and use a tier system that is employed by the military. For classification according to size, one can come up with the following sub-classes:

- Micro or Nano UAVs
- Mini UAVs
- Medium UAVs
- Large UAVs

Classification According to Range and Endurance

UAVs also can be classified according to the range as shown in Table 1 they can travel and their endurance in the air using the following sub-classes.

- Proximity range UAVs
- Close-range UAVs
- Less-range UAVs
- Medium-range UAVs
- long UAVs

Depending on its purpose, UAV varies according to various factors which is shown in Table. In particular, endurance and range play an important role in categorizing UAV based on its application in various fields.

Mostly, high endurance and ranged UAVs are used in the Military for lasting support and accurate outcome, whereas Medium is used in other real-time applications where it doesn't require the high category device. Low categorized UAV is used in a simple application which involves low cost used in photography.

Table 1. Category according to range and endurance

CATEGORY	ENDURANCE RANGE	RANGE
L	>24 Hours	>1500km
M	5-24 Hours	100-400 km
S	<5Hours	<100km

Classification by Maximum Altitude

Altitude is the maximum height covered by a UAV from ground level, which varies from drone to drone. The category on Altitude is shown in Table 2. The Low Categorized Drone lifts up to 1000m, and the medium ranges from 1000 to 10000m to the high-categorized vehicle flies higher than 10000m. All the working abilities differ in cost too.

Table 2. Category according to altitude

CATEGORY	MAXIMUM ALTITUDE
L_a	>10000m
M_a	1000-10000 m
S_a	<1000m

Classification by Wing Loading

Wing Loading is the ratio of the lift of a drone to its weight. To calculate the wing loading of a UAV, the total weight will be divided by the wing area. Usually, the essential equipment for flying, like a battery, rotor shares a high role in dealing with the wing loading, then the Equipment needed based on its application. It is mainly categorized into three, Listed out in Table 3. Low-less than 50 kg/m^2, Medium ranges from 50 to 100 kg/m^2, High- greater than 100 kg/m^2

Table 3. Category based on wing loading

CATEGORY	WING LOADING kg/m^2
L_l	**<50**
M_l	**50-100**
S_l	**>100**

APPLICATIONS

Unmanned Aerial Vehicles can be practiced in various fields, like Military, Agriculture, Geography mapping, Health care, Weather forecast, Disaster Management, Transportation, and delivery.

MILITARY

Being an efficient and successful device in the field, UAV plays a vital role in the military. It was initially built as a military device which works for reconnaissance and surveillance operations. Combat operations which risk human life can be substituted by UAV. Surveillance in the risk area operations costs human life several times. It can be prevented by using the drone where the vision of the area is clear and saves life. Recently, drones have been used for carrying IEDs to destroy enemy areas. In addition, prevention and destruction is possible by UAV. On the frontline, where transportation is a huge challenge because of the climatic conditions where soldiers suffer, the basic needs like food, medicine, ammunition and other essentials have been transported by drones, which is cost-efficient, eco-friendly and faster. Remote sensing has been done using drones for advancement of troops and Combat Vehicles. Aerial images produced by the drone are highly helpful for planning and advancement. It is more effective than satellite imagery because of its clear accuracy on the object on the earth. UAV is flexible for the operator for what exactly he should view, whereas the satellite imagery cannot achieve that efficiency. During times of calamities, the search and rescue becomes a handful for quick response by using drones.

AGRICULTURE

UAV's notable growth in a remote sensing platform facilitates various applications in agriculture in monitoring, storing data and mechanical work. Because of it, improved features in remote sensing have resulted in increased adoption of this technology in metropolitan areas. Drones can be used to collect data from ground sensors and forward it to control/base stations. Drones with sensors can be used to create an aerial sensor network for disaster management and environmental monitoring. Drones, remote sensing applications from plantation families, monitoring the quality of water/water bodies, detection of diseases, crop monitoring, yield predictions, and drought monitoring are just a few of the data sources. Some of the applications of drones in agriculture are:

- **Crop monitoring:** The crop fields are vast and challenging to monitor the unpredictable weather conditions, increasing the field risk and labor costs. Unmanned aerial vehicles equipped with high level cameras help to eliminate these challenges.
- **Precision agriculture:** Vegetation that focuses on crop diseases, nutrient deficiencies, and pest invasion in reduces productivity. Crop data is collected by UAVs and processed with AI techniques to address these challenges.

MEDICAL CARE

UAV has the potential to gather real-time data and deliver payloads at an affordable price, and it enhances the extension of various industrial, commercial, and civil applications. Drones are implemented for diagnosis and treatment, pre-planned inspection, and cost telemetering in rural locations. Microbiological and laboratory samples, drugs, vaccines, emergency medical supplies, and patient transportation can all be conveyed using drones. Drones have a variety of practical applications that have a lot of potential and are listed below:

- **Emergency supplies or medications on board:** EpiPens, poison antidotes, and oxygen masks are just a few types of life-saving kit.
- **Blood and tissue sample collection:** Drones may be able to provide goods and services while also allowing for speedier return transit to labs that are adequately prepared, eliminating human work and time.
- **Performing search and rescue missions:** People who are misled or are injured can be rescued at sea, in the mountains, in the desert, or in remote areas.
- **Accessibility to far-flung patients:** People are predominantly found in situations where the infrastructure for efficient emergency or continuity of care is lacking. Drones are being used to provide telemedicine, vaccinations, prescription drugs, and medical supplies to people in the place they present.
- **Integration of cloud and internet of things (IoT):** It dispenses a cost-effective way to connect heterogeneous devices and address rising data demands in healthcare applications, plus seamless application deployment and rendering service. One of the most exciting areas of drone development is within the healthcare industry. Such applications include delivering medicines, vaccines, blood and other medical supplies that are urgently needed in inaccessible areas. Supply challenges are frequently caused by poor transport networks, extreme weather conditions, natural disasters, or traffic congestion in urban areas. Delivery by drone could be a solution to such problems.

ROAD TRAFFIC CONGESTION CONTROL

Traffic congestion is a considerable global issue resulting from high population density, an increased increase in automobiles and their infrastructure. Researchers have outlined congestion in various forms. The standard definition of congestion in the state of traffic flow is when the travel demand exceeds road capacity. From the delay-travel time perspective, congestion occurs when the normal flow of traffic is interrupted by a high density of vehicles, resulting in excess travel time. Congestion can also be explained by the increment of the road user's cost due to the disruption of normal traffic flow. A variety of reasons are responsible for creating congestion in developed areas. For these reasons, congestion can be classified into recurring and non-recurring congestion. Recurring congestion occurs regularly, mostly due to the larger number of vehicles during peak times (Reed and Kidd, 2019). It also takes place in unpredictable events—weather, work zones, incidents, and special events—are the reasons for non-recurring congestion. Figure 1 represents the current procedure to measure congestion.

Figure 1. Current procedure to Measure Congestion (Afrin and Yodo, 2020)

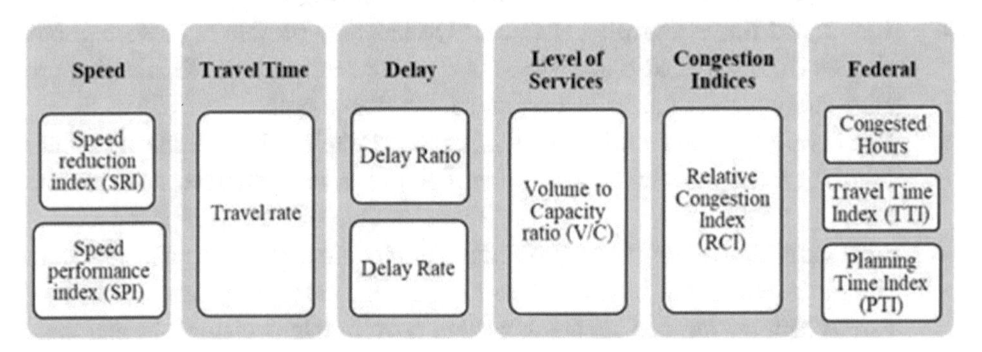

Speed

Speed Reduction Index (SRI)

It is the ratio of relative speed change between congested and free-flow conditions of 0 to 10. When the index value moves 4 to 5, it is considered as congestion and Values less than 4 it is determined as non- congested situation. (1) represents the mathematical representation of Speed Reduction Index

$$SRI = (1 - sac/sff) \times 10 \quad (1)$$

where SRI implies the speed reduction index, sac denotes the actual travel speed, and Sff determines the free-flow speed.

Speed Performance Index (SPI)

SPI had improved the regulation in the urban road traffic conditions. The value of SPI (ranging from 0 to 100) can be determined by the ratio between vehicle speed and the maximum permissible speed. To calculate the traffic state on the road with this index, classification criterion of the urban road traffic state is shown

Measuring the congestion level is for upgrading traffic management and improving control in urban areas. The decision-making steps which are listed are used towards a sustainable transportation system are highly dependent on the actual road traffic conditions. Thus, the measurement approach to quantify the congestion severity should be practical enough for the decision-makers to implement necessary steps to mitigate congestion promptly to achieve a sustainable and resilient transportation system.

Table 4. Transportation experts have suggested a range of attributes

Speed Performance Index	Congestion State Level	Description of Traffic State
(0,25)	Massive congestion	Low average speed, poor road traffic state
(25,50)	Mild congestion	Lower average speed, road traffic state bit weak
(50,75)	Smooth	Higher the average speed, road traffic state better
(75,100)	Very smooth	High average speed, road traffic state good

Travel Rate

It is rate of motion for a particular roadway segment or trip that can be represented by the ratio of the segment travel time by the segment length. The inverse of speed can also be employed to quantify the travel rate.

$$Tr = Tt / Ls \tag{2}$$

where, Tr denotes the travel rate, Tt is the travel time, and Ls indicates the segment length

Delay

Delay Rate

The delay rate is defines as rate of time loss for particular vehicles operating during congestion for a specific roadway segment or trip. It can be measured by the ratio between the actual travel rate and the acceptable travel rate as

$$Dr = Trac\text{-}Trap \tag{3}$$

where, Dr is the delay rate, Trac is the actual travel rate, and Trap is the acceptable travel rate.

Delay Ratio

The delay ratio can be measured by the ratio of delay rate and the actual travel rate. It is used to compare the relative congestion levels on different roadways

$$D = Dr / Trac \tag{4}$$

where D denotes the delay ratio, Dr is the delay rate, and Trac is the actual travel rate.

Measuring the congestion level is important for improving traffic management and control. The following decision-making steps towards a sustainable transportation system are highly dependent on the actual road traffic conditions. Thus, the measurement approach to quantify the congestion should be practical enough for the decision-makers to implement necessary steps to mitigate congestion promptly to achieve a sustainable and resilient transportation system.

Transportation experts have suggested a range of attributes that are often desired in a congestion measure, a good congestion measure should:

1. Be well-defined, easily comprehensible, and uncomplicated for non-technical users to interpret the results easily,
2. reflect the real level of service for any road types,
3. consider different system performances, such as travel time and speed,
4. provide a continuous range of values,
5. be able to be used in predictive and statistical analysis purposes,
6. offer comparable values to different road types, and be widely applicable for different road types

Table 5. Peak point of different congestion measures on a weekday

Day	Monday		Tuesday		Wednesday		Thursday		Friday	
Congestion Measures	a.m.	p.m.	a.m.	p.m.	a.m.	p.m.	a.m.	p.m.	a.m.	p.m.
SRI	-	5:31	-	5:21	-	2:10	-	5:31	-	4:11
SPI	-	5:31	-	5:21	-	2:10	-	5:31	-	4:11
Travel Rate	-	5:31	-	5:21	-	2:10	-	5:31	-	4:11
Delay Rate	-	5:31	-	5:21	-	2:10	-	5:31	-	4:11
Delay Ratior	-	5:31	-	5:21	-	2:10	-	5:31	-	4:11

Source (Afrin and Yodo, 2020)

Unmanned Aerial Vehicle Components

Various UAV is used in different types of areas, including dangerous material operation and control. The Quadcopter, which is an aerial vehicle with four rotors, has been involved in all recent research in the field of small independent drones. Its simple design is the key reason for its prominent role in small drones. It consists of two parts- rotating wings mounted at the top of a chasis. Vertical take-off and landings similar to traditional helicopters are achieved. Quadcopters should meet an appropriate time. Meeting the scheme should be fast, reliable and robust. No of propellers is one of the most important categories in drone making, usually termed as X-copter. Based on the number of propellers used, "X" was replaced with quad-, hexa- and octa-. The Table below gives the overview of the most acceptable components used in an quadcopter mounting to guarantee the endurance and reliability of the process. The table also gives the component description. Drones are used for preparing equipment, inspecting, transport, delivering in difficult terrain or distant land. They are used to collect information and things in areas where humans cannot live. Image recognition and mobile monitoring is possible (Arjomandi et al., 2006). The most used Components are listed below in the table 5 with its description.

Table 6. Components of UAV

Chasis	It is the construction that holds all the components together.
Motor	The design of motors is to provide rotatory motion for the propellers. Each rotor is controlled separately by a speed controller.
Speed regulator	Controls the speed of a rotor which directs it to the engine.
Rotor	A quadrotor holds four propellers, two "regular" propellers that rotate counter- clockwise, and two "pusher" propellers that rotate clockwise to dodge body spinning.
Flight controller	It is the brain of the quadrotor. It houses the sensors above-mentioned include the accelerometers and gyroscopes, which include how quickly each of the quadrotor motorsturns. (Sivakumar.M,&Malleswari.TYJ2021)
Radio Transmitter	It enables the regulation of the quadrotor, and it necessitates four channels for a basic quadrotor.
Power unit	Lithium polymer (LiPo) batteries are among the most frequent battery kinds for drones, as their size and weight benefit from high energy density with greater voltage per cell, which allows them to power drones on-board systems with less cells than other rechargeable systems. (Sivakumar and TYJ, 2021).
Telemetry module	It is practiced to regain flight information of the quadrotor on a computer to follow various aircraft parameters on the ground. (Sivakumar and TYJ, 2021).
Camera	It enhances the production of the quadrotor and adds value to its uses—the camera works as an attachment with a USB to observe the images (Sivakumar and TYJ, 2021).
Video Sender and Reciever	The transmitter transforms the information into a radio signal and outputs it to the imputed antenna, which later sends it out. The receiver operates to turn the radio signal into explicit videos (Sivakumar and TYJ, 2021).

DRONE SHAKTI

During the Union Budget 2021, an announcement was made on Drone Shakti. Industry got a great push after the relaxation given by the Government of India. As per the Regulations, use start-ups will be promoted for usage for use across the domain. The Kisan Drone is one of the drones promoted by the government to help farmers in the agricultural field in crop assessments, land data maintaining, Spraying pesticides. This revolution is expected to boost production in the Agri and farming sectors. Usage of drones for surveillance in crowded areas or public places

In India, drones were also deployed to deliver Covid-19 vaccines and other medical equipment during Pandemic 2020. This project, being led by the ICMR (Indian council of Medical Research), is highly used in the rural areas like Manipur, Nagaland and Andamans and Nicobar Islands. Drones are also being used for surveillance of Covid-19 hotspots and containment zones.

CONCLUSION

Unmanned aerial vehicles are highly versatile these days. This chapter tells the Features of Drones, Classification of Drones, and Various applications are discussed. It makes efficient use of scholars to get quick recap of the drone and its Multidomain applications, The study on various aspects on drone is discussed in this chapter. Drones plays a vital role in various domains like agriculture, military, Traffic management and medical field. This chapters discusses the important role and different way of usage of drone in civil applications. This Chapter pays ways to researchers to further research on the drone and its applications.

REFERENCES

ADDITIONAL READING

Afrin, T., & Yodo, N. (2020). A survey of road traffic congestion measures towards a sustainable and resilient transportation system. *Sustainability*, *12*(11), 4660. doi:10.3390u12114660

Aftabuzzaman, M. (2007, September). Measuring traffic congestion-a critical review. In *30th Australasian transport research forum* (pp. 1-16). London, UK: ETM Group.

Arjomandi, M., Agostino, S., Mammone, M., Nelson, M., & Zhou, T. (2006). *Classification of unmanned aerial vehicles. Report for Mechanical Engineering Class*. University of Adelaide.

Azevedo, C. L., Cardoso, J. L., Ben-Akiva, M., Costeira, J. P., & Marques, M. (2014). Automatic vehicle trajectory extraction by aerial remote sensing. *Procedia: Social and Behavioral Sciences*, *111*, 849–858. doi:10.1016/j.sbspro.2014.01.119

Hii, M. S. Y., Courtney, P., & Royall, P. G. (2019). An evaluation of the delivery of medicines using drones. *Drones (Basel)*, *3*(3), 52. doi:10.3390/drones3030052

Khan, M. A., Ectors, W., Bellemans, T., Janssens, D., & Wets, G. (2017). UAV-based traffic analysis: A universal guiding framework based on literature survey. *Transportation Research Procedia*, *22*, 541–550. doi:10.1016/j.trpro.2017.03.043

Khofiyah, N. A., Sutopo, W., & Nugroho, B. D. A. (2019). Technical feasibility battery lithium to support unmanned aerial vehicle (UAV): A technical review. In *Proceedings of the International Conference on Industrial Engineering and Operations Management* (*Vol. 2019*, pp. 3591-3601). Academic Press.

Kumar, S., & Kanniga, E. (2018). Literature Survey on Unmanned aerial Vehicle. *International Journal of Pure and Applied Mathematics*, *119*(12), 4381–4387.

Mancini, F., Dubbini, M., Gattelli, M., Stecchi, F., Fabbri, S., & Gabbianelli, G. (2013). Using unmanned aerial vehicles (UAV) for high-resolution reconstruction of topography: The structure from motion approach on coastal environments. *Remote Sensing*, *5*(12), 6880–6898. doi:10.3390/rs5126880

Reed, T., & Kidd, J. (2019). *Global traffic scorecard*. INRIX Research.

Sivakumar, M., & TYJ, N. M. (2021). A literature survey of unmanned aerial vehicle usage for civil applications. *Journal of Aerospace Technology and Management*, 13.

Chapter 2
A Review on the Applications of Unmanned Aerial Vehicles and Internet of Things Towards Smart Farming

Caprio Mistry
Brainware University, India

Mousumi Biswas
Brainware University, India

Ahona Ghosh
https://orcid.org/0000-0003-0498-285X
Maulana Abul Kalam Azad University of Technology, West Bengal, India

Arighna Basak
Brainware University, India

Bikalpa Bagui
University of Engineering and Management, India

ABSTRACT

In order to commercialize in the industry, various sensors and electrical gadgets are used to keep prices low in a few fields. Unmanned aerial vehicles (UAVs) can be utilized for surveillance, pesticide and insecticide application, and bioprocessing mistake detection to save money and improve the abilities of agricultural experts. Both single-mode and multi-mode UAV systems will perform admirably in this application. This chapter examines the constraints of the internet of things and UAV connectivity in remote areas, as well as smart agriculture application scenarios. In addition, the benefits and uses of employing the internet of things (IoT) and UAVs in agriculture were discussed. On the basis of several elements such as geographical, technological, and business, a system model has been presented. For various IoT applications, the architecture includes enabling technologies, scalability, intelligence, and supportability. Finally, interoperability issues are examined in depth in order to uncover the complications that arise during coordination between UAV and IoT components.

DOI: 10.4018/978-1-7998-8763-8.ch002

INTRODUCTION

Technology has gotten ingrained in every element of our life as a result of rapid technological growth and a reduction in human abilities. Agriculture and irrigation are two fields where man's potential can be fully realised. In order to commercialise in the industry, various sensors and electrical gadgets are used to keep prices low in a few fields. Agriculture is derived after two Latin words 'Ager' and 'Culture,' that indicate 'Land' and 'Cultivation,' respectively. It is a milestone in human evolution and one of the benchmark domains. Throughout human history, extensive revolutions have been planned to advance agricultural production with fewer assets and labour. Notwithstanding this, the population density has never allowed demand and supply to equal over time. Agriculture is vital to the survival of more than 60% of the world's populace. Based on the statements made by Food and Agricultural Organization of the United Nations, agricultural output covers around 12% of entire terrestrial zone (Zavatta, 2014). According to the projected scenario, the global population will spread up to 9.8 billion in 2050, representing a 25% increase over the current situation (Samir & Lutz, 2017). As a result, the emerging countries are expected to see nearly the same population growth as the developed countries (Le Mouël & Forslund, 2017).

On the other hand, urbanisation is expected to continue at a rapid pace, and nearly 70 percent of the population of whole world is expected to be built up by 2050 (Chouhan et al., 2020). In India, the population is estimated to be around 1.2 billion people. 50% people is employed in the agriculture sector, and approximately 61.5 percent of the population in India is mostly reliant on agriculture aimed at their living (FAO in India, 2017; Sawe, 2017).

Agriculture, on the other hand, has been undergoing the fourth revolution in recent decades as a result of the incorporation of Information and Communications Technologies into conventional agriculture (Sundmaeker et al., 2016). Machine Learning and Big Data Analytics, Remote Sensing, IoT, and UAVs are all promising technologies that might help agricultural systems innovate (Walter et al., 2017; Wolfert et al., 2017). Many agricultural restrictions, such as environmental circumstances, development state, soil status, irrigation water, pest and fertilisers, weed control, and greenhouse production environment, can be monitored in smart farming to increase crop yields, lower costs, and enhance process inputs (Nukala et al., 2016). Smart agriculture is a green technology practise since it reduces traditional farming's environmental footprint (Walter et al., 2017). Smart irrigation and minimal fertiliser and insecticide use in crops can further reduce leaching troubles and yields, as well as the influence of climate change in precision agriculture (Walter et al., 2017), (Wong, 2019). One of the most revolutionary technologies in modern wireless communications is the Internet of Things (Atzori et al., 2010). The main idea is to connect various physical items or devices to the Internet by employing certain addressing patterns. Transportation, healthcare, industry, cars, smart homes, and agriculture are all possible applications of IoT technology (Al-Fuqaha et al., 2015). To improve farming, IoT platforms deliver important data on a broad variety of physical restrictions in an agricultural system (Nukala et al., 2016). The importance of Wireless Sensor Networks (WSNs) to the IoT platform is undeniable, as most IoT applications in a variety of markets rely on wireless data transfer (Ghosh & Dey, 2021).

In the last two decades, the ubiquitous usage of the Internet has given public benefits to organisations and people all over the world. The ability to deliver real-time manufacturing and customer service was the key benefit of this achievement. By changing the operating environment, the IoT has recently claimed to give the same benefits with its revolutionary breakthroughs and boost customer awareness and capacities. IoT provides numerous solutions in the areas of health, shopping, traffic, defence, intelligent homes, smart cities, and agriculture. In agriculture, IoT implementation is viewed as the best answer because constant monitoring and control are required. The IoT is seen in the agricultural industrial chain on several times (Medela et al., 2013). The framework developed is much more intriguing when it comes to dealing with node problems and reconfiguring the network's weak connectivity ties on its own. (Zheng et al., 2016) proposes IoT management that tracks wind, soil, atmospheric, and water factors across a broad range of environments. In addition, based on its subdomains, IoT-based agricultural surveillance solutions have been developed. Soil tracking, air surveillance, disease, water, environment, temperature, insect monitoring, location, and fertilisation surveillance are among the sub-domains indicated (Hachem et al., 2015; Torres-Ruiz et al., 2016). Greenhouses, livestock, and precision farming are the most observed IoT applications of cultivation, which are all clustered under distinct surveillance domains. The wireless sensor network (WSN), which aids farmers in collecting essential data through detectors, keeps track of these applications using various IoT-based sensors/devices. Certain IoT-based settings use cloud providers to evaluate and process remote data, assisting scientists and farmers in making better decisions. Through the advancement of contemporary technology, environmental control systems now give extra management and decision-making capabilities. In hostile contexts, a specialised landslide risk management system has been built to enable quick deployments without user interference (Giorgetti et al., 2016). Additionally, data from various environmental indicators is communicated to the user through warnings or notifications to authorities (Liu et al., 2013).

Smart agriculture strives to boost output, yields, and profitability while reducing environmental impact through a variety of methods like effective irrigation, targeted and precise pesticide and fertiliser application, and so on. Smart farming is now possible because to the IoT and UAVs. As a crucial technology for intelligent agriculture, the Internet of Things adds value to data by automating collection, interpretation, and access by ensuring that data flows across diverse devices such as sensors, relays, and gateways. This improves the cost-effectiveness and timeliness of production and management activities in smart farms (Glaroudis et al., 2020).

Moreover, IoT reduces the effects of the natural environment by enabling for real-time responses to weed infestations, insect or disease diagnosis, weather tracking and prevention, soil conditions, and so on. As a result, UAV and IoT technology make it easier to effectively exploit resources such as water, fertilisers, and agrochemicals. Furthermore, these intelligent technologies have enhanced crop output and agricultural environmental consequences. Because (Osuch et al., 2020; Panchasara et al., 2021; Villa-Henriksen et al., 2020) are some of the main aspects of IoT and UAV-based smart farming:

- **Field monitoring:** By strengthening surveillance, special data gathering, and analysis, smart agriculture hopes to reduce crop waste. Animals that graze in open spaces in vast stables are detected by smart farming.
- **Views and tracking:** Technology aids in the evaluation of agricultural air quality and ventilation settings, as well as the detection of harmful waste fumes. Smart agriculture analyses microclimate conditions in greenhouses to improve fruit and vegetable yield and greenhouse quality.
- **Biomass management:** As a preventative strategy against fungus and other microbial pollutants, smart farming helps regulate humidity and temperature in crops such as straw, grass, and other grasses.
- **Offspring Care:** In animal farms, intelligent breeding maintains the conditions of offspring's upbringing and well-being.

Furthermore, UAVs have a wide range of applications, including those in the residential, military, commercial, and governmental sectors (Al-Fuqaha et al., 2015; Giorgetti et al., 2016; Hachem et al., 2015; Medela et al., 2013; Torres-Ruiz et al., 2016; Zheng et al., 2016). Environmental monitoring in the civilian sector (e.g., pollution, plant health, and industrial accidents) is an example. UAVs are used for surveillance and delivery applications in military and government zones to acquire information at disaster or epidemic sites and distribute medication or other vital materials; commercial applications transfer products and supplies in urban and rural areas. UAVs are considered part of the IoT since they rely on sensors, antennae, and embedded software to provide two-way communication for applications such as remote control and monitoring (Liu et al., 2013). The Internet of Things (IoT) generates a rapidly evolving cutting-edge ecosystem in which the main notion is the orchestration of a huge number of smart things. These can be used and triggered on a worldwide scale, either directly by users or through specific software that records behaviour and ideas. IoT allows things to engage in active everyday activities, which has a lot of potential in the "smart city" vision (Glaroudis et al., 2020). By 2022, it is estimated that there will be roughly 30 billion uniquely recognised objects in this worldwide society. With the arrival of 5G technology for UAVs, these projections are projected to skyrocket.

UAVs, on the other hand, aim to provide more views in intelligent farming, such as imaging analysis and agricultural monitoring (Kim et al., 2019). By patrolling a field of interest, UAVs encourage image analysis and agricultural field processing, as well as a full awareness of the situation (Mogili & Deepak, 2018). In addition, UAVs can use data transfer to deliver crucial information to the grounded tracking stations. UAVs are being used in a variety of agricultural applications, including insecticide and fertiliser prospecting and spraying, seed planting, weakening identification, fertility evaluation, mapping, and planting. These recent advancements in intelligent IoT and UAV-based agriculture aid the globe in achieving the '2030 Sustainable Development Agenda' targets, which aim to eliminate hunger by 2030 (Gubbi et al., 2013).

BACKGROUND

The load on agriculture has been greatly increased as the population has grown. With the advent of technology, this period has seen a shift away from traditional approaches and toward more inventive ones. Regardless of how people see agricultural advancement, the truth is that today's agriculture business is more data-driven, accurate, and intelligent than ever before. Almost every industry has been transformed by the rapid development of Internet-of-Things (IoT)-based technologies, including "smart agriculture," which has shifted from statistical to quantitative approaches. Such drastic changes are shaking current agricultural approaches and posing new and diverse challenges. Researchers, research institutions, academicians, and most nations throughout the world are passionate about developing and implementing joint projects to expand the field's horizons for the benefit of masculinity. The tech industry is vying for more cost-effective solutions. Incorporating IoT, cloud computing, big data analytics, and wireless sensor networks can provide enough capability to predict, analyse, and examine circumstances in real time, as well as recover actions. The perception of heterogeneity and device interoperability through the use of flexible, ascendable, and powerful methodologies models are also pioneering new areas in this sector.

Smart cities, smart homes, smart retail, and a variety of other applications can all benefit from the Internet of Things. It is critical to use IoT in agriculture repetitions. By 2050, the world's population will have reached a peak of 9.6 billion people, and the agriculture business wants to meet that need even faster. This is accomplished by the application of current technology, namely the Internet of Things (IoT). Farms with no workers are now a reality thanks to the Internet of Things. It can also be used to maintain cattle, greenhouse farming, and farm management, among other things. Sensors are the most important component of the Internet of Things. Sensors capture vital information, which is then analysed as planned. Sensors are mostly used in agriculture to collect data, determine NPK values, and detect illnesses and moisture content in the soil. The IoT is being studied and investigated to see how it may be used in the agricultural sector. Precision agriculture is the term specified to smart agriculture since it uses precise data to draw conclusions. It displays many sensors that aid IoT and agriculture, as well as their applications, problems, benefits, and drawbacks.

As a result of the rapid technological advancement and decline in human capacity, technology is now present at every stage of our lives. Agriculture and irrigation are two domains in which man's abilities can be put to use. In a few areas, various sensors and electronic gadgets are used to commercialise the industry while keeping costs down. To save money and improve agricultural workers' skills, unmanned aerial vehicles (UAVs) can be utilised for surveillance, pesticide and insecticide application, and bioprocessing fault detection. This application is applicable for both single-mode and multi-mode UAV systems. A network of UAV clusters coupled to ground infrastructures, GCS, or satellites can surpass the expertise of a single UAV system through good collaboration and synchronisation. As a result, the mobility model and specifications are the most efficient routing protocol for each agricultural application.

To comprehend the current situation, we must acknowledge that the farming trade is experiencing a new type of development. In most situations, farming has progressed beyond legacy decision support systems with pre-programmed time schedules. to a new generation of decision support systems. A new era of crop-growing systems that incorporate a variety of cutting-edge technology, like IoT, UAV, machine learning, etc. The majority of these are in a sample stage (not ready for profitable use), and they mainly deal with agricultural processes.

None of these systems integrate a set of agricultural operations or even the entire cultivation period's processes. As is to be expected, certain of the important empowering skills provide greater benefits in certain areas. Various cultivation procedures are more difficult to master than others; yet they are all tough to master. The implementation of IoT in numerous agricultural methods has enhanced total production in terms of yield and quality. Since it can support end-users, it has begun to disclose its potential benefits to them. There are numerous obstacles in the areas of technical difficulty, parameterization, system efficiency, performance, installation, and usability. Finally, addressing better farming techniques which bound the exact aims of agriculturalists will be a critical challenge for the widespread integration of IoT technology in cultivation.

Agriculture was the first industry to use UAV technology for remote observation. Though UAVs have a number of restrictions, most notably in terms of control independence and interaction competence, that have yet to be overcome, the advantages of using this technology have already been apparent. Furthermore, researchers, technologists, and farmers alike recognise the numerous benefits of using UAV technology in various parts of agriculture. To begin with, unmanned aerial vehicles (UAVs) play (and will continue to play) an essential role in weed detection and management. Because of the nature of this agriculture problem and the aerial capabilities of unmanned vehicles, end-users can efficiently manage weeds in cultivation.

The application of artificial intelligence-based methods to multi-spectral imagery record gathered from UAVs expands the capabilities of UAV technology even further. Second, the comparative advantage of using UAV technology and multi-spectral imagery to extract features of various vegetation indices highlighted the potential of such a technology in agriculture techniques. Field-level phenotyping is another important feature of the agricultural industry to which UAV technology can effectively contribute. Field-level phenotyping will be more efficient thanks to the use of UAV systems, which will allow farmers to evaluate overall plant growth and predict final yield more accurately. Once again, the comparative advantage of UAV technology in the field over the usage of static cameras has been demonstrated. Finally, the use of UAV technology in the field helps to address several complex agriculture issues (at an early stage).

Without a question, the agriculture industry, and thus the agricultural economy, is a highly complex and high-potential ecosystem of the global economy. As a result, main developing tools such as UAV and IoT are expected to play a significant part in the forthcoming. Weed management, field phenotyping, multi and hyper-spectral imagery to manage diseases, irrigation water, fertilisers, insecticides, growth of plant and yield, 3D mapping of herbal objects and management, crop value and amount enhancement, and other complex agricultural issues must all be addressed. Smart farming practises in this challenging environment can be contributed by UAV and IoT by meeting the restraints of scalability and simplicity of the system, user approachability, ease of connection, and increased profit. As a result, by meeting human requirements in town as well as rural settings, these skills may eventually change conventional farming practises into revolutionary farming environment.

UNMANNED AERIAL VEHICLE

In recent ages, the UAV has become a low-priced replacement of sensing technologies and data analysing methodologies. Remote sensing using electromagnetic energy, in its most basic form, measures the contents of aimed item since a distance and offers the benefits of comprehensiveness, flexibility, timeliness, non-invasiveness. Because of the complicated natures of farming manufacture, and soils, the remote sensing determines soil parameters in the farmhouse reserved from actual data (Berni et al., 2009; Elsenbeiss & Sauerbier, 2011; Ge et al., 2011). UAVs, according to academics, could be a viable option for gathering precise field data. It was recognized as a probable skill which can produce high spatial resolution imagery and at a temporal frequency sufficient for prompt reactions in the generation of relevant field and crop position information (Elarab et al., 2015). One of the key reasons for the UAV manufacturing exceeding market need is that it suits Low Altitude Remote Sensing that is not as much expensive than conventionally manned aircraft (Zhang et al., 2016).

However, an UAV is a kind of jet functioning without the use of a human pilot. UAVs come in a variety of shapes and sizes, and they are used for a variety of applications. Initially, the military used the technology for anti-aircraft target practise, gathering intelligence, and monitoring enemy territory. Technology, on the other hand, has expanded in importance in several sectors of human effort in recent years, far beyond its original aim. Unmanned aerial vehicles have become more adaptable as technology has advanced, allowing them to be used for a variety of applications. UAVs can be organised remotely by a ground station pilot or independently, with a pre-programmed flight plan and no onboard pilot. The potential for using UAVs in agriculture is considerable (Cano et al., 2017). One of these applications is ineffective, and evidence-based farming forecasts relies on spatial data obtained by the UAV. Farmers can also use UAVs to identify their farms from the air. This aerial perspective may disclose a variety of farming issues, including irrigation issues, soil variances, pest and fungal infestations, and more.

In addition, from a livestock standpoint, UAVs are utilised to conduct headcounts, observe animals, and examine eating habits and health-related behaviours. Farmers can use the information acquired to provide quick and efficient solutions for recognising difficulties and issues, better administration decisions, farm productivity recovery, and eventually improved revenue (Bacco, Berton, Gotta et al, 2018; Mukherjee et al., 2020).

The advantages of UAVs include the ability to capture images of a farmer's crop using a variety of camera filters that can provide multiple spectral imaging, sanction image processing and investigation, and provide better information about the health of their crop while also recognising areas of the crop that require specific types of attention. Furthermore, with minimal training, the small UAV can be quickly covered and conserved; production is a wonderful solution for agriculturalists looking to enhance their profession by combining agriculture with distant perception knowledge (Allred et al., 2020; Bacco, Berton, Gotta et al, 2018; Mukherjee et al., 2020).

TYPES OF UAV

A variety of factors, like size, maximum take-off weight (MTOW), range, and other factors, may be applied to classify UAVs (Wolfert et al., 2017). Furthermore, when it comes to UAV flight regulation, these will be taken into account. The classification was created with the wing type and autonomy degree in mind, as these may be the most acceptable standards for agricultural missions (Chapman et al., 2014; Sugiura et al., 2005; Sylvester, 2018; Vroegindeweij et al., 2014).

There are two types of wings in the wing category: rotary and fixed wings. The first group has jets and multi-rotor aircraft (usually named with drones). Multiple rotors form the airflow, which produces the appropriate belief to be lifted. Their key benefit is the ability to achieve flying flights, useful for aerial photography since it allows the cameras to gather more data for longer periods of time, compensating for poor lighting conditions. Furthermore, the airflow demonstrations perform admirably at low speeds and agree on low-altitude flights with the least amount of risk. Due to its mechanical simplicity, multi-rotors have become more popular than helicopters, which rely on a much more refined plate regulator instrument. As a result, unlike helicopters, multi-rotor flight is viewed through the variable velocity of many direct current motors, rather than motorized machinery.

Brushless motors, not needing any maintenance and have electronic supervisors, have also become an affordable alternative for numerous tasks in civil applications due to their lowered price. As a result, numerous drone builders have raised on the market, offering a diverse range of systems. Commercial drones' fundamental flaw is their lower payload capacity as compared to helicopters. However, drone builders have augmented the amount of rotors from four to six or eight, narrowing the difference. Drones having the highest payload capacity of 22 kg (Lv et al., 2019; Shafian et al., 2018; Torres-Sánchez et al., 2013; Zhang et al., 2018) are already accessible in the market. Not only the cargo capacity but also the safety increases with the increasing number of rotors. As a result, when one or more rotors fail, the aircraft may often fly in depreciate mode, allowing it to arrive in a safe manner. Fixed-wing aircraft, like planes, need to generate airflow to boost by moving fast their aerodynamic components like ailerons and wings.

As a result, the aircraft is unable to perform static flights. As a result, unlike rotary wings, the velocity cannot be reduced; instead, higher altitudes are required to achieve safe flight. Furthermore, rotary wings have a significant advantage over fixed wings in terms of manoeuvrability (e.g., immediate rotations on the vertical axis). Despite the fact that a secure wing aircraft's maximum range is greater than a rotary wing aircraft's due to regulatory rules for flying UAVs, it does not provide a meaningful benefit in most situations, and the aircraft nevertheless have a superior payload capacity for the most part. As a result, the investigation concluded that rotary-wing aircraft are preferred for agricultural missions over fixed-wing aircraft.

IOT FOR SMART AGRICULTURE

Smart cultivation is a modern farming idea that uses Internet of Things technologies to boost productivity. Farmers can successfully apply fertilisers to rise the value and amount of crops by using smart farming. Farmers, on the other hand, cannot be in the field continuously every day. Furthermore, they were unable to employ various instruments to determine the optimal ecological situations aimed at the crops. IoT delivers them an automated system that can operate without human intervention and inform them on how to deal with various types of challenges they may face while farming. Even if the farmer is not on the ground, it can spread and tell the farmer, allowing farmers to acquire more farmland and cultivate their productivity. By 2050, the global population is expected to reach 9 billion (Chouhan et al., 2020; Le Mouël & Forslund, 2017; Samir & Lutz, 2017; Sawe, 2017; Zavatta, 2014). As a result, IoT applications are required in farming to feed such a huge population and efficiently utilise farmland and other assets, which are scarce in some areas. Random weather events are affecting crops as a result of global warming, and farmers are losing money; hence, the IoT Smart Farming application will allow them to take immediate actions to avert this (Agarwal et al., 2019; Ayaz et al., 2019; Bacco, Berton, Ferro et al, 2018; Hunter et al., 2017).

However, major machineries of Agricultural IoT include sensor-based equipment, wireless communication technologies, internet connection, sensing and transferred data, and so on. It plays a critical part in the efficient configuration of IoT frameworks, classified by communication distance, spectrum, and application states. Figure 1 illustrated the IoT framework built on three layers: the physical layer in charge of sensing, network layer, in charge of data transport, and finally application layer, in charge of data storing and alteration (Villa-Henriksen et al., 2020).

Figure 1. IoT architecture

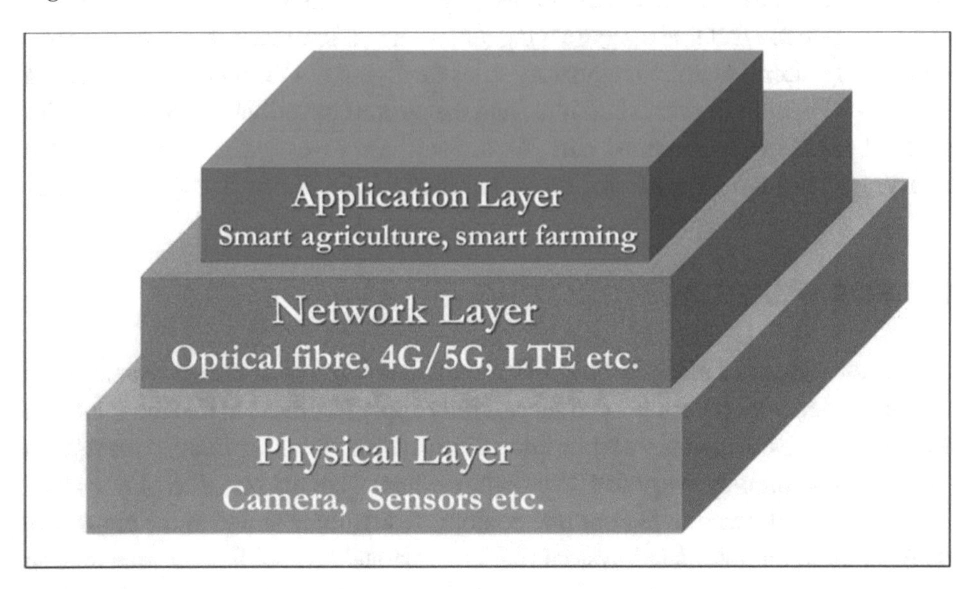

Physical Layer

Several terminal devices, cameras, sensors, Radio Frequency ID, Wireless Sensor Networks labels and readers, Near Field Communications devices, and other devices make up the physical layer (Tzounis et al., 2017). Sensors collect data on wind speed, temperature, humidity, nutrient levels, plant diseases, insect pests, and other factors for this layer. Embedded devices control the compiled data, which is then uploaded to a higher layer for further processing and examination. These sensors and terminal equipment are used to track, control, and identify agricultural and livestock crops. Wireless Sensor Networks are frequently used to govern and monitor storage and logistics services, for example (Tzounis et al., 2017).

Radio Frequency Identification technologies, on the other hand, are the most important pattern for networked devices. Radio Frequency ID tags, for example, store data using an Electronic Product Code, that is subsequently read, activated, and controlled by Radio Frequency ID readers. Radio Frequency Identification, and Near Field Communications technologies play a key part in the farming area by contributing to object detection, monitoring, control, and storing data on passive or active devices.

Network Layer

In the network layer, sensors and devices are required for connecting with gateway and adjacent nodes for creating the network. Sensor nodes of this layer collaborate and link with other nodes and gateways in a network to pass data to a tiny structure where it is stored, analysed, processed, and disseminated for usable information (Gubbi et al., 2013).

Application Layer

It is the topmost layer in the IoT infrastructure, where the IoT advantages are pushed outward. This layer contains a number of light stages or systems for controlling and monitoring soil conditions, water and nutrient levels, as well as plants and animals. These layers are also utilised to aid in the fast identification and organisation of illnesses and pests, infestation, and agricultural creation safety controllability, resulting in increased manufacturing efficiency.

COMMUNICATION MEANS APPLIED IN IOT BASED SMART AGRICULTURE

The technique for wireless communication was set by wireless protocols and standards. IEEE 802.15.4, for example, is a wireless standard for connecting internet-enabled gateways and end-nodes. ZigBee, Sigfox, EC-GSM, ONE-NET, Wireless HART, ISA100.11a, LoRaWAN, Bluetooth Low Energy (BLE), DASH7, and others are further examples (Suhonen et al., 2012). Figure 2 (Feng et al., 2019) shows how these standards are grouped in terms of transmission ranges.

Figure 2. Communication Technologies Classification for IOT based smart agriculture

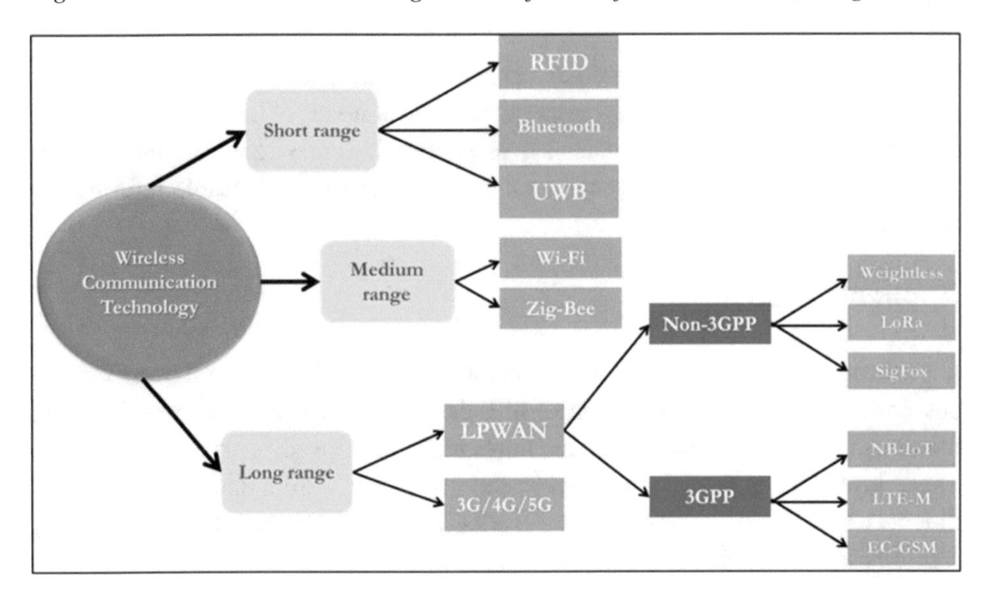

IEEE 802.15.4-based protocols are the best choice for short-distance wireless communication applications (Gubbi et al., 2013), and are expected to be used in Low-Power Wide-Area Networks. It has a data rate between 20 and 250 kbps, frequency bands like 2.4 GHz, 915 MHZ, 868 MHz, 433 MHz, and full outside Line of Sight series of 100 metres (Jain, 2016). On the other hand, IEEE 802.11 standards are ideal for medium-range connection scenarios. IEEE 802.11p is also a viable option for incorporating circumstances with a high level of mobility. Because of its great communication range, maximum permitted broadcast power of 1W, and less disturbed band having frequency of 5.9 GHz ISM (Bacco, Berton, Ferro et al, 2018), this issue appears to be receiving a lot of attention in agriculture.

Long-distance communication technologies, on the other hand, are the only solid and anticipated solutions when it comes to coverage. As a result, 3G, 4G, LTE, and 5G cellular connection technologies are the most suited and reliable standards for precision agriculture. For transmission and processing, a considerable amount of real-time data is required (Shi et al., 2019).

Furthermore, 5G communication technology is likely to deliver actual Device to Device communication, allowing vehicle localization. Furthermore, several devices per square kilometre can be maintained (Le et al., 2016). Unlike LTE, 5G is able to function with larger frequency bands and with larger channel bandwidths. 5G communication technology can provide new skills on-farm apparatus by contributing faster data rates and a larger transmission range under the model of actual connectivity, particularly in rural locations. However, the cellular network's availability and the economic viability of 5G technology in rural areas are also under doubt.

The most consistent IoT long-distance communication technologies are IEEE 802.11ah and LoRa or LoRaWAN. Furthermore, the first is a modification of the IEEE 802.11 family that was released in 2017 to aid IoT setups like smart monitoring (Bacco, Berton, Ferro et al, 2018). It uses 900 MHz bands, has a greater range of coverage, and uses less energy than IEEE 802.15.4. With a single access point, it can connect thousands of strategies within a one kilometre radius. On the other hand, LoRaWAN is a promising Low-Power Wide-Area Networks protocol that is being developed for a system of battery based wireless devices as shown in Figure 3.

Figure 3. Smart farming-based end to end communication (Ayaz et al., 2019)

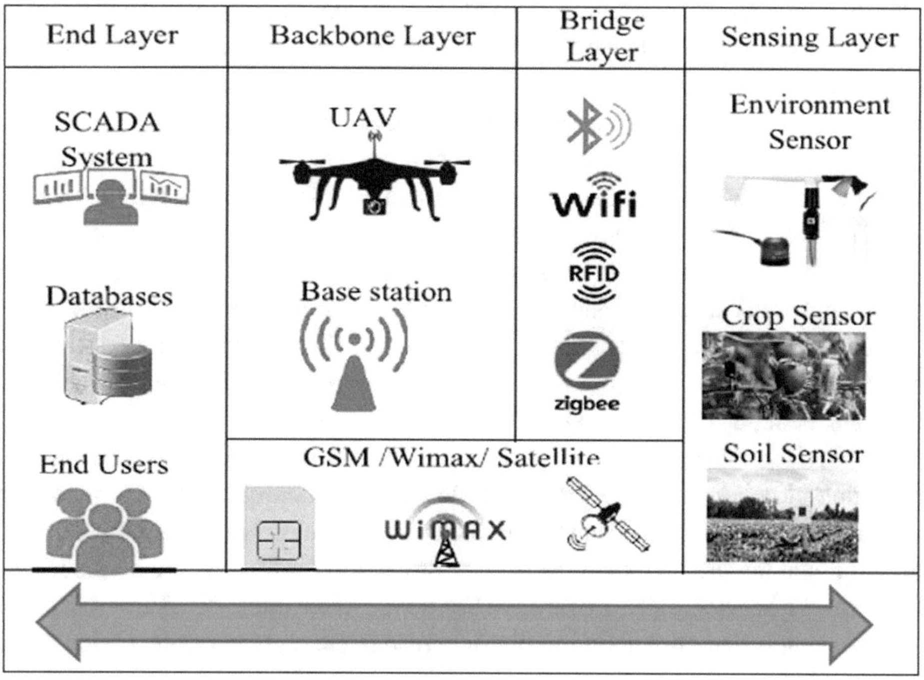

Based on the bandwidth requirement and purpose, cellular transmission systems ranging between 2G and 4G may be suitable; however, the dependability, and cellular network's availability in rural areas is the key concern. Another option for overcoming this is data transmission through satellite. Nonetheless, the cost of this communication technology is prohibitively high, making it unsuitable for small and medium-sized farms. Furthermore, the superior communication mode is determined by the needs of the application. Some farms, for example, require sensors that can operate at a low data rate but must run for extended periods of time and have a long battery life. In such cases, a new Low Power Wide Area Network is chosen as a superior resolution for cellular connectivity, not only in areas with long battery life but also in areas with a larger connectivity choice at acceptable rates (Beecham Research, 2016). Pasture and crop management are the most common applications for Low Power Wide Area Networks, although they can also be used in a variety of other farming-related applications.

In addition, the Micro Air Vehicle Link is a shared set of communicating rules that allows UAVs to connect with Ground Control Stations. It connects the calculating platform, the UAV monitoring platform, and the Ground Control Stations applications platform (Atoev et al., 2017). Directions, the location of a global navigation satellite system, and the velocity of the UAV are all transmitted via Micro Air Vehicle Link. The communication space amid UAV and Ground Control Stations is determined by the UAV's conditions. When the UAV is inside LOS, however, it can link up to 2 km (Salaan et al., 2019). UAVs are now scheduled to function in return-to-launch means, which allows them to automatically return to their initial location if transmission is disrupted. This strategy is used to prevent unintentional UAV mishaps or losses (Bhandari et al., 2018; Coombes et al., 2018; Kim et al., 2019). Other communication methods, like radio-frequency modules, ZigBee, etc. are available to communicate between Ground Control Stations and UAVs. Using various technology, such as phone apps, can extend the range of communication. Furthermore, the introduction of 5G cellular technology has the potential to significantly improve infrastructure and data processing speeds, both of which are important for high-definition charting (Alsalam et al., 2017; Saha et al., 2018).

In smart farming applications, cloud computing can be utilised for two purposes: I gathering and storing information transmitted from a remote client, and ii) processing the data and displaying the findings to the users. Visualization, data analytics, and decision-making are all examples of data processing. Infrastructure as a Service, Platform as a Service, and Software as a Service are three interrelated levels of cloud service models that correspond to physical resources, tools to create a wide range of applications, and Internet-based applications accessed by end-users, respectively. There are various considerations to consider while building and/or implementing a smart farming system, such as the associated charges and security standards.

UAV FOR SMART AGRICULTURE

The IoT has accelerated advances in a variety of industries, including agriculture. However, when it comes to agricultural, communication infrastructure such as base stations and Wi-Fi is woefully inadequate, stifling the growth of the IoT in this sector. One of the key steeplechases while introducing the IoT in the agriculture industry is that such communication arrangements and associated utilities are considerably worse in growing counties and rural areas. In the absence of a reliable communication setup, the data generated by the wireless sensors cannot be communicated. In such a case, unmanned aerial vehicles (UAVs) provide a viable alternative. To produce advanced processing and investigation data, the UAV system communicates with wireless sensors that cover broad areas. Robots are very valuable innovations in smart agriculture, and UAVs, sometimes called with drones, have been broadly applied (Mogili & Deepak, 2018; Muchiri & Kimathi, 2022). Agriculturalists are increasingly using drones or unmanned aerial vehicles (UAVs) to monitor and regulate farm development. Some UAVs get deployed for spraying water and insecticides in remote areas where human movement is difficult and crops remain at varied heights.

UAVs, often called with drones, are equipped with high-resolution cameras and precise sensors and can fly over thousands of hectares of farmland. The importance of investigation in all agricultural applications is enormous, especially in forestry and crop observation for large zones (Ammad Uddin et al., 2018). As a result, farm manufacturing requires a quick, affordable, practical, and wide surveillance system with precise data collection and broadcast capability. Currently, there are primarily two methods for obtaining aerial photos of a field or farm zone: satellites and aeroplanes. Both options are adequate for a macro view of the scenery, but they have serious issues with their quality in terms of P.C. views. The pictures are not of high-resolution and cannot provide the level of quality that is required throughout the investigation. Furthermore, the frequency of visits has issues, and it is difficult to collect images on a regular basis. Furthermore, another serious issue is that these controls extend above the cloud level, where there is a good chance that both will be crowded in terrible weather.

The application of multiple-UAV schemes to the smart agriculture could lead to a breakthrough in agriculture. Although still there exist different technical challenges to be resolved, the number of multiple-UAV schemes in smart framing has increased noticeably. Applying the circulated swarm algorithm, the authors of (Ju & Son, 2018) built a multi-UAV scheme aimed at agronomic areas. They evaluated the suggested system's performance and compared the results to those of a solo-UAV system. The results of the experiments showed that the multi-UAV scheme outperformed the single-UAV system. (Doering et al., 2014) We built and presented an autonomous precision agriculture system based on the usage of multiple-UAVs. Genetic Algorithm and Particle Swarm Optimization were integrated in (Zhai et al., 2018) to consider and solve the challenge of task planning of multiple-UAV schemes, that is a multi-target optimization issue. They developed a precision farming system made up of numerous components, agents, and drones that would work together to complete composite agricultural missions. The goal of this project was to maximise the use of restricted mechanised apparatus in smart agriculture. Lastly, in several agricultural settings, the integration of unmanned ground vehicles in UAVs was tested (Vu et al., 2018).

Figure 4. Different types of agricultural UAVs (Kim et al., 2019)

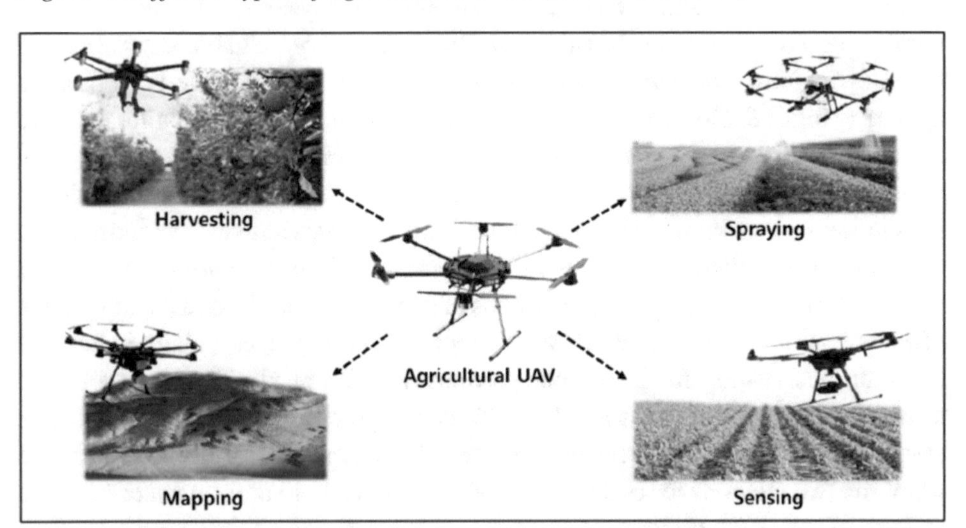

UAVs, on the other hand, are a platform that provides a "eye in the sky," which solves or eliminates the aforementioned issues with micro views as shown in Figure 4. The value of photos captured by UAVs is determined by the camera's resolution, which is typically many times higher than satellite images and, more importantly, varies according to application requirements. UAVs maintain faster and better NDVI to monitor agricultural conditions such as weed mapping, leaf assessments, and so on, and provide instant responses to farmers for timely moves. UAVs, on the other hand, are boosted in terms of frequency, even if wanted numerous times in a single day, and are unaffected by weather conditions unless it rains. UAVs have been chosen as the future of precision agriculture due to the aforementioned benefits.

In addition to incalculability, UAVs are required to transport heavy payloads in certain insecticide and fertiliser applications. As a result, optimum battery consumption becomes critical in such situations to spread out the flight time. Many parameters can be measured at this resolution to improve drone efficiency. Initially, special appropriate conditions, such as climate or air direction, must be met when flying. If you succeed, try to put together the best payload possible and place it correctly. For this situation, attributing the payload near the field, improving in smaller numbers, and refilling in its place for putting heavy quantities can be beneficial.

Furthermore, depending on the size of the area and the frequency of visits, optimal path selection plays an important role. For this resolution, numerous routing arrangements are planned particularly for the UAVs, so picking and achieving the suitable arrangement can give clear modification. New actions like tethering systems can help with pesticide application and UAV-based agriculture, when drones are required to fly with heavy payloads. In UAV tethering, an assembly is given that provides power via a lengthy cable, allowing it to fly as long as the farmers have power backup on the field; most importantly, it eliminates the need to transport hefty batteries. Agriculture is currently regarded as one of the most promising industries in which unmanned aerial vehicles (UAVs) can offer solutions to a variety of pressing and long-term issues. The following are some of the most important applications in which drones are already helping farmers through the system concept.

APPLICATIONS OF UAV INTEGRATED WITH IOT THROUGH A SYSTEM MODEL

UAVs integrated with IoT can be used to a variety of smart agriculture applications. Through a system paradigm, this section discusses some IoT and UAV applications in smart agriculture.

- **Smart crop monitoring:** Crop monitoring refers to the accurate detection of a farm's various constraints. One of the most important aspects of smart agriculture is automatic observation. Sensors placed at strategic locations can detect and transmit data to a gateway for further analysis and control. Sensors are utilised to control crop restrictions such leaf area index (Orlando et al., 2016), plant height, colour, form, and leaf size (Dadshani et al., 2015), among others. Controlling soil moisture (Sahota et al., 2011; Vellidis et al., 2007), farm water limitations such as pH level (Islam et al., 2020), and climate factors such as wind speed, wind direction, rainfall, radiation, air pressure, temperature, relative humidity, and so on are also possible with IoT and UAV (Crabit et al., 2011; Navulur & Prasad, 2017). Remote sensing is also present in a highly efficient manner. Remote sensors are connected in lower altitude UAVs due to simple sensors, allowing for efficient and cost-effective crop control. As a result, high-resolution records are obtained by removing various forms of abnormal situations, such as weather.
- **Smart pest management:** Pest control is usually based on three features: detecting, assessing, and treating. First, advanced infection and pest

appreciation methods are based on image processing, in which raw images are created utilising UAVs or remote sensing satellites through the crop zone. Remote sensors typically protect large areas and, as a result, provide higher productivity at a lower cost. On the other hand, UAV IOT sensors are capable of performing additional duties in data collection, such as environmental samples, plant health, and insect situations, at various angles throughout the crop cycle. Finally, IoT-based mechanical traps, for example, can capture, count, and illustrate insect types while also uploading statistics to the Cloud for a more thorough investigation, which is not possible with remote sensing.

- **Smart irrigation:** Agricultural UAVs equipped with cameras can provide incredible views of the exact areas of concern in the field. Farmers can use the cameras to control areas of low soil moisture, dehydrated crops, and saturated zones, as well as gain an understanding of the overall health of the farm's crops. With traditional farming, such precise observing was either impossible, incomplete, or prohibitively expensive due to the need to hire specialists to carry out the task and obtain satisfactory results. Currently, however, UAVs assist farmers by providing additional benefits to help them organise their own operations.
- **Livestock monitoring (Animal Husbandry):** Livestock monitoring is a difficult aspect of agriculture that requires a large amount of labour efficiency, resulting in cost-effectiveness issues. It is feasible to take remote surveillance on livestock monitoring with this IoT enabled 5g drone. As a result, a growing number of researchers are interested in using UAVs as a farming tool. The topic of organizing a cluster of UAVs for monitoring and tracking livestock like sheep or cattle in a pasture is investigated in this research. We presume that all beset creatures have been fitted using GPS collars, and that each animal's motion cannot be overlooked. We also accept that the amount of UAVs is adequate to cover the entire meadow, where our goal is to discover the best UAV placement to reduce the average distance between animal and UAV. We begin by describing a procedure for using UAVs to perform sweep coverage. The initial locations of all targeted animals can be acquired by deploying UAVs to achieve sweep coverage across the entire pasture. Then, using streaming k-means clustering with the initial and updated locations from the GPS collars, determine and update the UAVs' deployment.
- **Forecasting:** Smart agriculture's main feature is forecasting, that applies present and past data to estimate and calculate significant strictures. Machine learning and Scientific are two instances of forecasting apparatuses. For example, UAV provides various machine learning models such as the regression model for approximating phosphorus amount in the soil (Estrada-López et al., 2018), predicting soil moisture, plant ailment identification (Gao

et al., 2018), Artificial Neural Networks to predict field temperatures (Aliev et al., 2018), and so on (Ghosh & Dey, 2021).

System Model

The IoT is a cutting-edge technology dealing with multidisciplinary engineering frameworks for domestic and commercial automation libraries. If we consider three gateways as such, they are gateways 1, 2, and 3 as shown in Figure 5, the network layers will be more concentrated if we adopt the OSI model here. Three gateways are sufficient to optimise network traffic to and from the UAV band. After sensing data, it may be viewed through several portals with real-time readings after being relayed over the gateway to an online cloud server and from the client machine (mobile, tablet, PC).

Figure 5. System model

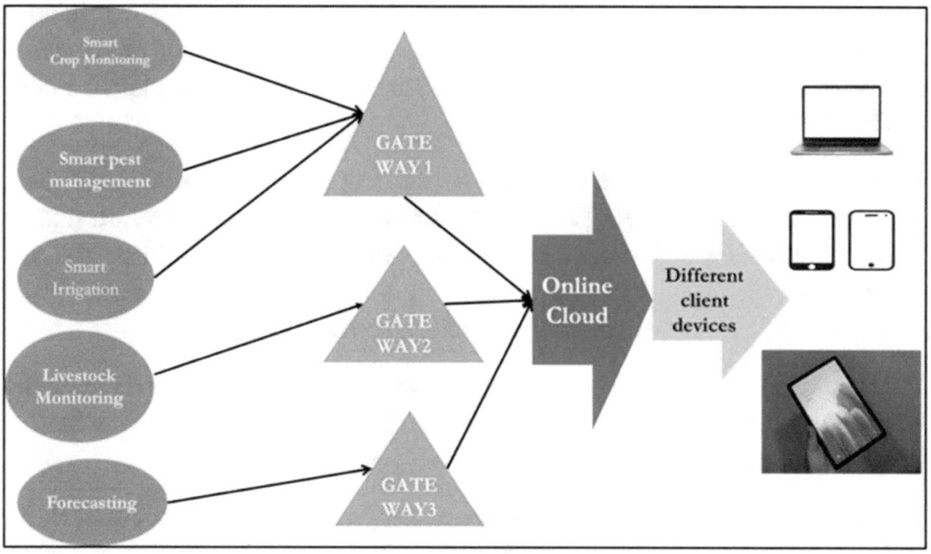

CONCLUSION

This chapter showcases a variety of communication technologies and highlights the challenges of IoT and UAV connectivity in remote places in terms of transmission range and communication technology. The goal of this chapter is to bridge the gap between IoT and UAV by identifying the technical and non-technical challenges that must be overcome. Furthermore, the chapter uses a system model to show how IoT and UAVs can be used for smart agriculture. By utilising the mobility of UAVs, the study provides a definitive answer to the question of IoT and UAV proximity. IoT scalability, heterogeneity, and network coverage constraints can all be considerably improved with a UAV-enabled IoT system.

REFERENCES

Agarwal, P., Singh, V., Saini, G. L., & Panwar, D. (2019). Sustainable Smart-farming framework: smart farming. In *Smart farming technologies for sustainable agricultural development* (pp. 147–173). IGI Global. doi:10.4018/978-1-5225-5909-2.ch007

Al-Fuqaha, A., Guizani, M., Mohammadi, M., Aledhari, M., & Ayyash, M. (2015). Internet of things: A survey on enabling technologies, protocols, and applications. *IEEE Communications Surveys and Tutorials*, *17*(4), 2347–2376. doi:10.1109/COMST.2015.2444095

Aliev, K., Jawaid, M. M., Narejo, S., Pasero, E., & Pulatov, A. (2018). Internet of plants application for smart agriculture. *International Journal of Advanced Computer Science and Applications*, *9*(4). Advance online publication. doi:10.14569/IJACSA.2018.090458

Allred, B., Martinez, L., Fessehazion, M. K., Rouse, G., Williamson, T. N., Wishart, D., Koganti, T., Freeland, R., Eash, N., Batschelet, A., & Featheringill, R. (2020). Overall results and key findings on the use of UAV visible-color, multispectral, and thermal infrared imagery to map agricultural drainage pipes. *Agricultural Water Management*, *232*, 106036. doi:10.1016/j.agwat.2020.106036

Alsalam, B. H. Y., Morton, K., Campbell, D., & Gonzalez, F. (2017, March). *Autonomous UAV with vision based on-board decision making for remote sensing and precision agriculture. In 2017 IEEE Aerospace Conference*. IEEE.

Ammad Uddin, M., Mansour, A., Le Jeune, D., Ayaz, M., & Aggoune, E. H. M. (2018). UAV-assisted dynamic clustering of wireless sensor networks for crop health monitoring. *Sensors (Basel)*, *18*(2), 555. doi:10.339018020555 PMID:29439496

Atoev, S., Kwon, K. R., Lee, S. H., & Moon, K. S. (2017, November). Data analysis of the MAVLink communication protocol. In *2017 International Conference on Information Science and Communications Technologies (ICISCT)* (pp. 1-3). IEEE. 10.1109/ICISCT.2017.8188563

Atzori, L., Iera, A., & Morabito, G. (2010). The internet of things: A survey. *Computer Networks*, *54*(15), 2787–2805. doi:10.1016/j.comnet.2010.05.010

Ayaz, M., Ammad-Uddin, M., Sharif, Z., Mansour, A., & Aggoune, E. H. M. (2019). Internet-of-Things (IoT)-based smart agriculture: Toward making the fields talk. *IEEE Access: Practical Innovations, Open Solutions*, *7*, 129551–129583. doi:10.1109/ACCESS.2019.2932609

Bacco, M., Berton, A., Ferro, E., Gennaro, C., Gotta, A., Matteoli, S., ... & Zanella, A. (2018). Smart farming: Opportunities, challenges and technology enablers. *2018 IoT Vertical and Topical Summit on Agriculture-Tuscany (IOT Tuscany),* 1-6.

Bacco, M., Berton, A., Gotta, A., & Caviglione, L. (2018). IEEE 802.15.4 air-ground UAV communications in smart farming scenarios. *IEEE Communications Letters*, *22*(9), 1910–1913. doi:10.1109/LCOMM.2018.2855211

Beecham Research. (2016). *An Introduction to LPWA Public Service Categories: Matching Services to IoT Applications*. Author.

Berni, J. A. J., Zarco-Tejada, P. J., Suarez, L., González-Dugo, V., & Fereres, E. (2009). Remote sensing of vegetation from UAV platforms using lightweight multispectral and thermal imaging sensors. *The International Archives of the Photogrammetry, Remote Sensing and Spatial Information Sciences*, *38*(6), 6.

Bhandari, S., Raheja, A., Chaichi, M. R., Green, R. L., Do, D., Pham, F. H., ... Espinas, A. (2018, June). Lessons learned from uav-based remote sensing for precision agriculture. In *2018 International Conference on Unmanned Aircraft Systems (ICUAS)* (pp. 458-467). IEEE. 10.1109/ICUAS.2018.8453445

Cano, E., Horton, R., Liljegren, C., & Bulanon, D. M. (2017). Comparison of small unmanned aerial vehicles performance using image processing. *Journal of Imaging*, *3*(1), 4. doi:10.3390/jimaging3010004

Chapman, S. C., Merz, T., Chan, A., Jackway, P., Hrabar, S., Dreccer, M. F., Holland, E., Zheng, B., Ling, T., & Jimenez-Berni, J. (2014). Pheno-copter: A low-altitude, autonomous remote-sensing robotic helicopter for high-throughput field-based phenotyping. *Agronomy (Basel)*, *4*(2), 279–301. doi:10.3390/agronomy4020279

Chouhan, S. S., Singh, U. P., & Jain, S. (2020). Applications of computer vision in plant pathology: A survey. *Archives of Computational Methods in Engineering*, *27*(2), 611–632. doi:10.100711831-019-09324-0

Coombes, M., Chen, W. H., & Liu, C. (2018, July). Fixed wing UAV survey coverage path planning in wind for improving existing ground control station software. In *2018 37th Chinese Control Conference (CCC)* (pp. 9820-9825). IEEE. 10.23919/ChiCC.2018.8482722

Crabit, A., Colin, F., Bailly, J. S., Ayroles, H., & Garnier, F. (2011). Soft water level sensors for characterizing the hydrological behaviour of agricultural catchments. *Sensors (Basel), 11*(5), 4656–4673. doi:10.3390110504656 PMID:22163868

Dadshani, S., Kurakin, A., Amanov, S., Hein, B., Rongen, H., Cranstone, S., Blievernicht, U., Menzel, E., Léon, J., Klein, N., & Ballvora, A. (2015). Non-invasive assessment of leaf water status using a dual-mode microwave resonator. *Plant Methods, 11*(1), 1–10. doi:10.118613007-015-0054-x PMID:25918549

Doering, D., Benenmann, A., Lerm, R., de Freitas, E. P., Muller, I., Winter, J. M., & Pereira, C. E. (2014). Design and optimization of a heterogeneous platform for multiple uav use in precision agriculture applications. *IFAC Proceedings Volumes, 47*(3), 12272-12277.

Elarab, M., Ticlavilca, A. M., Torres-Rua, A. F., Maslova, I., & McKee, M. (2015). Estimating chlorophyll with thermal and broadband multispectral high resolution imagery from an unmanned aerial system using relevance vector machines for precision agriculture. *International Journal of Applied Earth Observation and Geoinformation, 43*, 32–42. doi:10.1016/j.jag.2015.03.017

Elsenbeiss, H., & Sauerbier, M. (2011). Investigation of uav systems and flight modes for photogrammetric applications. *The Photogrammetric Record, 26*(136), 400–421. doi:10.1111/j.1477-9730.2011.00657.x

Estrada-López, J. J., Castillo-Atoche, A. A., Vázquez-Castillo, J., & Sánchez-Sinencio, E. (2018). Smart soil parameters estimation system using an autonomous wireless sensor network with dynamic power management strategy. *IEEE Sensors Journal, 18*(21), 8913–8923. doi:10.1109/JSEN.2018.2867432

FAO in India. (2017) Retrieved from https://www.fao.org/india/fao-inindia/india-at-a-glance/en/2017

Feng, X., Yan, F., & Liu, X. (2019). Study of wireless communication technologies on Internet of Things for precision agriculture. *Wireless Personal Communications, 108*(3), 1785–1802. doi:10.100711277-019-06496-7

Gao, J., Nuyttens, D., Lootens, P., He, Y., & Pieters, J. G. (2018). Recognising weeds in a maize crop using a random forest machine-learning algorithm and near-infrared snapshot mosaic hyperspectral imagery. *Biosystems Engineering, 170*, 39–50. doi:10.1016/j.biosystemseng.2018.03.006

Ge, Y., Thomasson, J. A., & Sui, R. (2011). Remote sensing of soil properties in precision agriculture: A review. *Frontiers of Earth Science, 5*(3), 229–238. doi:10.100711707-011-0175-0

Ghosh, A., & Dey, S. (2021). "Sensing the Mind": An Exploratory Study About Sensors Used in E-Health and M-Health Applications for Diagnosis of Mental Health Condition. In *Efficient Data Handling for Massive Internet of Medical Things* (pp. 269–292). Springer. doi:10.1007/978-3-030-66633-0_12

Giorgetti, A., Lucchi, M., Tavelli, E., Barla, M., Gigli, G., Casagli, N., Chiani, M., & Dardari, D. (2016). A robust wireless sensor network for landslide risk analysis: System design, deployment, and field testing. *IEEE Sensors Journal*, *16*(16), 6374–6386. doi:10.1109/JSEN.2016.2579263

Glaroudis, D., Iossifides, A., & Chatzimisios, P. (2020). Survey, comparison and research challenges of IoT application protocols for smart farming. *Computer Networks*, *168*, 107037. doi:10.1016/j.comnet.2019.107037

Gubbi, J., Buyya, R., Marusic, S., & Palaniswami, M. (2013). Internet of Things (IoT): A vision, architectural elements, and future directions. *Future Generation Computer Systems*, *29*(7), 1645–1660. doi:10.1016/j.future.2013.01.010

Hachem, S., Mallet, V., Ventura, R., Pathak, A., Issarny, V., Raverdy, P. G., & Bhatia, R. (2015, March). Monitoring noise pollution using the urban civics middleware. In *2015 IEEE First International Conference on Big Data Computing Service and Applications* (pp. 52-61). IEEE. 10.1109/BigDataService.2015.16

Hunter, M. C., Smith, R. G., Schipanski, M. E., Atwood, L. W., & Mortensen, D. A. (2017). Agriculture in 2050: Recalibrating targets for sustainable intensification. *Bioscience*, *67*(4), 386–391. doi:10.1093/biosci/bix010

Islam, N., Ray, B., & Pasandideh, F. (2020, August). Iot based smart farming: Are the lpwan technologies suitable for remote communication? In *2020 IEEE International Conference on Smart Internet of Things (SmartIoT)* (pp. 270-276). IEEE. 10.1109/SmartIoT49966.2020.00048

Jain, R. (2016). *Wireless protocols for iot part ii: Ieee 802.15. 4 wireless personal area networks*. IEEE.

Ju, C., & Son, H. I. (2018). Multiple UAV systems for agricultural applications: Control, implementation, and evaluation. *Electronics (Basel)*, *7*(9), 162. doi:10.3390/electronics7090162

Kim, J., Kim, S., Ju, C., & Son, H. I. (2019). Unmanned aerial vehicles in agriculture: A review of perspective of platform, control, and applications. *IEEE Access: Practical Innovations, Open Solutions*, *7*, 105100–105115. doi:10.1109/ACCESS.2019.2932119

Le, N. T., Hossain, M. A., Islam, A., Kim, D. Y., Choi, Y. J., & Jang, Y. M. (2016). Survey of promising technologies for 5G networks. *Mobile Information Systems.*

Le Mouël, C., & Forslund, A. (2017). How can we feed the world in 2050? A review of the responses from global scenario studies. *European Review of Agriculture Economics*, *44*(4), 541–591. doi:10.1093/erae/jbx006

Liu, Z., Huang, J., Wang, Q., Wang, Y., & Fu, J. (2013, June). Real-time barrier lakes monitoring and warning system based on wireless sensor network. In *2013 Fourth International Conference on Intelligent Control and Information Processing (ICICIP)* (pp. 551-554). IEEE. 10.1109/ICICIP.2013.6568136

Lv, M., Xiao, S., Yu, T., & He, Y. (2019). Influence of UAV flight speed on droplet deposition characteristics with the application of infrared thermal imaging. *International Journal of Agricultural and Biological Engineering*, *12*(3), 10–17. doi:10.25165/j.ijabe.20191203.4868

Medela, A., Cendón, B., Gonzalez, L., Crespo, R., & Nevares, I. (2013, July). IoT multiplatform networking to monitor and control wineries and vineyards. In *2013 Future Network & Mobile Summit* (pp. 1-10). IEEE.

Mogili, U. R., & Deepak, B. B. V. L. (2018). Review on application of drone systems in precision agriculture. *Procedia Computer Science*, *133*, 502–509. doi:10.1016/j.procs.2018.07.063

Muchiri, G. N., & Kimathi, S. (2022, April). A review of applications and potential applications of UAV. In *Proceedings of the Sustainable Research and Innovation Conference* (pp. 280-283). Academic Press.

Mukherjee, A., Misra, S., Sukrutha, A., & Raghuwanshi, N. S. (2020). Distributed aerial processing for IoT-based edge UAV swarms in smart farming. *Computer Networks*, *167*, 107038. doi:10.1016/j.comnet.2019.107038

Navulur, S., & Prasad, M. G. (2017). Agricultural management through wireless sensors and internet of things. *Iranian Journal of Electrical and Computer Engineering*, *7*(6), 3492. doi:10.11591/ijece.v7i6.pp3492-3499

Nukala, R., Panduru, K., Shields, A., Riordan, D., Doody, P., & Walsh, J. (2016, June). Internet of Things: A review from 'Farm to Fork'. In *2016 27th Irish signals and systems conference (ISSC)* (pp. 1-6). IEEE.

Orlando, F., Movedi, E., Coduto, D., Parisi, S., Brancadoro, L., Pagani, V., Guarneri, T., & Confalonieri, R. (2016). Estimating leaf area index (LAI) in vineyards using the PocketLAI smart-app. *Sensors (Basel), 16*(12), 2004. doi:10.339016122004 PMID:27898028

Osuch, A., Przygodziński, P., Rybacki, P., Osuch, E., Kowalik, I., Piechnik, L., Przygodziński, A., & Herkowiak, M. (2020). Analysis of the Effectiveness of Shielded Band Spraying in Weed Control in Field Crops. *Agronomy (Basel), 10*(4), 475. doi:10.3390/agronomy10040475

Panchasara, H., Samrat, N. H., & Islam, N. (2021). Greenhouse gas emissions trends and mitigation measures in australian agriculture sector—A review. *Revista de Agricultura (Piracicaba), 11*, 85.

Saha, A. K., Saha, J., Ray, R., Sircar, S., Dutta, S., Chattopadhyay, S. P., & Saha, H. N. (2018, January). IOT-based drone for improvement of crop quality in agricultural field. In *2018 IEEE 8th Annual Computing and Communication Workshop and Conference (CCWC)* (pp. 612-615). IEEE. 10.1109/CCWC.2018.8301662

Sahota, H., Kumar, R., & Kamal, A. (2011). A wireless sensor network for precision agriculture and its performance. *Wireless Communications and Mobile Computing, 11*(12), 1628–1645. doi:10.1002/wcm.1229

Salaan, C. J., Tadakuma, K., Okada, Y., Sakai, Y., Ohno, K., & Tadokoro, S. (2019). Development and experimental validation of aerial vehicle with passive rotating shell on each rotor. *IEEE Robotics and Automation Letters, 4*(3), 2568–2575. doi:10.1109/LRA.2019.2894903

Samir, K. C., & Lutz, W. (2017). The human core of the shared socioeconomic pathways: Population scenarios by age, sex and level of education for all countries to 2100. *Global Environmental Change, 42*, 181–192. doi:10.1016/j.gloenvcha.2014.06.004 PMID:28239237

Sawe, B. E. (2017). *Countries Most Dependent on Agriculture*. Retrieved from https://www.worldatlas.com/articles/countries most-dependent-onagriculture.html

Shafian, S., Rajan, N., Schnell, R., Bagavathiannan, M., Valasek, J., Shi, Y., & Olsenholler, J. (2018). Unmanned aerial systems-based remote sensing for monitoring sorghum growth and development. *PLoS One, 13*(5), e0196605. doi:10.1371/journal.pone.0196605 PMID:29715311

Shi, X., An, X., Zhao, Q., Liu, H., Xia, L., Sun, X., & Guo, Y. (2019). State-of-the-art internet of things in protected agriculture. *Sensors (Basel), 19*(8), 1833. doi:10.339019081833 PMID:30999637

Sugiura, R., Noguchi, N., & Ishii, K. (2005). Remote-sensing technology for vegetation monitoring using an unmanned helicopter. *Biosystems Engineering*, *90*(4), 369–379. doi:10.1016/j.biosystemseng.2004.12.011

Suhonen, J., Kohvakka, M., Kaseva, V., Hämäläinen, T. D., & Hännikäinen, M. (2012). *Low-power wireless sensor networks: protocols, services and applications*. Springer Science & Business Media. doi:10.1007/978-1-4614-2173-3

Sundmaeker, H., Verdouw, C., Wolfert, S., Freire, L. P., Vermesan, O., & Friess, P. (2016). Internet of food and farm 2020. In *Digitising the Industry-Internet of Things connecting physical, digital and virtual worlds*. River Publishers.

Sylvester, G. (Ed.). (2018). E-agriculture in action: drones for agriculture. Food and Agriculture Organization of the United Nations and International Telecommunication Union.

Torres-Ruiz, M., Juárez-Hipólito, J. H., Lytras, M. D., & Moreno-Ibarra, M. (2016, July). Environmental noise sensing approach based on volunteered geographic information and spatio-temporal analysis with machine learning. In *International Conference on Computational Science and Its Applications* (pp. 95-110). Springer. 10.1007/978-3-319-42089-9_7

Torres-Sánchez, J., López-Granados, F., De Castro, A. I., & Peña-Barragán, J. M. (2013). Configuration and specifications of an unmanned aerial vehicle (UAV) for early site specific weed management. *PLoS One*, *8*(3), e58210. doi:10.1371/journal.pone.0058210 PMID:23483997

Tzounis, A., Katsoulas, N., Bartzanas, T., & Kittas, C. (2017). Internet of Things in agriculture, recent advances and future challenges. *Biosystems Engineering*, *164*, 31–48. doi:10.1016/j.biosystemseng.2017.09.007

Vellidis, G., Garrick, V., Pocknee, S., Perry, C., Kvien, C., & Tucker, M. (2007, June). How wireless will change agriculture. In *Precision Agriculture '07–Proceedings of the Sixth European Conference on Precision Agriculture (6ECPA), Skiathos, Greece* (pp. 57-67). Academic Press.

Villa-Henriksen, A., Edwards, G. T., Pesonen, L. A., Green, O., & Sørensen, C. A. G. (2020). Internet of Things in arable farming: Implementation, applications, challenges and potential. *Biosystems Engineering*, *191*, 60–84. doi:10.1016/j.biosystemseng.2019.12.013

Vroegindeweij, B. A., van Wijk, S. W., & van Henten, E. (2014). *Autonomous unmanned aerial vehicles for agricultural applications*. Academic Press.

Vu, Q., Raković, M., Delic, V., & Ronzhin, A. (2018, September). Trends in development of UAV-UGV cooperation approaches in precision agriculture. In *International Conference on Interactive Collaborative Robotics* (pp. 213-221). Springer. 10.1007/978-3-319-99582-3_22

Walter, A., Finger, R., Huber, R., & Buchmann, N. (2017). Smart farming is key to developing sustainable agriculture. *Proceedings of the National Academy of Sciences of the United States of America, 114*(24), 6148–6150. doi:10.1073/pnas.1707462114 PMID:28611194

Wolfert, S., Ge, L., Verdouw, C., & Bogaardt, M. J. (2017). Big data in smart farming–a review. *Agricultural Systems, 153*, 69–80. doi:10.1016/j.agsy.2017.01.023

Wong, S. (2019). Decentralised, off-grid solar pump irrigation systems in developing countries—Are they pro-poor, pro-environment and pro-women? In *Climate change-resilient agriculture and agroforestry* (pp. 367–382). Springer. doi:10.1007/978-3-319-75004-0_21

Zavatta, G. (2014). Agriculture Remains Central to the World Economy, 60% of the population Depends on Agriculture for Survival. Expo2015, Milan, Italy.

Zhai, Z., Martínez Ortega, J. F., Lucas Martínez, N., & Rodríguez-Molina, J. (2018). A mission planning approach for precision farming systems based on multi-objective optimization. *Sensors (Basel), 18*(6), 1795. doi:10.339018061795 PMID:29865251

Zhang, J., Basso, B., Price, R. F., Putman, G., & Shuai, G. (2018). Estimating plant distance in maize using Unmanned Aerial Vehicle (UAV). *PLoS One, 13*(4), e0195223. doi:10.1371/journal.pone.0195223 PMID:29677204

Zhang, Y., Wang, L., & Duan, Y. (2016). Agricultural information dissemination using ICTs: A review and analysis of information dissemination models in China. *Information Processing in Agriculture, 3*(1), 17–29. doi:10.1016/j.inpa.2015.11.002

Zheng, R., Zhang, T., Liu, Z., & Wang, H. (2016). An EIoT system designed for ecological and environmental management of the Xianghe Segment of China's Grand Canal. *International Journal of Sustainable Development and World Ecology, 23*(4), 372–380. doi:10.1080/13504509.2015.1124470

Chapter 3
Recommendation of Crop and Yield Prediction by Assessing Soil Health From Ortho-Photos

J Dhalia Sweetlin
Anna University, Chennai, India

Visali A. L.
Anna University, Chennai, India

Sruthi Sreeram
Anna University, Chennai, India

Jyothi Prasanth D. R.
Anna University, Chennai, India

ABSTRACT

Agriculture is considered to be the driving force of the Indian economy. Production of crops is considered to be one of the complex phenomena as they are influenced by the agro-climatic parameters. From novice to experienced farmers, at times, fail to figure out the suitable crop for their lands, leading to financial loss. This is because of the dynamic change in soil nutrient levels and climatic conditions. Hence, it is important to predict crops according to the presence of the nutrients in a land. Recommending the crops to a farm after considering the nutrients levels of the soil and predicting the yield will largely help the landowner in taking necessary steps for marketing and storage in the future. These results will further assist the industries to plan the logistics of their business who are working in partnership with these landowners. In this work, pH and other soil nutrients are estimated from the input ortho images to recommend crops that can grow well under the given circumstances.

DOI: 10.4018/978-1-7998-8763-8.ch003

INTRODUCTION

India, being an agricultural land ranks second worldwide in agriculture outcome. But it is worth mentioning that the gross domestic product share is dropping. Agriculture sector contributes around 17% of the GDP and provides employment to more than 60% of the population as many of the rural households depend on agriculture (Arjun, 2013; Pawar & Chillarge, 2018). Hence the advances pertinent to technology in this era have been used to study the soil to introduce crops that can help increase the outcome from agriculture, as this can have an impact over GDP. Soil forms the uppermost layer of the earth and supports natural vegetation. For agriculture, soil supports the plants by acting as a preserver of water and nutrients. It is known for maintaining a balance in the ecosystem. All the properties of soil play a vital role in crop production. Out of the physical, chemical and biological properties of soil, the pH value forms the paramount criterion for crop sowing. This is because the nutrient levels of soil depend on the pH value (Khadka & Lamichhane, 2016). Plants consume macronutrients like Nitrogen (N), Potassium (K) and Phosphorus (P) along with a few other secondary nutrients.

The term 'precision agriculture' is simply the application of geospatial sensors and methodologies for detecting discrepancies in the field and fixing them with appropriate solutions. Unmanned aerial vehicles (UAV) are utilized as precious tools in precision agriculture as they have the ability to collect high resolution images. When these localized images collected by the drones are integrated after undergoing the rectification process, they form the ortho-photos or ortho-mosaics (Oliveira, 2018).

In this work, unsupervised image segmentation by backpropagation neural network is used for segmenting the soil section eliminating the trees and buildings from the ortho-photos. Multiple Linear regression models are developed to estimate the pH and nutrient value of the segmented soil images. Support Vector Machine (SVM) classifier is used to classify the nutrients levels into appropriate labels. Recommender System suggests crops that might be suitable for the land based on the similarity in the soil profiles.

BACKGROUND

Oliveira et al. (2018) proposed a method to detect areas that do not contain plant saplings in the coffee crops plantation using ortho-photos. To facilitate decisions, both the positions of failures and total failure length is obtained from the ortho-photos.

Ciriza et al. (2017) proposed a method to automatically detect uprooted orchards by using textural information from ortho-photos in order to reduce photo interpretation needed to update the agricultural database [AGDB]. Five textural features were selected based on GLCM and Wavelet planes. This methodology reduced the photo interpretation by 60-85% and achieved an accuracy of 85%.

Afrin et al. (2018) proposed a method to apply data mining techniques to predict the yields of staple crops of Bangladesh. Soil factors like pH, nutrients, organic substances and climatic data of the past six years were utilized to determine soil health. Tewari et al. (2013) proposed a method to estimate plant nitrogen content using digital image processing techniques. Observations include the average nitrogen consumption and the chlorophyll content. R,G,B and normalized R and G values were acquired and Regression models were built which had a correlation coefficient of 0.948.

Beycioglu et al. (2017) proposed a method to predict the pH of a solution by only using its images. Lee Nelan (2015) devised a method for identifying the soil characteristics from the ground images or soil maps through an image processing system. Pawar and Chillarge, (2018) proposed an application that can help farmers get literate about the soil in their land. J48 decision tree algorithm was utilized as a classifier. The devised system informs the farmers about the toxicity level of the soil, because of the environmental pollution due to fast industrialization. Boken(2016) examined the potential of the soil moisture estimates produced by satellite-based microwave sensors. This work concentrated on two areas namely Nebraska and California in the United States. He concluded that the soil moisture estimates are a vital source of the data that can be used for improvising the crop yield prediction models.

Veenadhari et al. (2014) has proposed a prediction model that can predict the crop yield well ahead of its harvest period, helping the farmers and other policy makers for taking necessary measures for marketing, storage and transportation. A software tool called 'Crop Advisor' has been developed along with a user-friendly interface for determining the influence of climatic parameters on the crop production.

Pudumalar et al. (2016) developed a crop recommendation system using data mining by deriving useful information from the agricultural data. Saeed Khaki and Wang (2019) developed crop yield prediction by taking into account the factors genotype, environment and their interactions. This system was developed using deep neural networks. Their model had an accuracy with RMSE 12% of the average yield. This method has outperformed Lasso, shallow neural networks and regression tree.

Chlingaryan et al. (2018) had reviewed the machine learning approaches for crop yield prediction and nitrogen estimation in precision agriculture. The work comprises remote sensing technologies for crop yield estimation and decision making along with the recent machine learning based techniques for crop yield prediction and nitrogen estimation.

Abhinav Sharma et al. (2021) presented a comprehensive review on how knowledge-based agriculture has the potential to increase product quality and productivity in the long run. The authors focused on soil factors such as carbon, moisture level, crop yield prediction, disease and weed detection in their review. The regression techniques, according to the author, serve as the foundation for soil qualities and meteorological parameters.

In this work, a crop recommendation model is developed to assess the quality of the soil by classifying the soil from the ortho images and recommending suitable crops.

SYSTEM FRAMEWORK

This section describes the modules involved in the proposed work. The modules include region of interest segmentation from orthophotos, prediction of pH and nutrient values, classification of the nutrient records and recommendation of the crops. Figure 1 shows the system framework of the crop recommendation and yield prediction system. Soil regions are segmented, eliminating the trees, weeds and buildings, through unsupervised image segmentation using backpropagation. The segmented regions are classified into different soil types such as Red, Black, Brown, Laterite and Coastal (National Portal of India,2021; Chandra Shekara et al., 2016).

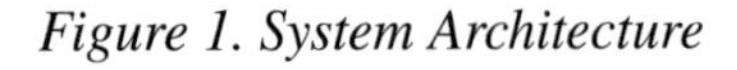

Figure 1. System Architecture

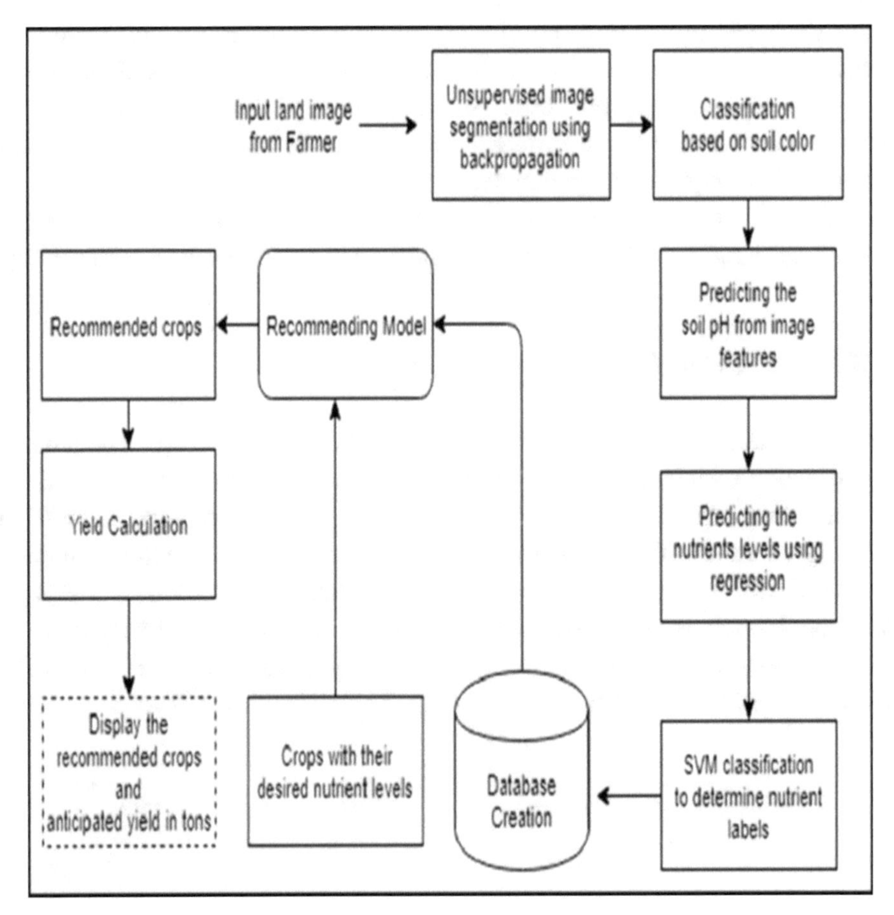

With respect to the boundaries marked as a result of segmentation, areas of the segmented regions are calculated to help find out the yield. Regions are labelled using different colors according to the category they fall in, and the mean red, green and blue values are extracted to calculate the pH index value. From the pH index, macronutrients like Nitrogen, Phosphorus and Potassium are estimated using linear regression.

Unsupervised Image Segmentation Using Backpropagation

Convolutional Neural Network is deployed to extract the pixel level features from the given ortho-images (Kanezaki, 2018). This is taken as the required output, and backpropagation is carried out by finding the error between the predicted cluster label for each pixel and the actual label.

Classification Based on Soil Color

A Convolution Neural Network is used to extract the pixel level features of the segmented ortho images. Images are classified as Black soil (type 0), Red soil (type 1), Brown soil (type 2), Laterite soil (type 3) and Coastal soil (type 4). Segment number, soil type, area, mean R, mean G, mean B are obtained from the segmented image. Every segment is given a segment id which can be used to plant the recommended crops. Area of each segment is also calculated to determine the yield that might be possible. The mean RGB values obtained are used to predict the pH values with the multiple linear regression model.

Prediction with Regression Models

An extension of simple linear regression that is used to predict an output variable 'y' on the basis of many other unique predictor variables 'x_1', 'x_2' etc. is multiple linear regression. A model with three predictor variables and one single response variable is given by the equation (1),

$$pH = b_0 + b_1R + b_2G + b_3B \quad (1)$$

Where b_0, b_1, b_2 and b_3 are the regression weights and R, G, B are the mean red, green and blue values (Kassambara et al., 2018). Simple linear regression models are used to find the relationship between pH and macronutrients. Multiple Linear Regression models are used to find the relationship between pH and the mean RGB values. pH was the response variable and the features extracted from the image were the predictors. The model was trained with the features of the image and their pH values in order to predict the pH for a query image. To interpret the multiple regression analysis, it is important to scrutinize the F-statistic and the associated p-value which is given by the summary of the model. The predictor variables are significantly related to the output variable if the p-value is less than 2.2 e-16 (Kumar et al., 2014). The features extracted from the image had a strong association with the pH value. Any attribute that is not considered worthy in prediction can be removed. For agricultural lands, the soil pH mostly varies from 6 to 8 (Chandra Shekara et al., 2016). The nutrient levels were predicted from the obtained pH values using simple linear regression (SLR). In SLR, the pH was the predictor variable and every macro nutrient N, P and K were considered the response variable. In this work all the macronutrients were positively correlated with the pH within the range of 4.0 to 8.5. This can be observed from the soil nutrients and pH availability chart suggested by Chandra Shekara et al. (2016)

The nutrients values dwindled in the extremes i.e., in the strong acidic and alkali regions highlighting that the lands are not suitable for cultivation. The overall quality of the model built can be assessed by the r-squared and the residual standard error (RSE). The value of r-square ranges from 0 to 1. In MLR, this value represents the coefficient of correlation between the observed values of pH and the fitted values. RSE gives a measure of the error of the prediction. The model is more reliable with a less RSE. The error rate of the model is determined by dividing the RSE by the mean outcome variable.

Classification of Nutrients

SVM is a supervised learning model that is associated with learning algorithms that analyze data used and can be made use in classification problems (DataCamp Community, 2019). Radial basis kernel function is used to classify the nutrient records. The performance of the model can be increased by tuning its parameters. All the pH and the corresponding nutrient levels with labels are fed to the classifier. In this work, 223 soil instances with pH ranging from 4 to 9 and nutrients values in mg/kg were given. The classifier is proved reliable by analyzing the hit and miss count in the test phase. This labelled nutrient database is used to build the recommendation module.

Building a Recommendation Model

Crops based collaborative approach is a model-based algorithm for recommending crops. Model based algorithms consist of building a model based on the dataset on crop yields. This approach offers both speed and scalability. In this algorithm, the similarity values between crops are calculated by observing all the lands that have given a good yield of these crops. This is obtained using cosine similarity measures. The similarity values are used to predict the crops for new queries (Kawasaki & Hasuike, 2017).

Cosine based similarity, also known as the vector-based similarity (Han et al., 2012) is computed using equation (2). Any two crops and their yields are taken as vectors and the similarity between them is defined as the angle between these vectors.

$$sim(i,j) = \cos\cos(\vec{i}.\vec{j}) = \frac{\vec{i}.\vec{j}}{\|\vec{i}\|^2 . \|\vec{j}\|^2} \quad (2)$$

All the crops that are similar to the target crop are acquired from the similarity matrix and prediction is carried out. From this set, the crops for which the land has given a good yield alone are taken. The weight for each crop is given by the similarity between the crop in the list and the target crop. A threshold yield value is set for those crops that don't have more yield scores.

Crops that satisfy most of the state's needs are collected from the past crop production dataset (National Portal of India,2021). The different nutrient levels that these crops require are collected and arranged in a form accessible for analyzing. Crop specific details like their common name, scientific name, and desired nutrient levels to give optimal growth, number of heads that could be planted in per meter square and number of grains that can be extracted, which are notable attributes were acquired. Having identified the nutrient levels, the yield attribute helps in conveying on what nutrient level the yield is better, which acts like a score. If crop A and B had given high yields to land L1 and L2, then crops like A and B (similar in nutrient requirements), can be recommended to lands similar to L1 and L2. This is a crops-based collaborative recommendation. By doing this, possible crops for the land are identified and also the ones that might give high yields, thus helping to work on the profits.

IMPLEMENTATION AND RESULTS

Segmentation and Feature Extraction

Segmentation was done on the ortho-photos collected through the USGS satellite imagery for selected areas in Tamil Nadu (Kanezaki, 2018). The network used for performing the unsupervised image segmentation consists of a block of 2D convolution, ReLU, and batch normalization layers repeated for M times followed by a block of 1D convolution, batch normalization, and Argmax classification to classify each pixel into the corresponding cluster. When the query image is given as input to the model, the algorithm predicts the cluster labels for each pixel with fixed network parameters and trains the CNN model with the cluster labels identified through super-pixel refinement using SLIC. The network is trained such that all the pixels in a superpixel are allocated to the same cluster. Three major soils are widely found in the state according to the Tamil Nadu Agricultural University and they are Red, Black and Laterite Soils (National Portal of India,2021). Parts of the images have vegetation and few other noises like trees and buildings, which have to be extracted from the region of interest. As the segmentation method is unsupervised, the ground truth values are given by the simple linear iterative clustering algorithm that clusters super-pixels, which are a group of pixels that share similar characteristics (Kanezaki, 2018). The backpropagation network makes use of the ground truth to classify the pixel into the right cluster. The segments are correctly marked when the error between the predicted and actual cluster label is as shown in Figure 2 and Figure 3.

Figure 2. Segmented Image

Figure 3. Input Ortho-photo

A classifier model labels the type of the segmented regions based on the features extracted. It helps in classifying the regions as red (type 0), black (type 1), laterite soils (type 2), coastal (type 3) and others. If a segment fails to fall in type 0, 1, 2 and 3 it is put into the others folder (trees, weeds and buildings). The quantitative information collected from the image is illustrated in TABLE 1.

Table 1. Values extracted from the region of interest

Sl No	Color Assigned	Area	R	G	B
1	Dark green	65.225	178	162	149
2	Violet	139.561	117	95	90
3	Dark Pink	173.713	152	120	110
4	Orange	606.661	131	118	111
5	Light green	963.654	181	175	161

Every segment is marked with a segment id and its type in order to traverse back to have an idea of where the recommended crops are to be planted. Area of the segment is also calculated to find the yield that might be possible after having sowed the offered crops. Dots per inch are taken from the properties of the image to find out the area. Orthophotos considered in this work, possessed 96 dpi. One pixel is measured to be 0.264 mm. So, the product of the number of pixels in the area and the area of one pixel gives the total area of the segment. Using the mean red, mean green and mean blue values taken from the region, the multiple linear regression model predicts the pH value.

Prediction using Regression

Soil samples with their measured pH values from the laboratory were taken for the purpose of prediction (Kumar et al.,2014). The MLR model was deployed to learn the relationship between the red, green and blue values (independent variable) and the pH of the sample (dependent variable). The training and the test set was split in an 80-20 ratio. TABLE 2 shows the actual and the predicted pH values of the test set. Multiple r-squared and adjusted r-squared values were 0.9463 and 0.9428 respectively.

Table 2. Actual and Predicted pH values

Index	Actual Values	Predicted Values
1	7.318242	7.29
2	6.982465	7.11
3	6.821710	6.85
4	7.427179	7.44
5	7.642593	7.59
6	7.330362	7.32
7	6.816855	6.90

The soil nutrients dataset as in TABLE 3, collected from 'PANGEA' contains the nutrient levels analyzed for over 600 soil samples (Neelan, 2015). Notable attributes included the soil pH, Nitrogen, Potassium and Phosphorus levels in mg/kg. This helped to identify the nutrient levels based on the pH value. The dataset was labelled to help classification, based on the guidelines given in the farmer's handbook by the Organization of Ministry of Agriculture and Farmers welfare namely deficient, insufficient, sufficient, excess and toxic (Chandra Shekara et al.,

2016). Figure 4 shows the relationship between the pH and the nitrogen values of the samples considered. It can be seen that the pH and nitrogen are positively correlated in the range of 4 to 9. Nitrogen samples range from 1 mg/kg to 25 mg/kg. Figure 5 shows the relationship between the pH and Phosphorus. pH and Phosphorus have a positive correlation and the multiple r-squared value was 0.9 proving the prediction can be a reliable one. Similarly Figure 6 shows how the pH is associated with the Potassium values. All the macronutrients are positively correlated within the pH range 4 to 9 and they have negative correlation in the strong acidic (less than 4) and the strongly basic (above 9) regions.

Figure 4. Plot b/w pH and N

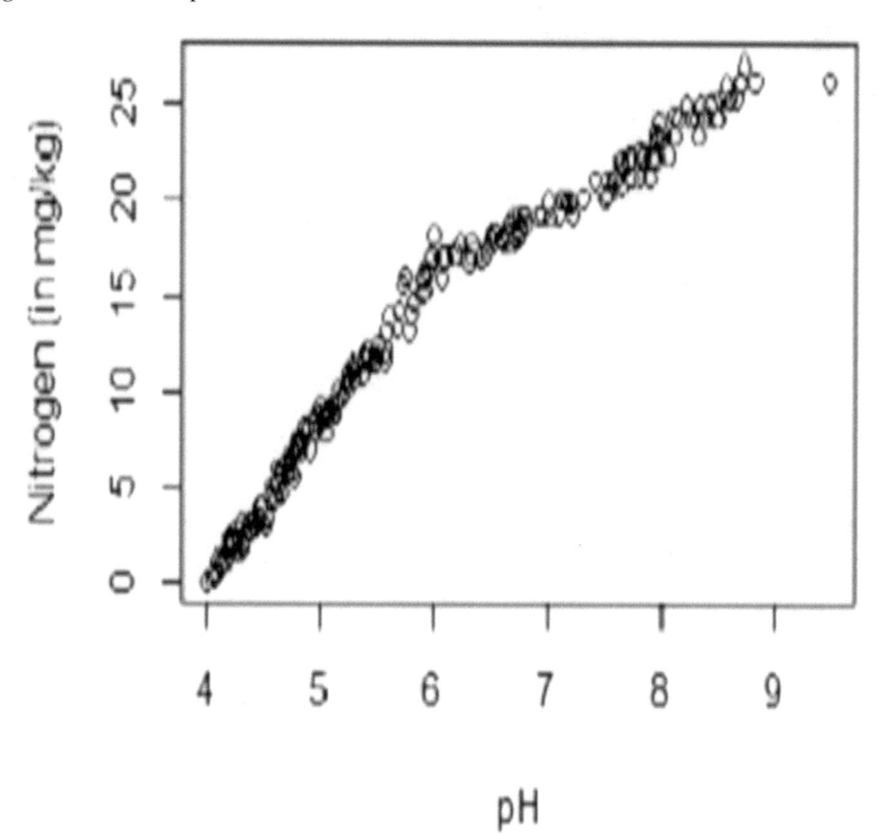

Figure 5. Plot b/w pH and P

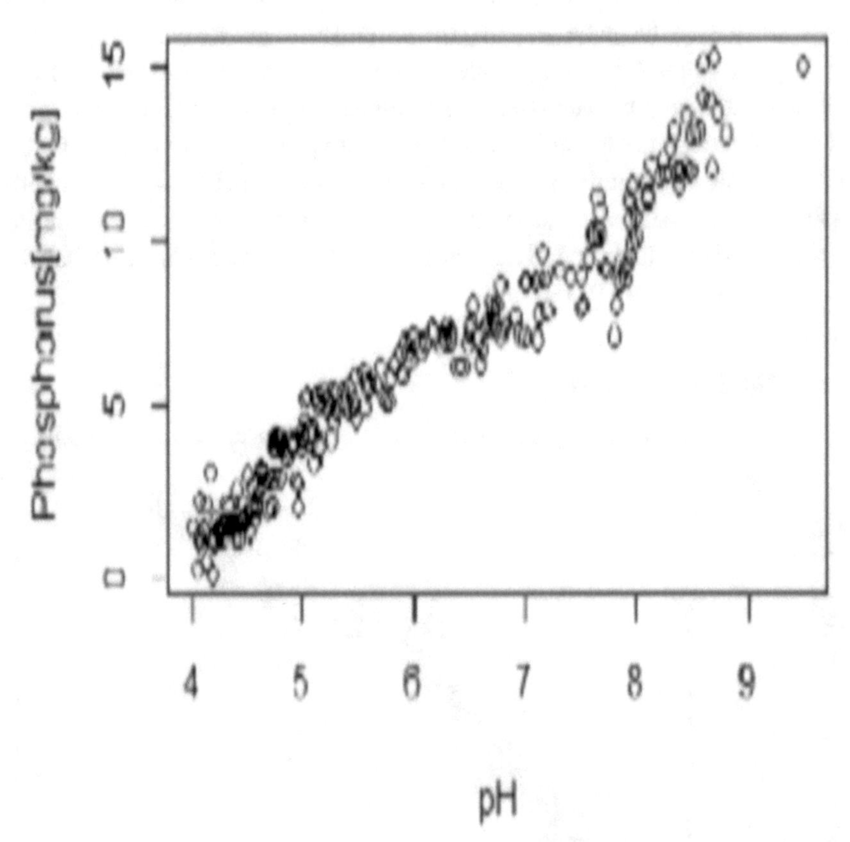

Figure 6. Plot b/w pH and K

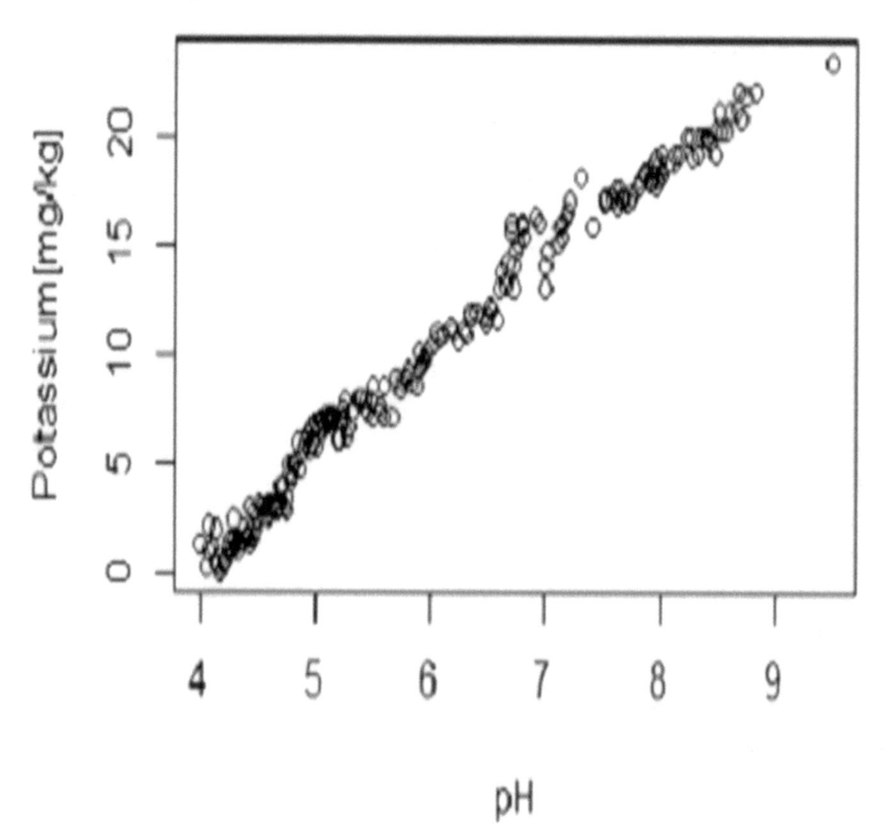

Table 3. Actual vs Predicted values in classification

Actual / Predicted	Deficient	Insufficient	Sufficient	Excess	Toxic
Deficient	35	0	1	0	0
Insufficient	0	5	0	0	4
Sufficient	0	0	22	0	0
Excess	0	1	0	6	0
Toxic	0	0	0	0	0

For all the segments identified, nutrients levels are estimated so as to create a nutrient database. It consists of the segment number, pH value, macronutrients levels and type of the soil as shown in Table 4.

Table 4. Nutrients database

S.No	AREA	pH	N	P	K	Label
1	0.042	7.68	22.26	10.136	17.9	S
2	0.058	7.01	18.69	8.5114	14.4	I
3	0.03	7.6	18.69	9.9426	17.1	S
4	50.27	7.55	21.83	9.8255	16.9	S
5	12.66	6.38	21.58	6.9927	11.6	I

The nutrients records are classified using the SVM. The class label was determined by the model using the pH and the macro nutrient levels. 62 support vectors were constructed in order to separate the records 12,21,12,12 and 5 for deficient, insufficient, sufficient, excess and toxic respectively. Radial basis kernel function was used and the gamma value was set to 0.25. A classification accuracy of 91.89% was achieved when the test set was given to classify.

Recommendation by Collaborative Filtering

Collaborative models make use of the given data by looking into the dataset how the crops have responded with yields to the different nutrient levels. It takes in the information about the crops-nutrient levels and their yields in tons. If the crop is able to grow in one or more nutrient levels, then it is recommended for the land that has nutrient level for which the crop has given better yields. Hence this model aims at collaboratively recommending crops based on the yields. Few other crops with guidelines can be recommended, so that the farmer can change the soil pH by adding organic manures to prepare the land to give out profits. Top 3 crops are offered by the model which is presented according to the yield levels. Yield is calculated as given in equation (3) which is suggested by Pawar and Chillarge (2018).

$$\frac{Yield\ Y(AHS)}{104} \tag{3}$$

where A is the area of the land in m^2, H is the number of saplings that could be planted per m^2 and S is weight of the seed in mg. The result is given in tons/hectare and is recorded in Table 5.

Table 5. Crops offered along with yield prediction

Segment_id	Crop_name	Yield
1	Dry Ginger Cardamom Brinjal	283.17968 14.3868 0.169068
2	Crab-apple Kiwi Peanut	0.06791 0.01480 0.01258
3	Pineapple Apple Corn-sweet	22.1224 0.0665 0.1256
4	Tomato Onion-Dry Banana	99.7731 2.76170 0.01456
5	Tomato Tamarind Kiwi	99.7731 7.47845 0.01480

The recommendation model built is evaluated using a 5-fold cross validation method. The dataset is split into five sets and each set acts as both a testing and training set. This helps to know the effectiveness of the recommendation system. Table 5 shows the crops recommended by the system developed along with the estimated yield.

CONCLUSION AND FUTURE WORK

It is possible to affirm that the methodology proposed is feasible and can be used for filtering out the crops most suitable for the land. Unsupervised image segmentation using backpropagation helps to extract the region of interest completely without any sorts of noise. It is very costly to determine the pH and soil nutrients from soil testing laboratories and so existing datasets were requested from similar works to estimate nutrient levels. Recommending crops collaboratively will help to match the demand by giving out more new crops according to the demand. Best three crops with yields are produced and it is left for the cultivators to choose their own as they wish depending on other criteria. Recommendations can be made giving importance to any of the attributes like yields, demands, climatic conditions. The work can be implemented in the form of mobile applications to make it available for farmers. The yield calculation is approximate and can assist the industries to know the yield that can be expected from the farm. Future work can be to extend the recommendation phase by adding more vital parameters to the model to train, contributing significant advancement in the agriculture sector.

REFERENCES

Afrin, S., Khan, A. T., Mahia, M., Ahsan, R., Mishal, M. R., Ahmed, W., & Rahman, R. M. (2018). Analysis of Soil Properties and Climatic Data to Predict Crop Yields and Cluster Different Agricultural Regions of Bangladesh. *IEEE/ACIS 17th International Conference on Computer and Information Science (ICIS).* 10.1109/ICIS.2018.8466397

Arjun, K. M. (2013). Indian Agriculture - Status, Importance and Role in Indian Economy. *International Journal of Agriculture and Food Science Technology*, *4*(4), 343–346.

Beycioglu, A., Comak, B., & Akcaabat, D. (2017). Evaluation of pH value by using image processing. *Acta Physica Polonica A*, *132*(3-II), 1142–1144. doi:10.12693/APhysPolA.132.1142

Boken, V. K. (2016). Potential of soil—moisture-estimating technology for monitoring crop yields and assessing drought impacts-case studies in the United States. *IEEE Technological Innovations in ICT for Agriculture and Rural Development (TIAR).*

Chlingaryan, A., Sukkarieh, S., & Whelan, B. (2018). Machine learning approaches for crop yield prediction and nitrogen status estimation in precision agriculture: A review. *Computers and Electronics in Agriculture*, *151*, 61–69. doi:10.1016/j.compag.2018.05.012

Ciriza, R., Sola, I., Albizua, L., Álvarez-Mozos, J., & González-Audícana, M. (2017). Automatic Detection of Uprooted Orchards Based on Orthophoto Texture Analysis. *Remote Sensing*, *9*(5), 492. doi:10.3390/rs9050492

Han, J., Kamber, M., & Pei, J. (2012). *Data mining: concepts and techniques*. Elsevier.

Kanezaki, A. (2018). Unsupervised Image Segmentation by Backpropagation. *IEEE International Conference on Acoustics, Speech and Signal Processing (ICASSP).*

Kassambara, L., Sara, M. F., & Kassambara, V. (2018, March 10). *Multiple Linear Regression in R.* STHDA. http://www.sthda.com/english/articles/40-regression-analysis/168-multiple-linear-regression-in-r

Kawasaki, M., & Hasuike, T. (2017). A recommendation system by collaborative filtering including information and characteristics on users and items. *IEEE Symposium Series on Computational Intelligence (SSCI)*. 10.1109/SSCI.2017.8280983

Khadka & Lamichhane. (2016). The Relationship between Soil pH and Micronutrients, Western Nepal. *International Journal of Agriculture Innovation and Research*, *4*(5).

Khaki, Saeed, & Wang. (2019). Crop yield prediction using deep neural networks. *Frontiers in Plant Science, 10.*

Kumar, V., Vimal, B. K., Kumar, R., Kumar, M., & Kumar, M. (2014, January). Determination of soil pH by using digital image processing technique. *Journal of Applied and Natural Science*, *6*(1), 14–18. doi:10.31018/jans.v6i1.368

Lee Nelan, A. (2015). *Image Processing system for soil characterization*. Academic Press.

National Portal of India. (2021). https://www.india.gov.in/topics/agriculture

Oliveira, H. C., Guizilini, V. C., Nunes, I. P., & Souza, J. R. (2018). Failure Detection in Row Crops From UAV Images Using Morphological Operators. *IEEE Geoscience and Remote Sensing Letters*, *15*(7), 991–995. doi:10.1109/LGRS.2018.2819944

Pawar & Chillarge. (2018). Soil Toxicity Prediction and Recommendation System Using Data Mining In Precision Agriculture. *3rd International Conference for Convergence in Technology (I2CT).*

Pudumalar, S., Ramanujam, E., Rajashree, R. H., Kavya, C., Kiruthika, T., & Nisha, J. (2017). Crop recommendation system for precision agriculture. *Eighth International Conference on Advanced Computing (ICoAC)*. 10.1109/ICoAC.2017.7951740

Sharma, Jain, Gupta, & Chowdary. (2021). Machine Learning Applications for Precision Agriculture: A Comprehensive Review. Institute of Electrical and Electronical Engineers.

Shekara, Balasubramani, & Shukla. (2016). *Farmer's Handbook on Basic Agriculture*. Agricultural Department and Sciences.

Support Vector Machines in R. (n.d.). *DataCamp Community*. https://www.datacamp.com/community/tutorials/support-vector-machines-r

Tewari, V. K., Kumar, A. A., Kumar, S. P., Pandey, V., & Chandel, N. S. (2013). Estimation of plant nitrogen content using digital image processing. *Agricultural Engineering International: CIGR Journal*, *15*(2), 78–86.

Veenadhari, S., Misra, B., & Singh, C. (2014). Machine learning approach for forecasting crop yield based on climatic parameters. *International Conference on Computer Communication and Informatics*. 10.1109/ICCCI.2014.6921718

KEY TERMS AND DEFINITIONS

Agriculture: It is the art of growing crops. Many families in India depend on agriculture.

Classification: Categorization of objects into groups based on the existing ground truth values. Artificial Intelligence based algorithms classify objects easily.

Collaborative Filtering: This method can be used to predict the plantation of a suitable crop that can yield profit to the requesting user by collecting and analyzing relevant agricultural data in different types and from different sources.

Crop Recommendation: Suggestions about crops that are given to agriculturalists based on the pH values, types of soil, the nutrient contents present in the soil and other characteristics. The crops planted based on the suggestions yield maximum profit.

Image Segmentation: This technique is a pre-processing activity that is carried out in image processing. This partitions the image into multiple segments and identifies the different bounded regions present in the image using computer algorithms.

Soil Color: Soil color refers to the composition of the soil. It might be due to the minerals present in the soil. Soil exhibits different colors like black, red, brown, etc.

Soil Nutrients: The three important nutrients present in the soil are Nitrogen (N), Phosphorous (P), and Potassium (K) which are popularly quoted together as NPK values.

Unsupervised Segmentation: Algorithms that understand the different bounded segments that are present in an image by itself without referring to the ground truth.

Chapter 4
Design and Implementation of an Amphibious Unmanned Aerial Vehicle System for Agriculture Applications

Arun Kumar Manoharan
GITAM University, India

Mohamed Ismail K.
Agni College of Technology, India

Nagarjuna Telagam
https://orcid.org/0000-0002-6184-6283
GITAM University, India

ABSTRACT

Today, drone systems have become an emerging technology for agriculture applications as an unmanned aerial vehicle (UAV). They help the farmers in crop monitoring and production. They are used to reduce human resources and to control pollution in the agriculture field. In real-time, drones are suitable for working in the agriculture field during strong winds and even in various climate conditions. This chapter proposes an amphibious unmanned aerial vehicle (UAV) system design and implementation for agriculture applications. Drones are useful to avoid deforestation in India. The estimated simulation results are used to calculate the drones' efficiency using their weight, flying time, and power consumption. In this chapter, three different UAV system phases have been discussed (i.e., design of drones, the building of payload, and evaluation of drone using the software). This chapter helps the beginners understand the necessary calculations of the drone design along with thrust values, select the propellers sizes, and calculate the drone's flying time, stability, and power consumption.

DOI: 10.4018/978-1-7998-8763-8.ch004

INTRODUCTION

In present days, Unmanned Aerial Vehicles (UAVs) are the emerging technology in the modern technological world. The UAVs are also called Drones, and an aircraft system operated without a human pilot. The drones are smaller than the commercial aircraft system, which can be performed in an autopilot mode (Tezza and Anjudar, 2019). A ground-based controller controls the onboard electronic components of the drones via wireless communication. The drones can be operated either through remote control by a human or autonomously by programmed computers. Self-automated drones are employed in recent advancements which carried out missions without human involvement like surveillance systems, artificial intelligence, cloud computing, machine, and deep learning. In earlier days, drones were designed and implemented in military applications (Mogili and Deepak, 2018). UAVs have been used in various commercial and agricultural applications in recent days. Future UAV-based networks are needed to supply high data rates, security, range, and dynamicity. During this context, operative UAVs victimization the coming tactile web surroundings and low latency 5G networks will solve network coverage and data rates (Hayat et al., 2016).

Furthermore, the employment of computer code outlined Networking (SDN), Network performs Virtualization (NFV), and Intent-based Networking (IBN) will solve the difficulty of dynamic network management? The applications of UAVs are speedily increasing in most civilian domains. Air taxis, Food drones, drones for medication delivery are several UAVs' main civilian applications. It's imperative to verify the UAVs' genuineness and operations in such applications that Blockchain technology may be an excellent resolution. UAVs' power constraints may be self-addressed by recharging batteries on the go with solutions like star panels or wireless charging. Besides, higher algorithms may be enforced to create the UAV computations additional energy economical (Mozaffari et al., 2016). Most sensible applications of UAVs would usually need a swarm of the many drones instead of one drone. Correct management, cooperation, and autonomy of such hives would need numerous computing and Machine Learning algorithms. The geographic vary of operations, clearance of access to civilian and military airspaces, network coverage, the period of flight, security needs, autonomy, among several others, are several the problems that should be self-addressed before such applications become a reality. Broadly, these are the class of measurability and security. This special issue explores application-specific UAV platforms that use novel techniques for multiple new applications that are scalable and secure (Alzenad et al., 2017).

The UAVs are real-time systems that require a fast response to changing the sensor information. Therefore, UAVs depend on single board programmed computers for their computational necessities. The agricultural applications of drones are weather analysis, monitoring crops' healthiness, spraying fertilizers, sowing seed balls, irrigation systems, etc. The drones are classified into four types based on the type of aerial platform.

Multi-Rotor Drones

The multi-rotor drones are commercially used, easy to manufacture, and more economical. The speed and flying time of these drones are limited. It is used for standard applications like photography and video surveillance systems. The flying time of these drones is around 30 minutes. Hence these drones are not suitable for larger-scale applications.

Further, these drones are classified into four types depending on the number of rotors used. The four types of Tricopter have three rotors, Quadcopter has four rotors, Hexacopter has six rotors, and Octocopter has eight rotors. Among these types, quadcopters are widely used for commercial purposes (Mogili and Deepak, 2018)

Figure 1. a. Tricopter; b. Quadcopter; c. Hexacopter; d. Octocopter

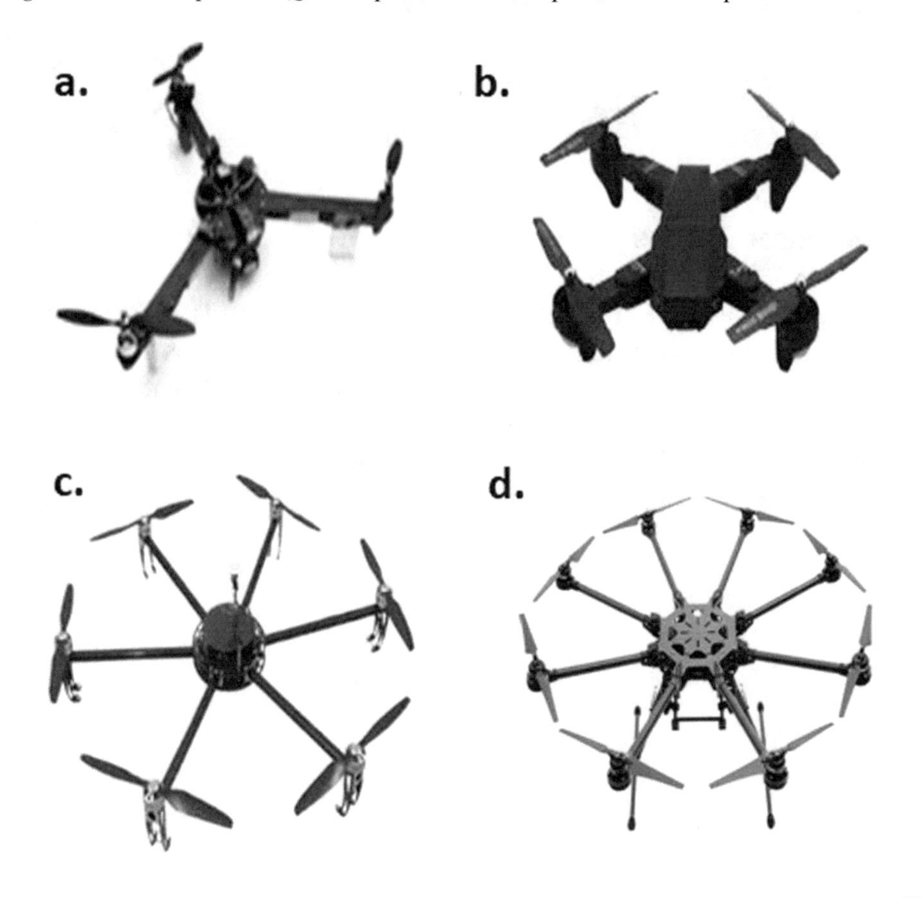

Fixed Wing Drones

The fixed-wing drones are operated remotely by humans or onboard computers. They are usually used to carry heavier loads to a longer distance. These drones are capable of flying around for a few hours. They are flying at higher altitudes with more stability. The typical applications are field surveys, surveying forest areas, topographical maps, etc. The fixed-wing drones are operated with solar energy, in which the solar panels are placed over the surface of the two wings (Pobkrut et al., 2016).

Figure 2. Fixed-wing drone

Single Rotor Drones

The design and structure of these drones are like helicopters. The drone's size is enormous compared to the multi-rotor and fixed-wing drones. Gas engines can also power single rotor drones. The flying time is high and has more efficiency than multi-rotor drones. The design of these drones is much more complex and more economical. Training is required to lift and fly them. These drones have one oversized rotor for spinning wings and a small rotor for direction and stability (Murugan et al., 2017).

Figure 3. Single rotor drone

Hybrid VTOL

It is a hybrid version of a drone that combines the benefits of fixed-wing and rotor-based models. In recent days, these types of drones are used for delivering packages due to their capability of longer flying time. These drones have gyros and accelerometers used to maintain the drone's stability in the air. They can operate either by remote or automated mode (Puri et al., 2017).

Figure 4. Hybrid VTOL

Further, the drones are categorized with three other parameters: size, range, and endurance:

By Size

- The Nanosize of the drone is up to 50 cm.
- Small – the size is less than 2 meters in length.
- Medium – smaller aircraft had a lighter weight.
- Large – Aircraft for military applications

By Range

- Close-range: coverage area is around 3 miles with an average 20-30 minutes flying time.
- Short-range: coverage area is around 30 miles away from the users and has flying time ranges from 1 hour to 6 hours.
- Mid-range: operated around 90 miles surface with up to 12 hours flying time.

By Endurance

These drones are used for surveillance systems and military applications. More endurance drones can continuously fly up to 3 days with a coverage area of around 400 miles.

BASIC MATERIALS AND METHODS TO DESIGN THE DRONE

Drones are advanced machines with different components connected to carry out the work. Each element in the drone plays another function. The material selection for each element is being considered to minimize the weight and enhance the performance. To increase the flying time and performance, the drones must be light-weighted. Low-density materials are an excellent choice to reduce the overall mass of drones. The parts of the drones are frames, motors, propellers, batteries, cameras, and sensors (Saari et al., 2017).

The drones must generate enough thrust compared to their weight to fly in the air. Hence, the selection of materials plays a vital role in drone design. The commonly used materials are carbon fiber-reinforced composites, Aluminium, Lithium-ion batteries, Thermoplastics such as polyester, nylon, polystyrene, etc. The weight of each component is considered to improve efficiency.

Software Used to Design the Drone Structure

The computer-Aided Three-Dimensional Interactive Application (CATIA) is the software used to develop the computer-aided design, computer-aided manufacturing, computer-aided engineering, and 3D formats. We can realize how to configure or reuse the working information accessible from previous plans without any preparation. It is highly used in the aerospace and defense industries for electrical and electronic system design and mechanical engineering. The structure of the drone and payload is varied depending upon the applications. This software is used to create the drone's 3D frame structure, including the top plate, bottom plate, mainframe, payload, etc. figure 5 shows the drone's isometric view (Jawad et al., 2019).

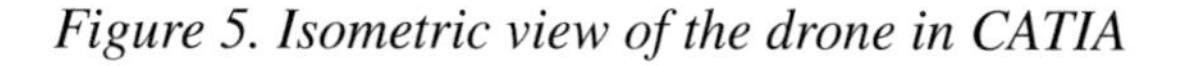

Figure 5. Isometric view of the drone in CATIA

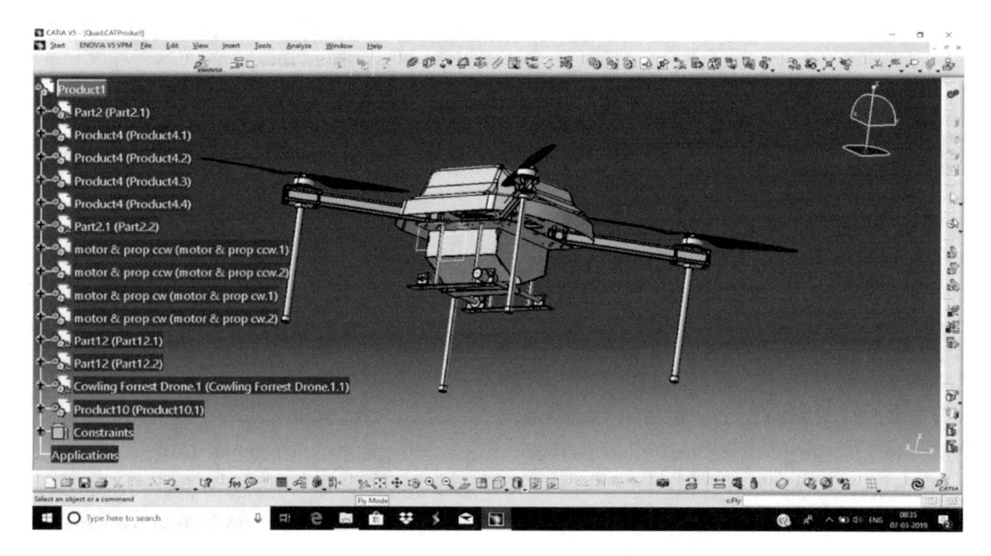

A Computer Numerical Control (CNC) router is a system programmed cutting machine that is highly used to cut various materials such as wood, composites, aluminum, plastics, glass fibers, steel, and glass. The drone parts are designed using plastics, carbon, and glass fiber because the weight should be minimum. Hence, CNC routers are widely used to effectively cut the various parts of the drones. Figure 6 shows the bottom plate design of the drone (Telagam et al., 2017)

Figure 6. Bottom plate design in CNC router

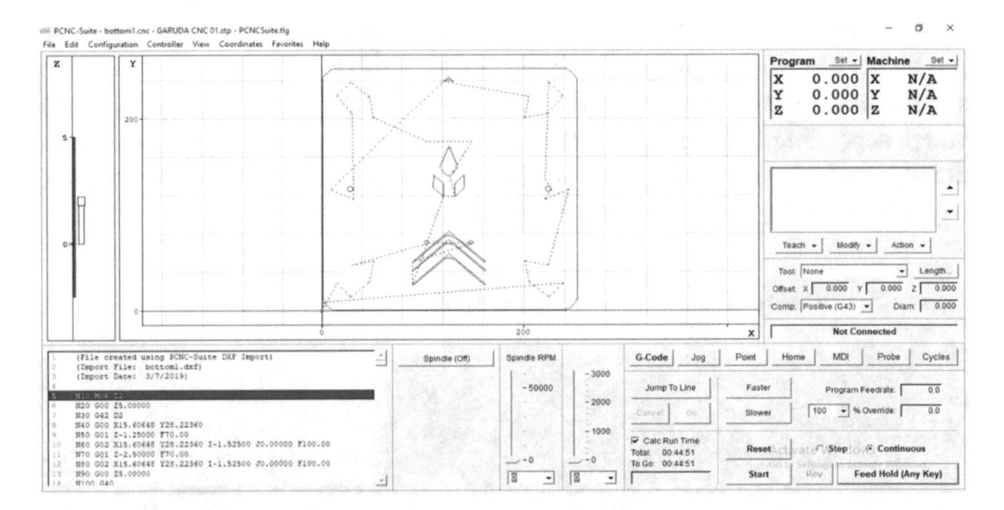

Parameters Used to Design the Drone

The drones' design parameters are the frame, motors, propellers, batteries, cameras, and sensors. The structure gives mechanical strength to the drones. They are holding the various parts of the drone. The frame should be more assertive as well as less weight. Thermoplastics and carbon fiber-reinforced composites are suitable materials with good strength and high stiffness with low density. The motors and propellers are used to lift the drone from the ground. The choice of motor and propeller plays a vital role in determining the performance. The engine consists of winding of coils and magnets. The motor generally dissipates more heat. It is encapsulated with proper heat dissipating material like aluminum. The choice of propellers is another parameter that should withstand high speed.

The batteries are the power hub of the drones. The energy efficiency of the batteries should be more. The higher capacity supports more flying time in the air. Lithium-ion batteries are the right choice, offering good energy and power with lesser weight. In recent days, supercapacitor technology has been incorporated in better-performance drones (Moribe et al., 2018). The sensors are considered the nervous system of drones. They are doing various roles in the internal and around the drones. The sensors are used to maintain balance when flying in the air. The commonly used sensors are gyroscopic, accelerometers, tilt, magnetic, airflow, and GPS (Liang and Delahaye, 2019). The cameras are attached to the mainframe of the drones. High-quality cameras with lesser weight are more preferred. Advanced microchip technology is installed in the drones to make them intelligent and fly without humans' help. Semi-automated drones are highly preferred in industries that require less human supervision (Murugan et al., 2016).

ARCHITECTURE DESIGN OF THE DRONE

The drone's architecture consists of three main categories: a radio control section, a flight control section, and a payload section. The radio control section includes the wireless transmitter, receiver, and servos. The transmitter section unit controls the drones' speed, direction, and sensor operation by sending the command signal at a specified radio frequency to the receiver. Figure 8 shows the architecture of the drone (Agarwal et al., 2018).

Figure 7. The architecture of the drone system

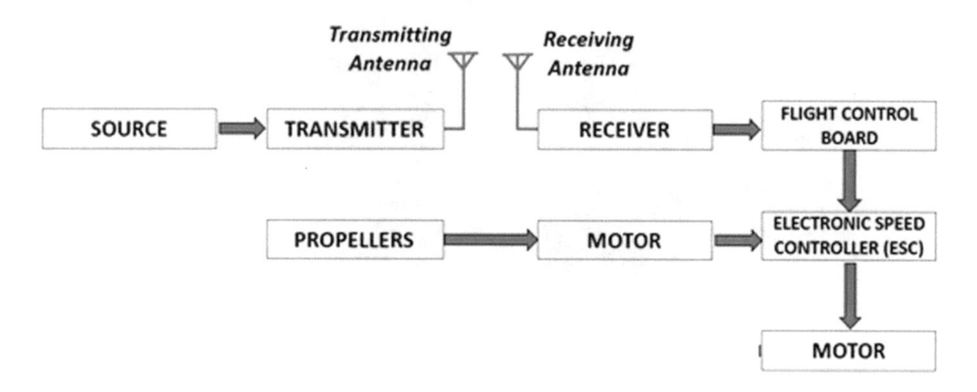

The flight control system takes the responsibility to control and maintain the drones' stability in the air. It is acting like a human pilot in unmanned aerial vehicles. Finally, the payload structure is designed accordingly to the specific operation of the drones. There is no fixed payload structure in the design of drones. Generally, the payload is an additional part attached to the drones to carry out the specified applications (Saha et al., 2018).

Transmitter and Receiver of the System

The transmitter and receiver system of the drones uses the omnidirectional antenna to communicate. The transmitter sends the control information to the drone in radio waves at a frequency of 2.4 GHz. The receiver unit is placed at the top of the drones.

The receiver unit shares the power backup from the drones. The transmitter unit is energized with the help of separate batteries.

The effective communication range between the transmitter and receiver is generally around one kilometer. The receiver's sensitivity is made too high because the drone's flying and specific operation are commanded by the human or automated computer via the remote control. Figure 8 shows the RC transmitter (Adamu et al., 2020).

Figure 8. RC Transmitter

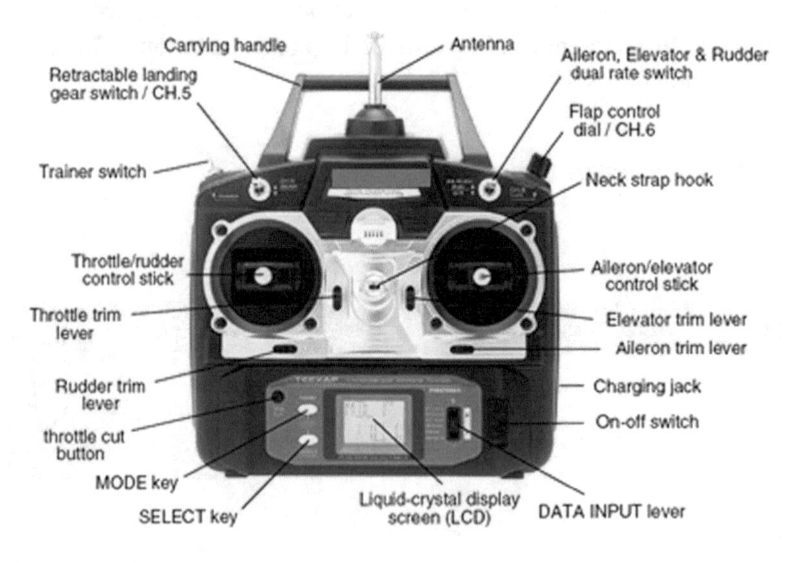

Figure 9. RC Receiver

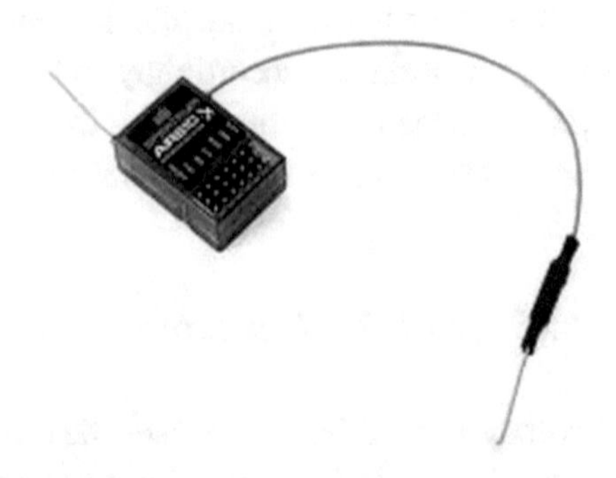

The RC transmitter primarily uses four ways: Roll, Pitch, yaw, and throttle to move the drone. The Roll is used to move the drone to the left or right direction, Pitch is used to move the drone in the forward and backward movement, The Yaw is used to rotate the drone in a counter-clockwise or clockwise direction the air. The throttle controls the power delivered to the drone motors from the battery, making the flying drone go faster or slower. Figure 9 shows the RC receiver (Telagam et al., 2017).

Pulse Width Modulation (PWM)

The pulse width modulation technique is commonly used in wireless communication protocol, such as radio wave communication in drones and flight controllers. The pulse width range in radio controllers ranges from 1000 to 2000 microseconds at a frequency range from 40 to 200 Hz. The flight controller's information to the RC receiver is in the form of a pulse width modulated signal. The pulse width modulated movement uses a separate wire for each channel (Patel et al., 2015).

The PWM is highly preferred to achieve the operation at the load. The load was not affected because the PWM switching frequency is very high, which results in the switching at load can be done smoothly. The advantage of using PWM is that the loss of power at the device is significantly less. Hence, modern drones prefer PWM because when the switch is on, the power is delivered to the drone's motors is high, the button is off, and there is no current supplied to the engines. The controlling of the drone's speed is achieved through PWM with minimum loss of power. Figure 10 shows the duty cycle duration used in designing (Fujimori et al., 2018).

Figure 10. The duty cycle duration of CNC router

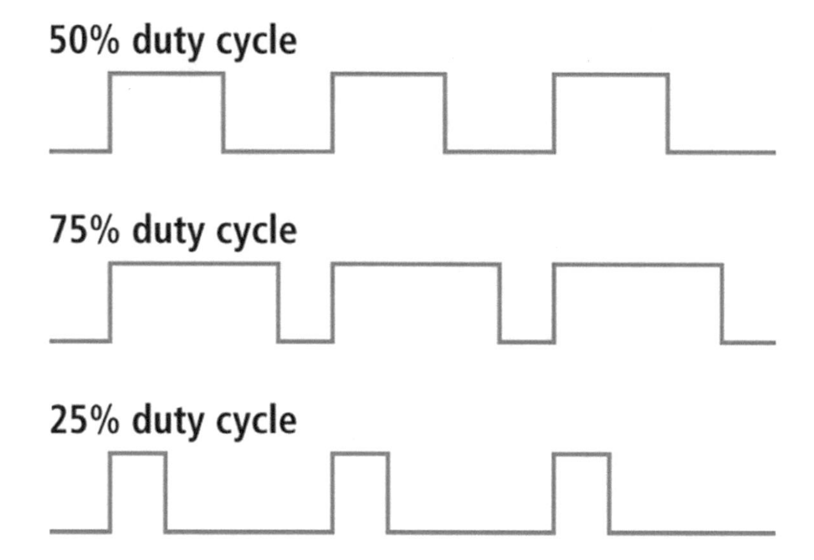

Electronic Speed Controller (ESC) and Motors

The Electronic Speed Controller (ESC) is an electronic device that plays a significant role in controlling and regulating an electric motor's speed. ESC is also used to change the direction of the engine and dynamic braking. ESC is mainly used in remote control applications such as electric vehicles, boats, airplanes, model trains, helicopters, and quadcopters. Nowadays, it is an essential component of the present quadcopters that gives the electric motors higher power, frequency, and resolution of three-phase AC power. ESC provides a wide variation in the speed of the engine. It controls the speed of the motor, which is always necessary for a quadcopter to fly. The electronic speed controller is shown in figure 11.

Generally, ESC is categorized as the different current ratings. It is also called the current speed controller. The power given to the motor is rapidly switched ON and OFF. The MOSFET transistor inside the controller is used as a switch, around 2000 times per second. Hence, the speed of the electronic motor is effectively controlled by the controller. If the MOSFET is ON, the larger current flows through the motor winding, rotating the engine at high speed. If the MOSFET is OFF, the motor coils' current flow is blocked, reducing the motor's speed. The larger current rating devices are used in airplanes (Bristeau et al., 2011).

Figure 11. Electronic Speed Controller

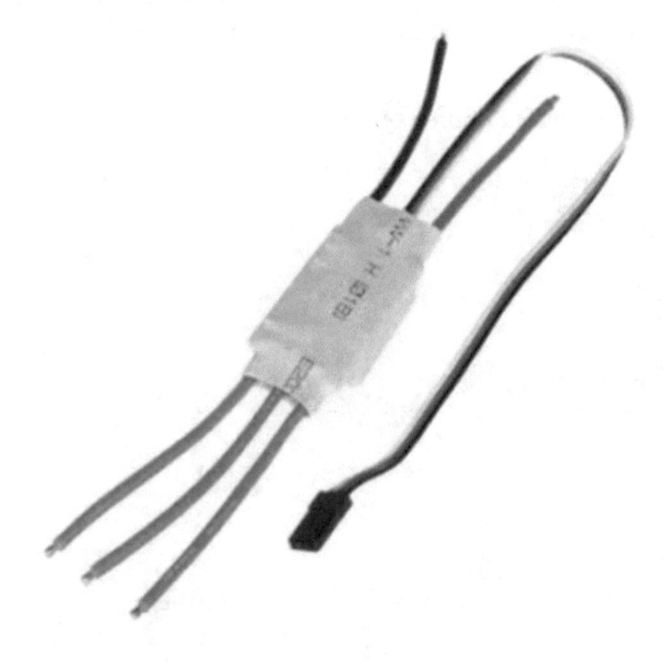

The pulse width modulation signal with pulse width varies from 1 ms to 2 ms is used as an input to operate electronic speed controllers. The duration of the pulse width controls the speed of the motor. If we give smaller pulse width, the engine will rotate at full speed. The electronic speed controller has two types:a brushed electronic speed controller and a brushless electronic speed controller. The brushed ESC is commonly used for remote control applications. It carries small powers, and the performance is quietly low. The brushless ESC is used nowadays because it offers high power performance compared to the brushed ESC. It is widely used in quadcopter applications because of its long-lasting capability (Somanaidu et al., 2018).

A Brushless Direct Current Motor (BLDC) is an electronically commutated motor. It is an asynchronous DC motor. Compared to the brushed DC motor, the BLDC motor has a high power to weight ratio, high speed, reasonable speed control, longer lifetime, high efficiency, and low-cost maintenance. The rate of the BLDC motor is controlled by a specific closed-loop controller called Electronic Speed Controller. The controller delivers current pulses to the engine's winding, which manages the required speed and torque. Figure 12 shows the BLDC motor (Telagam, et al., 2017).

Figure 12. BLDC Motor

The rotating permanent magnet surrounds the fixed armature of the brushless motor. The electronic controller regularly switches the current phase to the engine to regulate the required speed and direction. Too much heat due to high input power affects the magnetism of the permanent magnet inside the motor. BLDC motors' applications are electric vehicles, industrial engineering, motion control systems, positional and actuation systems, radio-controlled cars, and aeromodelling. The modern quadcopters use BLDC motors, which are available in various power-to-weight ratios and are more reliable (Telagam et al., 2019).

Battery Elimination Circuit (BEC)

The battery eliminator circuit is a voltage regulator. The modern drone uses high voltage batteries to supply its various electric and electronic parts. Hence, a device is needed to step down the high voltage into smaller voltages. The drones use the BEC to properly step out the single source's power into various parts. The drones cannot carry multiple batteries to deliver the energy for their various electric and electronic components. Hence, the recent drones effectively utilize the single energy source with the help of battery eliminating circuit (Telagam et al., 2019).

Figure 13. Battery Elimination Circuit

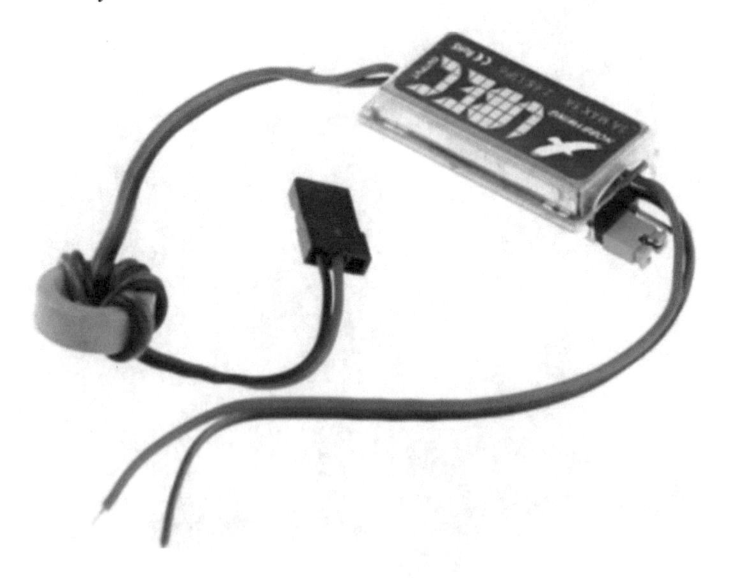

Rotor-Propellers for the Flying Drone

The propellers in the drones effectively transform the rotary motion into thrust. They create the difference in air pressure between the drones' top and bottom surfaces, sufficient to lift the drones into the air. The propellers are arranged in pairs that can spin either clockwise or counter-clockwise to balance the drones in the air. The propellers' rotating speeds are controlled by varying the voltage applied to the motors through electronic speed controllers. The battery elimination circuit and pair of propellers are shown in Figures 13 and 14.

The rotor's power is positively related to the size of the propellers. Smaller size propellers require minimum energy to lift the drones because these have less inertia, which is easier to control and change the speed. The longer propellers lift the drones to a greater height than the shorter propellers, but it requires more power from the motor (Torres et al., 2020).

Figure 14. Pair of Propellers

The propellers having more blades provide a more comfortable lift of the drones into a more considerable height. Increasing the number of edges in the propellers leads to the issue of faster battery drain. Hence, the drone propellers can be constructed with two, three, or four blades. The blades are made from plastic or carbon fiber to give the drone more flexibility with lesser weight (Al Murad et al., 2020).

THRUST ESTIMATION OF THE DESIGN

Thrust is the vital force to lift the drone from the ground. Thrust calculation is necessary for the drones to fly in the air. The drone's weight is known to choose the number of motors required to lift the drone. The weight is split into three main components: drone weight, battery weight, and equipment weight, and weight, including the weight of the main structure like frames, motors, propellers, and landing gears. The equipment weight includes the importance of the detachable equipment such as sensors, cameras, and payload.

The thrust-to-weight ratio is an important parameter to determine the optimal thrust of the drone motors. The drone motors' manufacturers display the thrust value, the weight that the single motor can lift from the ground. In general, the thrust-to-weight ratio is maintained at 2:1 for the drones' effective operation. The high thrust-to-weight ratio is advisable for easier control of the drones.

Throttle Measurement

The throttle is used to control the upward and downward motion of the drones in the air. The positive throttle lifts the drone upward, and the negative throttle will pull down the drone to the ground. The flying height of the drone is not similar for all applications. The drone's size in the air is chosen according to its operation. Hence, throttle control plays a vital role in maintaining the drone's altitude in the air for its specific function.

Flying Mechanism of the Drone

The rotors of the drone are used for propulsion and control. The rotor rotates the propellers at high speed, which creates a difference in the air pressure between the drones' upper and lower surfaces. It refers that the basic principle behind lifting the drone from the ground. The rotors can rotate in the clockwise as well as counter-clockwise direction. It allows our drones to change their direction of flying in the air. The faster rotation of the rotors lifts the drone to a more considerable height.

The drone lifting process includes the rotors' net thrust greater than the gravitational force. The weight of the drones is positively related to the thrust estimation. The gyroscope stabilization ensures the smooth flying capability of the drones in the air. It provides the essential and required information to the flight controller to maintain the stability of the drones.

The flying of drones in the air was controlled by manual or auto mode. Manually, the drones in the air are maintained remotely by human. There are four significant controls of the drone are Roll, Pitch, yaw, and throttle. The Roll is used to change the direction of the flying drone to the left or right direction. For example, they push the roll stick to the left results in the propellers, allowing the air to the right and forcing them to fly towards the left (Ganeshamurthy, et al., 2020).

Pitch is used to move the drones in the air to the forward and backward direction. While pushing the transmitter's pitch stick, it sends the command to the rotors that change the drones' motion. Yaw is another flight control tool used to rotate the drones clockwise or counter-clockwise in the air.

Figure 15. Flying mechanism of the drone.

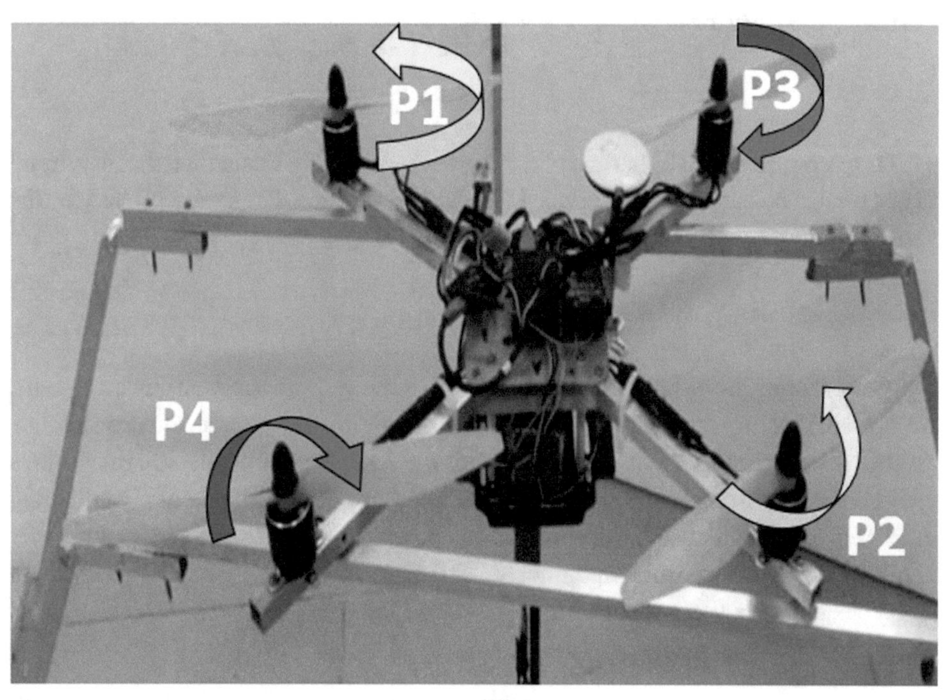

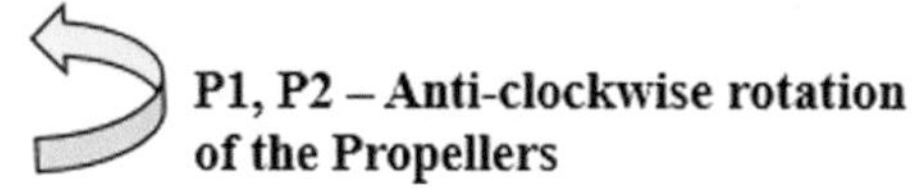

The yellow color arrow mark shows the propellers' P1, P2 -Anticlockwise rotation, and the Blue color indicates the P3, P4- Clockwise rotation of the propellers. Figure 16 shows the flying mechanism of the drone.

Fly Time Calculation

The maximum time of the drones or UAVs in the air without refueling or recharging is called flying time. The flying time of the small drones ranges from 20 to 30 minutes. The larger drones can fly around in a few hours. Solar-powered drones are used in the coastal and forest areas, flyingfor up to three days. The weight of the drone and the battery's capacity determine the flying time. The importance of the payload is also kept at a minimum to increase the drones' flight time (Kumar et al., 2020).

The battery delivers the power to the drones to fly. The smaller batteries provide lesser weight with lower energy, and the larger battery provides higher energy. The drones with the solar system have more flying time than the battery-powered drones.

$$Maximum\ Flight\ time\left(min.\right) = \frac{Battery\ Storage\ Amp\ Rating\left(mAh\right)\text{x}60}{Current\ Draw\ of\ Drone\left(Amps\right)} \tag{1}$$

The recommended flight time of the drones is relatively lesser than the maximum flight time. The safety factor reduces the total flight time. This time is used for the safe landing of the drone.

Power Consumption

The power supply is the fundamental unit of the drone system. In recent advancements, we are still facing difficulties in finding a unique power source for drones. The drones' power management system depends on various parameters like the drone's weight, the importance of the payload, the number of sensors operated and flying time. Hence, we can't use the drone's unique power source for commercial and other applications (Siriborvornratanakul et al., 2018).

Types of the power supply unit of the drone.

a. Batteries
b. Fuel Cells
c. Hybrid Energy Sources

Batteries

Bat has energized most drones used for commercial purposes in recent mysteries. These batteries give power to the drones for a maximum of two hours. Hence, we find difficulties in using battery-powered drones for more extensive operations than commercial applications. Cascading batteries is not the solution because it increases the weight and space complexities. Different approaches are made for battery-powered drones, which allow them to operate the drones for long hours through specific energy managing skills like swapping, laser recharging, and tethering.

The drones can recharge or replace their depleted batteries during the mission via ground-based power stations in the swapping mechanism. The swapping of batteries can be done in an autonomous mode or by humans (Mohan and Poobal, 2017). Wireless recharging is also one of the solutions to recharge the batteries of the operating drones. Laser energy is used to recharge the battery of the drones in a wireless mode (Patel et al., 2015). The drones are connected to the ground power supply (Galkin et al., 2018). Figure 16 shows the wireless recharging methods which are used in the drone.

Figure 16. Battery-powered drones. (a) Wireless recharging through lasers (b) Tethered drone

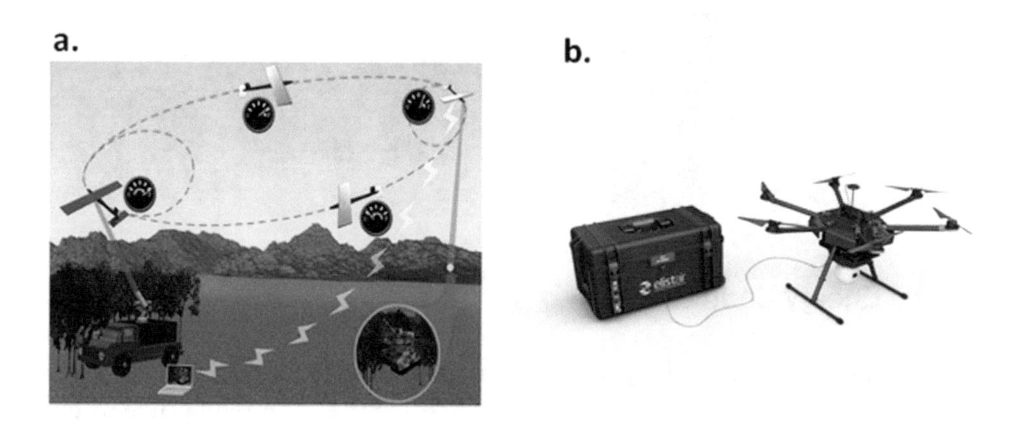

Fuel Cells

Fuel cells are used in drones alternate to the battery sources. Instant refueling is the main advantage of fuel cells compared to battery sources. But fuel cells have lower energy density because the size of the fuel tank is limited. In addition to that, the fuel cells have a maximum efficiency level of 60%, which is lower than that of lithium batteries having a top efficiency level of around 90% (Galkin et al., 2018). Figure 17 shows the fuel cell-powered drones.

Figure 17. Full cell-powered drones.

Hybrid Energy Sources

Solar energy is the right energy resource for drones having fixed wings. The solar cells can be placed in the drones' wings, Which Increases the flying time of the drones (Muttin, 2011). In recent days, Supercapacitors are used in drones, which overcome battery energy sources (Saad et al., 2014). A supercapacitor should have high charge storage capacity, stability, fast charging, and slow discharging capability. Hence, an Energy Management System is needed to effectively couple the fuel cell, battery, and supercapacitor in a single drone. This Hybridization produces good performance and increases the flying time of the drones (Muttin, 2011). The solar energy-based drone is shown in figure 18.

Figure 18. Solar energy-powered drones.

PAYLOAD DESIGN

The payload is the excess weight or unmanned aerial vehicle (UAV). It usually carries the drones' additional supported components, such as sensors, cameras, or deliver packages. The large drones use payload as a necessary unit for commercial and specific purposes. The payload plays a different role for different drones. The payload's weight plays a significant role, starting from the design, battery capacity, flying time, etc. To achieve the drone's longer flying time, the importance of the payload is kept small. The payload structure is not defined, which is designed according to its specific applications. In the Agricultural field, the payload carries the fertilizers with a spray system, plant cutting blades, cameras to monitor crop health, and seed balls (Saad et al., 2014). Figure 19 shows the conveyer belt, and the payload mechanism is shown in figure 20.

Figure 19. Belt conveyer model for payload

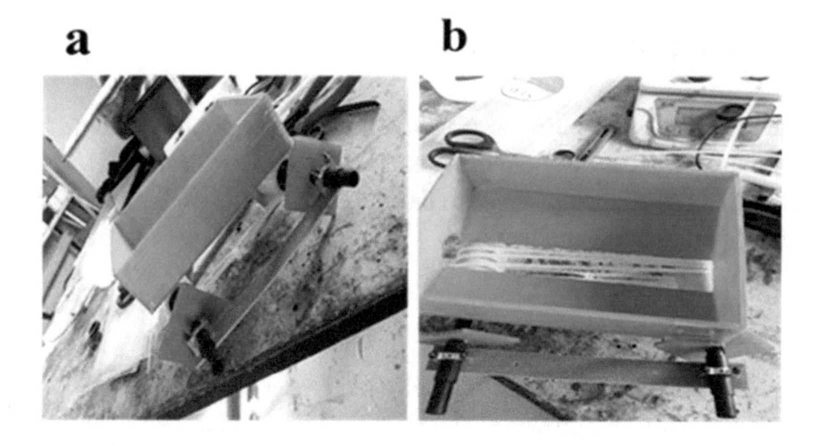

Figure 20. Payload section (a) Metal Blade (payload) (b) Metal Blade connected to the drone as a payload

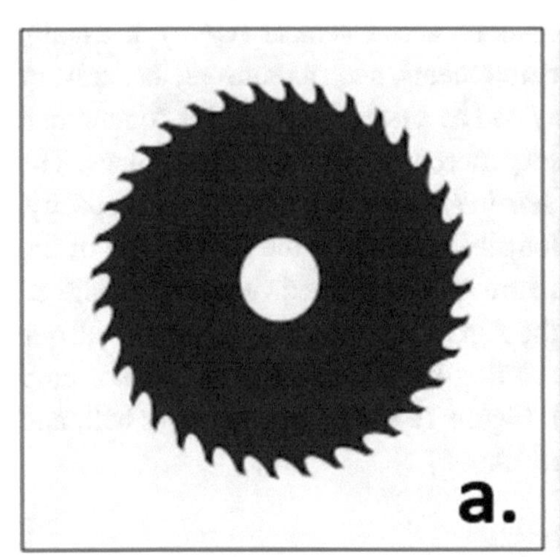

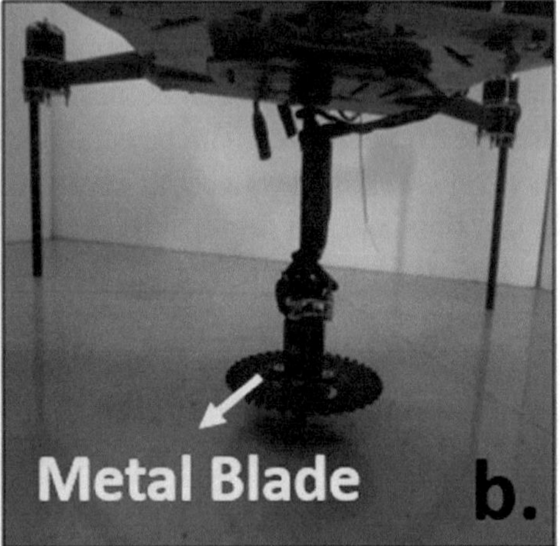

PRACTICAL ASPECTS OF THE UAV

Unmanned aerial vehicles are the best solution to carry out the allotted work in the dangerous and most dirty environments. The usage of drones is increasing in day-to-day life. The challenges need to be addressed while implementing drones in real-time for commercial applications.

Regulations

The usage of drones in regular life is rapidly increasing. The demands for drones in various industries are also growing. These drones are used for their specific military, defense, delivering goods and medicines, photography, agriculture, navigation, geographic mapping, and telecommunication. Everyone can buy a drone in the online market and use it for their purpose. Privacy is the central issue in day-to-day life while using drones. The loss of controls in the operation of drones may lead to severe accidents. Hence, the government authorities may decide to overcome these issues in the future. The standard rules and regulations are framed and followed to avoid the above challenges (Dricus, 2015).

Fully Autonomous Drones

The drone operation's autonomy is needed for those drones having more flying time in the forest and coastal areas. Humans cannot continuously operate drones during the whole day. Hence autonomy is needed in drone technology. Today, drones are performed in an autonomous mode with human supervision. There are more challenges in the design and operation of fully autonomous drones. The drone ensures and meets the local laws, regulations developed by government bodies, air traffic conditions, and safety systems to achieve this (Aneke and Wang, 2016).

The Drone of IoT in 6G Wireless Communications

The growth of drones enhances the business opportunities in mobile communication networks. It is necessary to increase the communication efficiency between the drones and the ground station. The drone systems may integrate with sixth-generation networks to enhance the data transmission rate, speed, and system capacity. It offers higher performance in various applications such as agriculture, smart parking, home, industrial automation, etc. The IoT builds drones with artificial intelligence, and 6G wireless communications are the key factors that make them invaluable to the various commercial and industrial applications (Wang et al., 2011).

Traffic Management System

The drone traffic management system is necessary to ensure the air safety of unmanned aerial vehicles. It provides information and notifications to the drone operators. The report includes weather alerts, the availability of air space, and the other flights in the area. The traffic management system has a specificcontroller that monitors the drones within the specified limit. It allows the drone operators to operate at low altitudes during multi drone operations (Fotouhi et al., 2019).

APPLICATIONS

Drones are machines capable of flying in the air without a human pilot. The drones are built to reduce human beings' efforts and time in commercial and industrial areas. Nowadays, drones are used in various industries to enhance their productivity with less effort and minimized cost. They are having numerous applications in the field of military and defense. The drones are used for commercial purposes like delivering goods and medicines, archaeological surveys and agricultural areas, etc. (Telagam et al., 2017).

Bomb Detection in Military

Bomb detection is one of the challenging and high-risk tasks. To detect the bomb by humans is a time-consuming process. Hence, the detection of bombs at an earlier time is necessary to save the life of humans. In recent days, drones are used to detect bombs in various places in the military field. These drones are smaller in size and have high-quality wireless cameras. They save the lives of numerous people in highly sophisticated areas. Bomb detection of the drone is used in the military is shown in figure 21.

Figure 21. Bomb Detection using drone

Surveillance in Defense

Drones are effectively used in the surveillance system, collecting the images and videos day and night. They are well equipped with modern technologies that trap the suspects' phone calls and use GPS locations to collect field information. The drones are also used for regular patrol services. Figure 22 shows the surveillance of dronesoperated in airports (Gantala et al., 2017).

Figure 22. Surveillance example of drone

Agriculture

Drones are used in the agricultural field for many applications, which reduces the farmers' burden and improves productivity. The drones with high-quality imaging technologies such as thermal, hyperspectral, and multispectral are used to give specific information to the formers regarding the soil wet, crop health and fungal infections, etc. The drones with suitable payloads are used for sowing seeds and spraying fertilizers. The drones with cutting blades are used for trimming the plants in the gardening and highways. Figure 23 shows the drone used in agriculture applications (Telagam et al., 2017).

Figure 23. drone in agricultural applications

Healthcare

Drones are used in healthcare systems, such as delivering medical kits, vaccines, drugs, blood and collecting biological samples in remote areas. They have played a vital role in the recovering everyday life of humans in disaster areas. In the COVID pandemic, drones are used to spray sanitizers to control the spread of COVID viruses. Figure 24 shows the drone applications in healthcare (Gantala et al., 2017).

Figure 24. Drone in healthcare applications

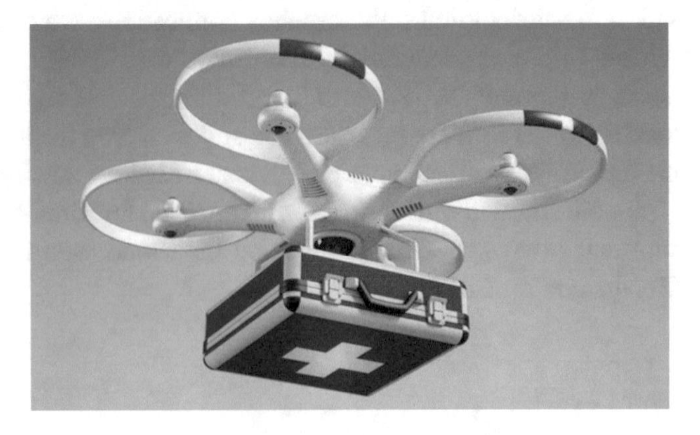

Aerial Photography

In the modern world, the usage of drones to capture images and footagehas increased. Drones play a significant role in the cinematography industry. These autonomy-operated drones are used in sports, forest, and real estate photography. Further, it is also effectively used in journalism for live broadcasting. Figure 25 shows the aerial photography taken by drone (Thotakuri et al., 2017).

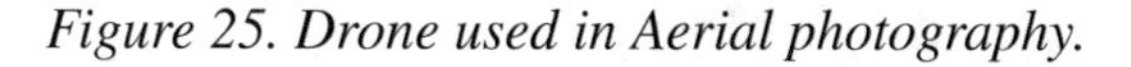

Figure 25. Drone used in Aerial photography.

Geographic Mapping

Geographic information systems (GIS) have many industrial applications and technological advancements that have enhanced GIS data. Drones are capable of capturing high-quality images with the use of high-end cameras. The high-quality captured images of coastlines, mountaintops, and islands are used to generate 3D maps and supports to give locations to reach those places in cloud-based applications.

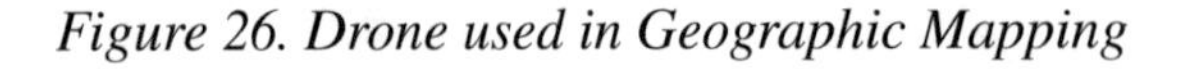

Figure 26. Drone used in Geographic Mapping

Disaster Management

Drones give their maximum support to society to collect or gather information after a natural or human-made disaster. High-definition cameras, sensors, and radars provide accurate information to disaster management rescue teams. The unmanned helicopters are used to rescue humans and animals in disaster area.

Figure 27. Drones are used in disaster management

Weather Monitoring

Weather monitoring is necessary, especially in areas where regular violent storms occur. Hence, drones are effectively used to monitor dangerous and unpredicted weather. It gives information about the air quality, temperature, pressure, wind speed, and wind direction to the weather scientists and the public. The drones capture the information through various sensors and send it back to the ground in real-time.

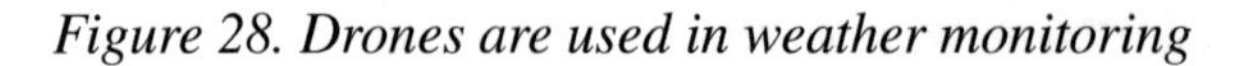

Figure 28. Drones are used in weather monitoring

Telecommunications

In telecommunications, drones are used in maintenance to monitor the installed equipment's working condition at the top of the cell towers. Solar-powered drones are used to provide internet services in remote areas such as coastal, forest, and hills. They are used to finishing autonomous missions, data uploading to the clouds, and enhanced performance of telecom providers.

Figure 29. Drones are used in telecommunications

FUTURE ENHANCEMENT

Drones are becoming the most powerful advanced machines in the future. Much research is carried out to fine-tune and enhance the present drones' performance to their industry needs.

Personal Transportation

In the future, fully autonomous drones may be used for personal transportation. The drone's weight and the flying time are significant issues to these drones' design and development. This kind of vehicle is not the cheap one. It will take more time to standardize the regulations and commercial utilization. Safety measures are the essential key while implementing personal transportation using drones. Figure 30 shows the private transportation application of drones.

Figure 30. Personal transportation example drone.

Delivering Goods

The next revolution of drones in the future is in delivering parcel services. Drones are delivering goods and parcels in a very quick time. It reduces the human resources and delivery time in the parcel services. Amazon tested its unmanned delivery services in the UK. The supervision of multiple drones in remote locations is a significant issue. Another issue is the safety of delivering goods and drones.

Figure 31. Drone as Delivery agent

Fire Fighting

The advancement technologies introduced in drones may be used in firefighting to control the forest and domestic areas. These kinds of drones are used for emergencies to maintain the spreading of fire. They are stored in small packages of dry chemical powder, released over the area along the specified route. This will control the spread of fire in the wild and domestic (Telgam et al., 2020).

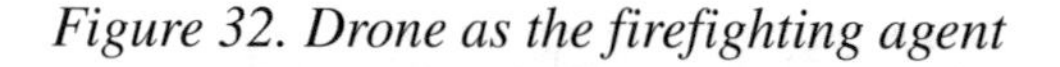

Figure 32. Drone as the firefighting agent

Rescue Missions

All the drones used for commercial and industrial applications are equipped with GPS systems. This allows the play of drones in the search and rescue operation in the future. The nano drones are smaller in size and can fly in hard-to-reach locations. These drones are operated in emergency services for searching the victims or missed persons in any environment (Radha et al., 2021).

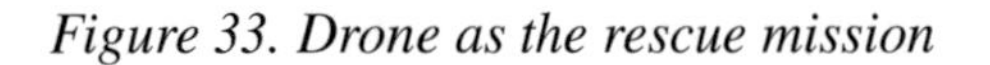

Figure 33. Drone as the rescue mission

CONCLUSION

The drone receives the input from the setting, which is then processed by the neural network, and the most appropriate action is chosen. The drone executes the action (after making confident that it poses no danger to itself or its surroundings). This successively changes the setting's state, and the drone receives the new form of the ground, and the method is recurrent.With the event of mobile net networks (MIN) and the internet-of-things (IoT), there are tremendous challenges for the long run B5G/6G wireless Communication networks to support huge property and seamless affiliation beneath a restricted spectrum. To handle the higher than challenges, physical layer technologies like huge multiple-input multiple-output (MIMO), non-orthogonal multiple access (NOMA), mm-wave (mm Wave), little cell networks (SCN), satellite communication, and pilotless aerial vehicles (UAV). UAV communication has attracted goodish attention from educational and trade among these technologies', additionally generally referred to as drones or remotely piloted aircraft, have found numerous applications these days. At first, UAVs square measure primarily employed in the military to defend against enemies and cut back the utilization of pilots. At present, regardless of within the civil or military field, the appliance of UAVs may be aforementioned to be all over like environmental detection, traffic watch, agricultural plant protection, disaster relief, etc. With the speedy development of computing, caching, sensing, communications, and management, UAV communications networks play a non-negligible role for the B5G/6G networks, which may be employed in various applications. UAVs may be acted as aerial base stations, access points (APs), relays, or flying mobile terminals among a cellular network to boost coverage, dependableness, and energy potency of wireless networks. UAVs meet challenges to introduce paradigms associated with the B5G/6G industrial situations.

This chapter discusses drone's types, structure, design, and methodology. They are easy to operate. The drone technology is incorporated to reduce the human efforts in agricultural, medical, surveillance, civil, military, and commercial applications. The drones are designed to be easy to handle. Data security and privacy are the major issue in recent days, which can be overcome in the future. The current standard and regulations are not allowing drones to fly beyond the line of sight. The weather conditions are also considered, which reduces the flight time and causes control failures. Achieving longer flight time is also a significant challenge in drone technology.

REFERENCES

Adamu, A. A., Wang, D., Salau, A. O., & Ajayi, O. (2020). An integrated IoT system pathway for smart cities. *International Journal on Emerging Technologies*, *11*(1), 1–9.

Agarwal, A., Singh, A. K., Kumar, S., & Singh, D. (2018, December). Critical analysis of classification techniques for precision agriculture monitoring using satellite and drone. In *2018 IEEE 13th International Conference on Industrial and Information Systems (ICIIS)* (pp. 83-88). IEEE.

Al Murad, M., Khan, A. L., & Muneer, S. (2020). Silicon in horticultural crops: Cross-talk, signaling, and tolerance mechanism under salinity stress. *Plants*, *9*(4), 460. doi:10.3390/plants9040460 PMID:32268477

Alzenad, M., El-Keyi, A., Lagum, F., & Yanikomeroglu, H. (2017). 3-D placement of an unmanned aerial vehicle base station (UAV-BS) for energy-efficient maximal coverage. *IEEE Wireless Communications Letters*, *6*(4), 434–437. doi:10.1109/LWC.2017.2700840

Aneke, M., & Wang, M. (2016). Energy storage technologies and real life applications–A state of the art review. *Applied Energy*, *179*, 350–377. doi:10.1016/j.apenergy.2016.06.097

Bristeau, P. J., Callou, F., Vissière, D., & Petit, N. (2011). The navigation and control technology inside the ar. drone micro uav. *IFAC Proceedings Volumes*, *44*(1), 1477-1484.

Dricus. (2015). *Top 8 solar powered drone (UAV) developing companies*. https://sinovoltaics.com/technology/top8-leading-companiesdeveloping-solar-powered-drone-uav-technology

Fotouhi, A., Qiang, H., Ding, M., Hassan, M., Giordano, L. G., Garcia-Rodriguez, A., & Yuan, J. (2019). Survey on UAV cellular communications: Practical aspects, standardization advancements, regulation, and security challenges. *IEEE Communications Surveys and Tutorials*, *21*(4), 3417–3442. doi:10.1109/COMST.2019.2906228

Fujimori, A., Ukigai, Y. U., Santoki, S., & Oh-hara, S. (2018). Autonomous flight control system of quadrotor and its application to formation control with mobile robot. *IFAC-PapersOnLine*, *51*(22), 343–347. doi:10.1016/j.ifacol.2018.11.565

Galkin, B., Kibilda, J., & DaSilva, L. A. (2019). UAVs as mobile infrastructure: Addressing battery lifetime. *IEEE Communications Magazine*, *57*(6), 132–137. doi:10.1109/MCOM.2019.1800545

Ganeshamurthy, A. N., Kalaivanan, D., & Rajendiran, S. (2020). Carbon sequestration potential of perennial horticultural crops in Indian tropics. In *Carbon Management in Tropical and Sub-Tropical Terrestrial Systems* (pp. 333–348). Springer. doi:10.1007/978-981-13-9628-1_20

Gantala, A., Nehru, K., Telagam, N., Anjaneyulu, P., & Swathi, D. (2017). Human tracking system using beagle board-xm. *International Journal of Applied Engineering Research: IJAER*, *12*(16), 5665–5669.

Gantala, A., Vijaykumar, G., Telagam, N., & Anjaneyulu, P. (2017). Design of Smart Sensor Using Linux-2.6. 29 Kernel. *International Journal of Applied Engineering Research: IJAER*, *12*, 7891–7896.

Hayat, S., Yanmaz, E., & Muzaffar, R. (2016). Survey on unmanned aerial vehicle networks for civil applications: A communications viewpoint. *IEEE Communications Surveys and Tutorials*, *18*(4), 2624–2661. doi:10.1109/COMST.2016.2560343

Jawad, A. M., Jawad, H. M., Nordin, R., Gharghan, S. K., Abdullah, N. F., & Abu-Alshaeer, M. J. (2019). Wireless power transfer with magnetic resonator coupling and sleep/active strategy for a drone charging station in smart agriculture. *IEEE Access: Practical Innovations, Open Solutions*, *7*, 139839–139851. doi:10.1109/ACCESS.2019.2943120

Kumar, M. A., Telagam, N., Mohankumar, N., Ismail, M., & Rajasekar, T. (2020). Design and implementation of real-time amphibious unmanned aerial vehicle system for sowing seed balls in the agriculture field. *Int J Emerg Technol*, *11*(2), 213–218.

Liang, M., & Delahaye, D. (2019, October). Drone fleet deployment strategy for large scale agriculture and forestry surveying. In 2019 IEEE Intelligent Transportation Systems Conference (ITSC) (pp. 4495-4500). IEEE. doi:10.1109/ITSC.2019.8917235

Mogili, U. R., & Deepak, B. B. V. L. (2018). Review on application of drone systems in precision agriculture. *Procedia Computer Science*, *133*, 502–509. doi:10.1016/j.procs.2018.07.063

Mohan, A., & Poobal, S. (2018). Crack detection using image processing: A critical review and analysis. *Alexandria Engineering Journal*, *57*(2), 787–798. doi:10.1016/j.aej.2017.01.020

Moribe, T., Okada, H., Kobayashl, K., & Katayama, M. (2018, January). Combination of a wireless sensor network and drone using infrared thermometers for smart agriculture. In 2018 15th IEEE Annual Consumer Communications & Networking Conference (CCNC) (pp. 1-2). IEEE. doi:10.1109/CCNC.2018.8319300

Mozaffari, M., Saad, W., Bennis, M., & Debbah, M. (2016). Unmanned aerial vehicle with underlaid device-to-device communications: Performance and tradeoffs. *IEEE Transactions on Wireless Communications*, *15*(6), 3949–3963. doi:10.1109/TWC.2016.2531652

Murugan, D., Garg, A., Ahmed, T., & Singh, D. (2016, December). Fusion of drone and satellite data for precision agriculture monitoring. In *2016 11th International Conference on Industrial and Information Systems (ICIIS)* (pp. 910-914). IEEE.

Murugan, D., Garg, A., & Singh, D. (2017). Development of an adaptive approach for precision agriculture monitoring with drone and satellite data. *IEEE Journal of Selected Topics in Applied Earth Observations and Remote Sensing*, *10*(12), 5322–5328. doi:10.1109/JSTARS.2017.2746185

Muttin, F. (2011). Umbilical deployment modeling for tethered UAV detecting oil pollution from ship. *Applied Ocean Research*, *33*(4), 332–343. doi:10.1016/j.apor.2011.06.004

Patel, K. D., Jayaraman, C. S., & Maurya, S. K. (2015). Selection of BLDC Motor and Propeller for Autonomous Amphibious Unmanned Aerial Vehicle. *WSEAS Transactions on Systems and Control, 10*, 179–185.

Pobkrut, T., Eamsa-Ard, T., & Kerdcharoen, T. (2016, June). Sensor drone for aerial odor mapping for agriculture and security services. In *2016 13th International Conference on Electrical Engineering/Electronics, Computer, Telecommunications and Information Technology (ECTI-CON)* (pp. 1-5). IEEE. 10.1109/ECTICon.2016.7561340

Puri, V., Nayyar, A., & Raja, L. (2017). Agriculture drones: A modern breakthrough in precision agriculture. *Journal of Statistics and Management Systems*, *20*(4), 507–518. doi:10.1080/09720510.2017.1395171

Radha, D., Kumar, M. A., Telagam, N., & Sabarimuthu, M. (2021). *Smart Sensor Network-Based Autonomous Fire Extinguish Robot Using IoT*. Academic Press.

Saad, E. W., Vian, J. L., Vavrina, M. A., Nisbett, J. A., & Wunsch, D. C. (2014). *Vehicle base station*. Academic Press.

Saari, H., Akujärvi, A., Holmlund, C., Ojanen, H., Kaivosoja, J., Nissinen, A., & Niemeläinen, O. (2017). Visible, very near IR and short wave IR hyperspectral drone imaging system for agriculture and natural water applications. *The International Archives of the Photogrammetry, Remote Sensing and Spatial Information Sciences*, *42*(W3), 165–170. doi:10.5194/isprs-archives-XLII-3-W3-165-2017

Saha, A. K., Saha, J., Ray, R., Sircar, S., Dutta, S., Chattopadhyay, S. P., & Saha, H. N. (2018, January). IOT-based drone for improvement of crop quality in agricultural field. In *2018 IEEE 8th Annual Computing and Communication Workshop and Conference (CCWC)* (pp. 612-615). IEEE. 10.1109/CCWC.2018.8301662

Siriborvornratanakul, T. (2018). An automatic road distress visual inspection system using an onboard in-car camera. *Advances in Multimedia, 2018*. doi:10.1155/2018/2561953

Somanaidu, U., Telagam, N., Nehru, K., & Menakadevi, N. (2018). USRP 2901 based FM transceiver with large file capabilities in virtual and remote laboratory. *iJOE, 14*(10).

Telagam, N., Kandasamy, N., & Nanjundan, M. (2017). Smart sensor network based high quality air pollution monitoring system using labview. *International Journal of Online Engineering*, *13*(08), 79–87. doi:10.3991/ijoe.v13i08.7161

Telagam, N., Kandasamy, N., Nanjundan, M., & Arulanandth, T. S. (2017). Smart Sensor Network based Industrial Parameters Monitoring in IOT Environment using Virtual Instrumentation Server. *Int. J. Online Eng.*, *13*(11), 111–119. doi:10.3991/ijoe.v13i11.7630

Telagam, N., Lakshmi, S., & Nehru, K. (2019). Ber analysis of concatenated levels of encoding in GFDM system using labview. *Indonesian Journal of Electrical Engineering and Computer Science*, *14*(1), 80–91. doi:10.11591/ijeecs.v14.i1.pp77-87

Telagam, N., & Manoharan, A. (2020). Multi user based performance analysis in upcoming 5G Techniques. *Journal of Critical Reviews.*, *7*(12). Advance online publication. doi:10.31838/jcr.07.12.152

Telagam, N., Nanjundan, M., Kandasamy, N., & Naidu, S. (2017). Cruise Control of Phase Irrigation Motor Using SparkFun Sensor. *Int. J. Online Eng.*, *13*(8), 192–198. doi:10.3991/ijoe.v13i08.7318

Telegam, N., Lakshmi, S., & Nehru, K. (2019). USRP 2901-based SISO-GFDM transceiver design experiment in virtual and remote laboratory. *International Journal of Electrical Engineering Education*, 0020720919857620.

Tezza, D., & Andujar, M. (2019). The state-of-the-art of human–drone interaction: A survey. *IEEE Access: Practical Innovations, Open Solutions*, *7*, 167438–167454. doi:10.1109/ACCESS.2019.2953900

Thotakuri, A., Kalyani, T., & Vucha, M., MC, C., & Nagarjuna, T. (2017). Survey on Robot Vision: Techniques, Tools and Methodologies. *International Journal of Applied Engineering Research: IJAER*, *12*(17), 6887–6896.

Torres, M., Llamas, I., Torres, B., Toral, L., Sampedro, I., & Bejar, V. (2020). Growth promotion on horticultural crops and antifungal activity of Bacillus velezensis XT1. *Applied Soil Ecology*, *150*, 103453. doi:10.1016/j.apsoil.2019.103453

Wang, Y., Chen, K. S., Mishler, J., Cho, S. C., & Adroher, X. C. (2011). A review of polymer electrolyte membrane fuel cells: Technology, applications, and needs on fundamental research. *Applied Energy*, *88*(4), 981–1007. doi:10.1016/j.apenergy.2010.09.030

KEY TERMS AND DEFINITIONS

Battery Eliminating Circuit (BEC): Voltage regulator. It is designed to drop a big voltage down to a smaller voltage. As modern RC airplanes use high voltage batteries, it allows you to run your receiver, servos, and other accessories from your main battery without using a separate lower voltage one.

Electronic Speed Controller (ESC): Devices that allow drone flight controllers to control and adjust the speed of the aircraft's electric motors. A signal from the flight controller causes the ESC to raise or lower the voltage to the motor as required, thus changing the speed of the propeller.

Propellers: A mechanical device for propelling a boat or aircraft, consisting of a revolving shaft with two or more broad, angled blades attached to it.

Pulse Width Modulation (PWM): Modulation technique that generates variable-width pulses to represent the amplitude of an analog input signal. PWM is widely used in ROV applications to control the speed of a DC motor and/or the brightness of a lightbulb.

Unmanned Aerial Vehicle (UAV): Military aircraft that is guided autonomously, by remote control, or both and that carries sensors, target designators, offensive ordnance, or electronic transmitters designed to interfere with or destroy enemy targets.

Chapter 5
Scope of UAVs for Smart Cities:
An Outlook

Jyoti Singh
Sant Hirdaram Girls College, Bhopal, India

ABSTRACT

New technologies are always remarkable for the sustainability of human beings and for their enhancement. Unmanned aerial vehicles (UAVs) are highly used on a wide range of commercial as well as defense purposes. UAVs are also called drones. With the phenomenon scope of application, drones have reached a very high level in each and every field with smart cities being no exception. UAVs provide many services for the development of smart cities like traffic control, natural disaster management, monitoring, transportation, infrastructure, mapping, air quality, and many other parameters. UAVs with high resolution cameras and advanced techniques have many properties (i.e., less time consuming, highly efficient, data to collect and analyze, etc.). Collection of data is very fast and accurate; even analyzation of any task given is very authentic. They are also considered as an aid to surveillance for security purposes.

INTRODUCTION

The interest in smart cities is growing by the day, especially in the aftermath of the global financial crisis. The global population is growing and is expected to double by 2050. As a result of these estimates, cities and communities face new difficulties and possibilities. There is a greater desire to concentrate on employing ICT services and

DOI: 10.4018/978-1-7998-8763-8.ch005

advanced technologies in long-term smart city development. The design proposals of such smart cities necessitate extensive and complete integration of ICT and related tendencies. UAVs help to achieve these objectives effectively and efficiently. As a result, UAVs are used in a broad variety of applications and functions in smart cities. These applications range from traffic flow monitoring to measuring and forecasting floods and natural calamities utilizing wireless sensors. Additionally, potential for UAVs and their applications in smart cities will continue to grow rapidly. As urban populations grow, so will expenditure budgets to fulfil these cities' needs. According to a Cloud Services Corporation report, smart city technology expenditures hit approximately $80 billion in 2020, with analysts forecasting that investment will reach $435 billion by 2025. According to a McKinsey and Company estimate, global spend on building and infrastructure is around 2 trillion US dollars per year, with ICT spending accounting for 1.5 to 2% of that total. However, improvements in cloud computing, wireless sensors, networked unmanned systems, big data, open data, are projected to continue in the future decade. Furthermore, billions of gadget s will be linked together.

Figure 1. Unmanned Aerial Vehicle

Conventional cities will confront several problems in the future decades, affecting their development and sustainability. Among the difficulties are: the rapidly expanding population which has polarized economic growth, necessitating more inhabited regions. Environmental pollutants and the prerequisites for sustainability will result in significant requirements` for UAVs to be used in Smart Cities. In terms of development, transforming any city into a smart city is the current global trend. It will construct and build a connected and maintainable Smart City by using emerging technologies such as the Internet of Things and cloud computing. Indulgent the phrase "Smart City" is not totally etched in stone because there has only been limited research on the subject. However, as technology, information systems, and communications develop, some crucial basic components that are necessary to adequately grasp and describe the notion of a smart city may be identified. Smart city efforts are influenced by eight major elements. Integration of controlling and managerial aims is required aimed at smart cities towards function successfully also productively. Smart city is built on a set of smart technologies like computing etc. that are useful to vital infrastructure capabilities and services. Smart trending computing leads to a fundamental range of included (HSNT) hardware, software, and network technologies that give Information technologies systems by concurrent knowledge of the existent domain. Unmanned aerial vehicles (UAVs) can provide an array of programs and possibilities that might enhance smart cities. Advanced analytics increasing sustainable users in making better informed judgments. Governance entails the execution of procedures with constituents that share information in accordance with instructions and principles in order to attain goalmouths and purposes. For efficient smart city administration, several variables such as teamwork, communiqué, governance, and statistics interchange are essential. Policy context: Awareness on how to use information systems appropriately requires an understanding of the policy environment. It primarily defines technical and non-technical built-up challenges besides establishes situations conducive to urban growth. Citizens of the city can become active participants in the government and management of the city through smart city projects. If they are essential actors, they may be able to participate in the effort to such a degree that they can impact its success or failure. Economy is one of the key drivers of smart city projects, and a city with a high level of economic competitiveness is regarded to have one of the smart city attributes. Many factors influence smart city growth, like that unproven measurement scaling of new and recent technologies, new machinery challenging the surviving status of association and successively conventional cities, technology attentiveness among city sectors, developed smart city resolutions, and the complication of how smart cities are worked, financed, regulated, and planned (Mohammed et al., 2014). The consequences are primarily business formation, occupation creation, employment services, and productivity gains. The development of an ICT infrastructure is critical to the development of a smart city and is dependent on various variables linked to its performance and availability. Because of the global economy's unpredictability, governments have begun to turn parts of their towns into smart cities through leveraging advances in ICT to boost efficiency and tolerance power, cut costs, and develops excellence quality of life sustainability. While a variety of developing technologies will be considered, it is anticipated that the usage of unmanned aerial vehicles (UAVs, or drones) will make significant and diversified contributions to the growth of smart cities. According to a new Science Direct research, unmanned aerial vehicles (UAVs) may be utilized for anything from environmental monitoring to traffic control to deliver cost-effective assistance to local administrations. But how would the use of unmanned aerial vehicles (UAVs) affect the design and administration of smart cities overall?

The Role of UAV's

Though originally designed for the military and aerospace industries, drones have now forayed into the limelight because of the efficiency, efficacy and enhanced levels of safety and precision. Drones have changed how we perceive the world. UAVs may be utilized for a variety of purposes in cities, including traffic and crowd surveillance, civic security, item delivery, infrastructure inspection, and more, with minimal technological and security breakthroughs. The widespread deployment of UAVs in day-to-day municipal operations may give benefits that achieve the precise purpose of what a smart city seeks to achieve improvement of the lives of its citizens (Liu et al., 2015). UAVs exist in a variety of shapes and sizes. They are classified into three types: safety regulation, scientific research, and commercial uses. However, in order to develop a well-designed UAV application, precise information support is required, which is required for a successful system (Mohammed et al., 2014). It is commonly known that unmanned aerial vehicle (UAV) applications have gotten involved in a variety of industries ranging from agriculture to oil and gas production and transportation. A typical UAV architecture comprises of four basic components: the control system, the monitoring system, the data processing system, and the landing system.

Figure 2. Parts of Unmanned Aerial Vehicle

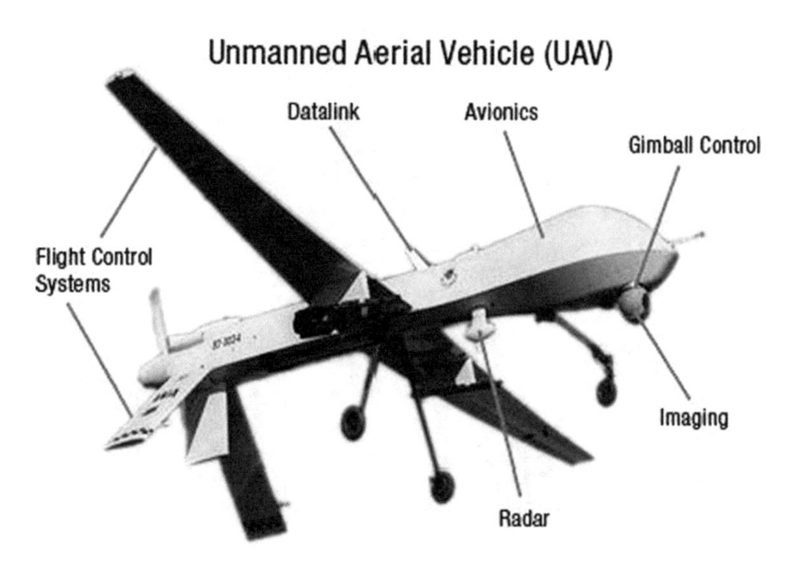

An internal system performs a variety of activities ranging from navigation to transmitting data to the ground. The UAV industry is still expanding, and UAVs are being used in new activities and solving new challenges daily. Many organizations are interested in creating unmanned aerial vehicles (UAVs) in order to minimize the cost of connected services. To date, some of the impediments to deploying UAVs in many civilian applications include the expense of procuring these devices, developing the necessary apps, and developing the operating systems. UAVs are simple to deploy, with the ability to do challenging jobs, enable high-resolution photography, and cover remote locations (Idries et al., 2015).

Figure 3. Role of UAV For Smart Cities

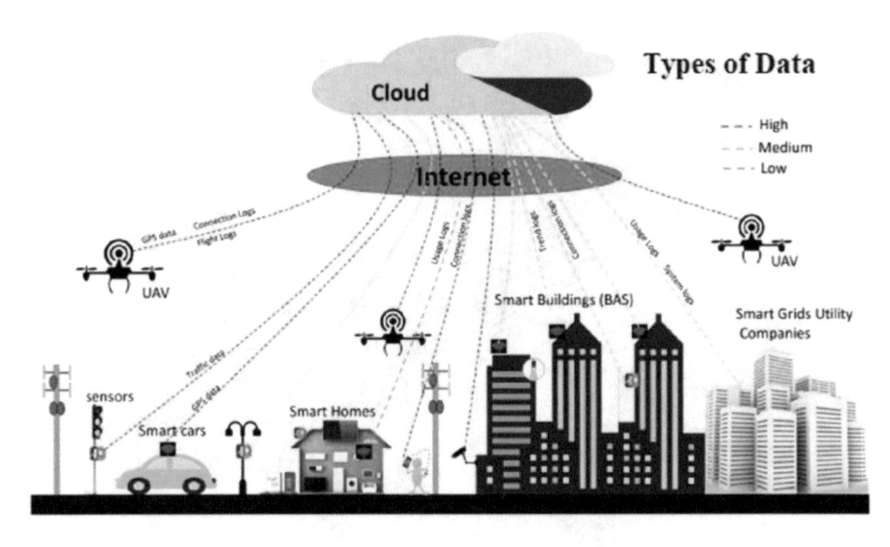

UAVs were first recognized for their military applications, which gave some individuals a restricted view of this technology. When UAVs were authorized to be used in civil applications, their image altered, providing the media with a positive picture and opinion of UAVs. In addition, UAVs were used for humanitarian purposes, such as monitoring hurricane-affected areas. In Nepal, for example, UAVs were used to safeguard animals. The initiative's Non-Governmental Organization (NGO) instructed the guards on how to utilize UAVs to preserve animals, which managed to prevent certain problems. NGOs in Japan utilize unmanned aerial vehicles (UAVs) to monitor unlawful Japanese whales in the southern hemisphere. That is what created a favourable perception of UAVs in the academic and technological areas and encouraged their adoption. A technology with such capabilities must have certain ethical and legal consequences. Some jurisdictions have statutes and regulations governing privacy and data protection. However, the majority of UAV applications have been deployed in the military and security domains. The difficulties that were studied in general were safety, privacy, and ethics, which will be priority for security services.

Smart Transportation and Traffic Management

For decades, cities throughout the world have been plagued by traffic problems ranging from unexpected gridlocks to the ever-present rush hour. To find an acceptable solution to traffic problems, cities must first understand the causes of congestion, where regions get the most crowded, the state of the roads in the most crowded regions, and so on. Static cameras can only provide so much intelligence when gathering this data. This is where unmanned aerial vehicles (UAVs) may play an immediate and critical role in traffic management inside smart cities.

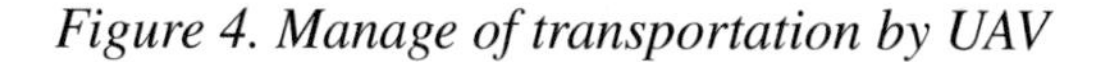

Figure 4. Manage of transportation by UAV

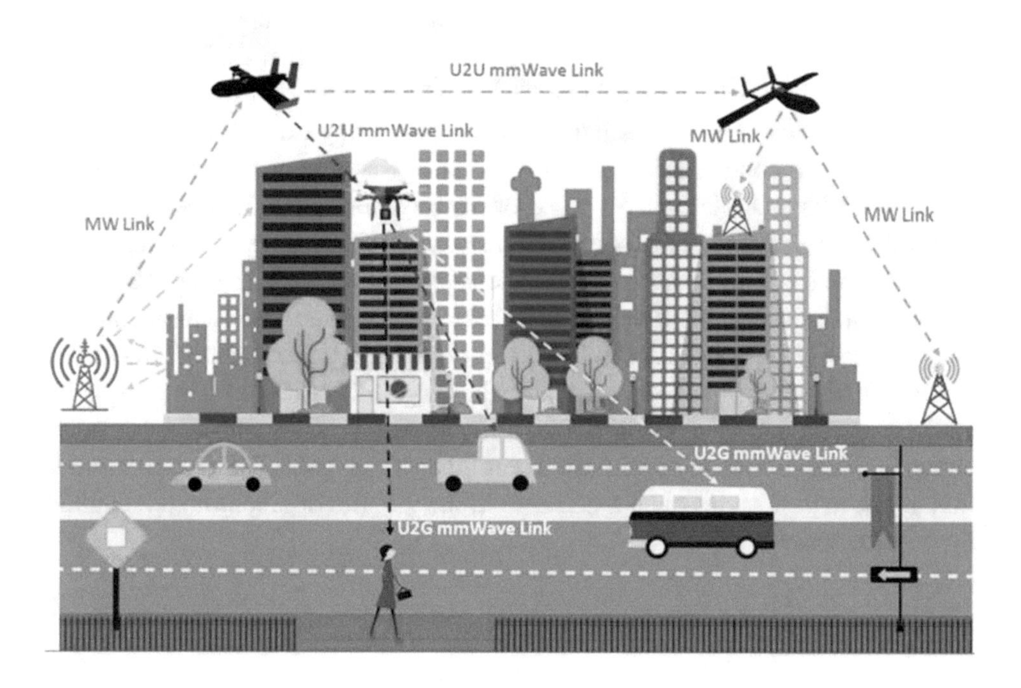

UAV sensors can be used to gather and send real-time data on traffic congestion. Because of their great visibility and sleek mobility, UAVs can provide live feeds of traffic congestion, allowing other technology or human personnel to remedy the problem or shift traffic to prevent even more congestion. Furthermore, when combined with other ICT or Internet of Things advancements, UAVs might assist drivers in finding parking, reducing vehicle miles driven while circling for a place. UAVs may also be used to assist map out metro developments, design more efficient bus routes, and even locate the optimum locations for bike lanes and other types of public and green transportation. UAVs can effectively monitor present traffic while also preventing future jams. This potential will be extremely useful to any smart cities that use UAVs for economic development and growth.

Spatial and Mapping Activities

Using UAVs in geospatial surveying is a new trend in UAV civil applications in smart cities. The primary architecture of a smart city necessitates the optimization of data flows given by wireless sensor networks, as sensors are the primary component of any autonomous system, including those using UAVs. Because portable devices and wireless sensors are recognized for their low power consumption and excellent performance, the system also requires real-time procedures that are integrated with the accessible information repository. This can act as a tool for the technological foundation of smart cities. This integration of technology expands a range of new applications and opportunities, such as fire management in open areas, where the usage of UAVs and micro-UAVs is extremely advantageous. The potentials span from a diverse set of accessible solutions and technologies that are rapidly emerging. However, the challenges and hurdles to the deployment of UAV systems are more closely related to political and cultural concerns than to cost and benefit considerations. Because most UAV designs are reliable, integrating such technologies allows for the installation of wireless sensors on-board, allowing UAVs to be used in geospatial, land surveying, and Geographic Information System (GIS) applications in smart cities, in addition to being useful for environmental analysis. These opportunities may result in cost savings and a decrease in the number of labour hours required for such operations.

Figure 5. Process of Spatial and Mapping

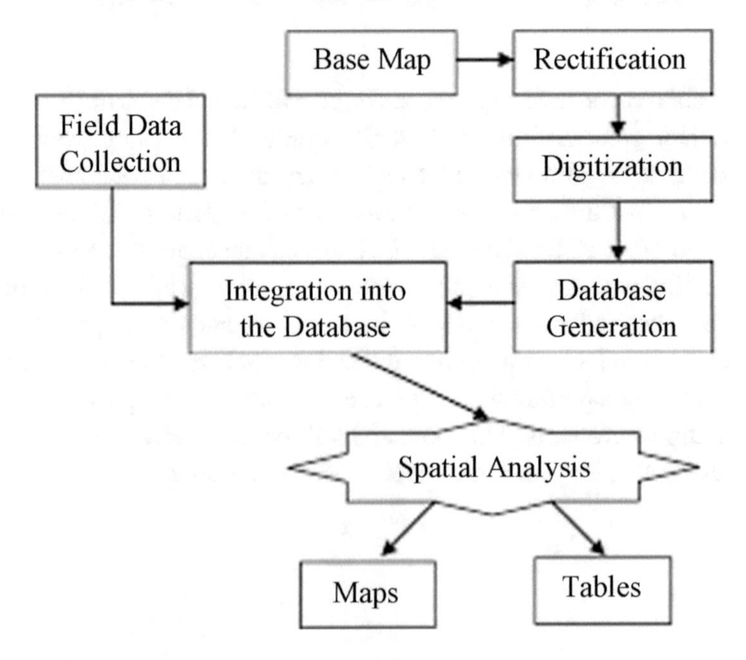

Control of Public Protection

The integration of UAV systems with M2M, RFID, LTE, and live video streaming expanded UAVs' role in public protection. However, the movement toward intelligence and data mining allows UAVs to participate in civil security operations such as providing security services for smart cities. This new tendency will shift city management employees from being reactive to proactive and data driven. Furthermore, the use of unmanned aerial vehicles (UAVs) in surveillance missions will lower costs while increasing operational efficiency.

Urban Security Management

Security management is one of the UAV's prospects in smart cities (Urban Security). The use of UAVs in such areas will allow the city to deploy a rapid operations room that is kept up to date with efficient data flow and will allow the city to handle large public events with large crowds while also providing complete technological coverage.

Processing of Large Datasets

The integration of technologies can be achieved, or at least to make simpler, by using UAVs since information transfer among them may be rapid and precise. Coordination of information from UAVs can operate as a third-party technology to coordinate information from disparate systems. As they are controlled from a base station, once they acquire information, the ground system to send a command to the UAV to transfer the data to another system or UAVs. Big data processing systems for smart infrastructure applications would necessitate the use of several technologies, including as

1. Time series data processing integration with GIS
2. Structure of equipment use
3. Modelling and simulation are used in combination.

Monitoring Natural Disaster and Health Emergencies

Natural environment: one of the primary aims of a smart city is to improve natural resource management and sustainability. Furthermore, the conservation of natural resources and accompanying infrastructure is critical (Mohamed et al., 2020). Due to their regular and extensive surveillance, UAVs can also assist authorities in taking preventative actions ahead of natural catastrophes. They may be utilized as dependable, adaptable, and safe instruments to swiftly disseminate messages and do real-time situational analysis, and they can even control circumstances within natural catastrophes that are too dangerous for people. The use of unmanned aerial vehicles (UAVs) in catastrophe circumstances such as fires, floods, and earthquakes would assist authorities in swiftly and successfully controlling such emergency situations. Because UAVs can reach locations that people cannot, they will correctly examine the issue and assist in acting appropriately in some hazardous scenarios In the figure depicts a firefighting unmanned aerial vehicle (UAV). UAVs can be utilized to help offer and deploy health emergency services more quickly. Whether it's bringing medical items or services to patients quickly or employing UAVs as operating ambulatory services to transport crucial life support equipment and medical supplies, the use of UAVs can substantially enhance the way cities respond to public health requirements.

CONCLUSION

Strengths, Drawbacks, and Possibilities

UAVs have the potential to provide several benefits to smart cities, ranging from obvious benefits such as smart transportation and crowd control to hidden ones like as tourism support and product order fulfilment. However, these intelligent solutions and broad uses are dependent on several problems that must be overcome before UAVs can become a city staple. There is presently no method to smoothly incorporate UAVs into smart cities because to license and certification constraints, as well as privacy and security concerns. Leaders must manage the hazards connected with UAVs to unleash the potential of a "smart city" for widespread use in order to promote a greater quality of life in cities supported by UAV applications. UAV development can be costly because to technical challenges, deployment challenges, training, and system integration. It is also costly to design a UAV for a certain service since it must perform effectively. UAV have limited capacities such as it cannot think or communicate like humans. The legal restrictions: it is not allowed to fly them in specific locations. There are fears that they will distract, scare, or injure humans and birds. Training is required since they must be operated by someone who is knowledgeable about them.

UAVs may have an impact on airplanes and their route guidance. Companies face a hurdle in introducing UAVs to manage specific elements of their organization since this necessitates the expenditure of additional resources. However, if UAVs are deployed, they may be extremely beneficial to the organization in terms of gaining strategic benefits. The smart city strategy aims not only to preserve inhabitants' and tourists' quality of life, but also to improve living by utilizing IT infrastructure and innovative communication technologies. A smart city is an example of efficiency, creativity, and ubiquitous access to a variety of services. Smart living, smart economics, smart tourism, smart environment, smart transportation, and smart governance are some of the axes along which a city's smartness may be measured. For example, the United Arab Emirates is presently concentrating on constructing Smart Dubai, an effort that will totally redefine and alter its people's quality of life. The desire of being knowledgeable has only become stronger since winning Expo 2020. It is an excellent opportunity to create UAV applications that will aid in the achievement of these objectives. Several of these prospects include intelligent road traffic, privacy and security while utilizing smart devices, monitoring of the environment and ecosystems, creation of smart malls, and more. All these possibilities may be realized with UAVs in smart cities.

Irrespective of what you call them – UAVs, Miniature pilotless aircraft or flying mini robots, drones are entering the mainstream in a big way. Not only limited to defence organizations and tech savvy consumers, but everybody is also starting to see the potential drones may possess. Drones are cost effective solutions for many businesses, replacing humans in highly dangerous jobs. Their use in emergency services has seen an upward spike as they are far more effective at spotting things with a Hawkeye view. They help in surveillance, tracking and other security purposes. They do efficient work in little time, things that take humans weeks or months to do, thus making lives easier. Drones have made the aerial view common place and accessible to all, from news reports to geographic surveys to wedding photography.

Chronic congestion is major concern in the cities – drones help planners to understand the geographic data better and implement data driven improvements.

Witnessing the word through the eyes of drones can be both a boon as well as bane. Boon as it shows accurate and precise details and on the flip side it subjects people to unwelcome and uninformed surveillance and scrutiny. As it reaches every nook and corner of our lives – a deeper understanding of the ethics of aerial vision is essential.

REFERENCES

Idries, A., Mohamed, N., Jawhar, I., Mohamed, F., & Al-Jaroodi, J. (2015, March). Challenges of developing UAV applications: A project management view. In 2015 International Conference on Industrial Engineering and Operations Management (IEOM) (pp. 1-10). IEEE. doi:10.1109/NTMS.2014.6814041

Liu, B. L., Wang, L., Lee, S. J., Liu, J., Qin, F., & Jiao, Z. L. (Eds.). (2015). *Contemporary Logistics in China: Proliferation and Internationalization*. Springer.

Mohamed, N., Al-Jaroodi, J., Jawhar, I., Idries, A., & Mohammed, F. (2020). Unmanned aerial vehicles applications in future smart cities. *Technological Forecasting and Social Change*, *153*, 119293. doi:10.1016/j.techfore.2018.05.004

Mohammed, F., Idries, A., Mohamed, N., Al-Jaroodi, J., & Jawhar, I. (2014, May). UAVs for smart cities: Opportunities and challenges. In *2014 International Conference on Unmanned Aircraft Systems (ICUAS)* (pp. 267-273). IEEE.

Mohammed, F., Idries, A., Mohamed, N., Al-Jaroodi, J., & Jawhar, I. (2014, March). Opportunities and challenges of using UAVs for dubai smart city. In *2014 6th International Conference on New Technologies, Mobility and Security (NTMS)* (pp. 1-4). IEEE.

Chapter 6
Urban Intelligence and IoT-UAV Applications in Smart Cities:
Unmanned Aerial Vehicle-Based City Management, Human Activity Recognition, and Monitoring for Health

Prince R.
https://orcid.org/0000-0002-6004-4000
Quilon, India

Navneet Munoth
https://orcid.org/0000-0003-2704-1403
Maulana Azad National Institute of Technology, Bhopal, India

Neha Sharma
https://orcid.org/0000-0003-1302-9593
Mindtree, India

ABSTRACT

The objective of this chapter is to propose a model of an automated city crime-health management that can be implemented in future smart cities of developing countries. The chapter discusses how a suitable amalgamation of existing technologies such as IoT, artificial intelligence, and machine learning can output an efficient system of unmanned city management systems, thereby facilitating indirect engendering of innovative scopes for technology workers and researchers and alleviating the

DOI: 10.4018/978-1-7998-8763-8.ch006

living standards within the city fabrics, catalyzing infrastructure development. In this chapter, the authors have structured an ideal UAV-matrix layout for city fabric surveillance built over the scopes of artificial intelligence. Succinctly, this chapter provides a platform that would galvanize the possibilities and that could be reimagined to structure a more resourceful working model of new emerging smart cities and enlighten the settings of existing ones.

INTRODUCTION

In the present scenario of urban habitats, the cities have become chaotic. By 2050, 68% of the population would be residing in the cities. (United Nations, 2019) If things go according to the current scenario of city development, the city would not be able to sustain this tremendous amount of human population. A 'City' is a culmination of the holistic aspirations of its inhabitants. A city should cater to the demands and necessities regarding victuals, security, healthcare, navigation, and rights. After-all the evaluation of the city design is based entirely on the forward and backward interaction between the city and its users. However, every single task within delivering services as per the demands of its users is not a fluent task, instead, it is laborious and requires rational decision-making units. Hence, to manage and accomplish such tremendous number of tasks that all together build up to a happy and smart city, the management is exposed to challenging aspects like Precision, Duration, Accuracy, Timeliness, and Consistency. But can humans alone tackle these obstacles to build such an ideal city? Or is it smart enough to waste much of human resources in just the maintenance of the city fabric?

In the modern times, the age of Artificial Intelligence, IoT, and robotics, implementation of such technology in the management systems of the city fabrics could drastically solve the challenging aspects in delivering each of its citizens, their demands. These multitasking technologies have proved to be both veracious and timely. Why are AI and robotics more than the existing technologies in city management? They simply work by inputting, processing and outputting, but through the implementation of IoT, and machine learning-the system now can review itself and correct the existing algorithms.

This chapter is an endeavor to give a thought regarding future degree, potential conceivable outcomes and issues concerning instalment of UAV (Unmanned Aerial Vehicles) Systems in the city fabric and how might it get change the lives of the general city population sooner rather than later in significance to create a better bond between the city and its users. The UAV system is simply a 3D matrix over the city fabric managed and directed by its CPU unit, performing tasks like monitoring, reporting, actions, and learning. The scale and complexity of the purpose can vary from basic food and package delivery to health and crime monitoring. Currently, the governing bodies of certain countries have framed a set of permits and privacy laws. Many works have been done on UAV-based Traffic Analysis, UAV-based Package delivery (Amazon Drone Delivery: Prime Air), UAV surveying, etc. Scopes like UAV based healthcare and crime monitoring using motion tracking and human activity recognition will be examined and discussed in this chapter.

Heinous Activities that are Remote and Inaccessible by the Supervisors and Inceptors by a Specific Time

What happens if a region of the city is at the moment with the absence of human at night, and someone has had a heart attack or dehydrates to faint and is helpless? How do we notice such events and solve the case? What happens if a Crime is occurring in dark corners of a city, where human presence is absent? How can be such activities recognized and controlled with necessary action? What if there is a city where such aspects are well supervised and maintained? Won't the users get a subconscious feeling of safety? Won't there be a drastic decrease in crime rate itself, if the users know that they are being watched? Won't there be fewer health breakdowns of the citizens if they are diagnosed immediately?

TRADITIONAL SURVEILLANCE

The traditional surveillance methods include the implementation of a security camera or a network of cameras, a control room for monitoring and humans working in the control room to take the required decisions. Even though the technology proves to be effective in crime management, the challenging aspects like precision, duration, accuracy, timeliness, and consistency are not efficiently tackled.

CCTV Surveillance

Closed Circuit TV, a self-contained surveillance system comprising cameras, recorders, and displays for monitoring activities in a constrained volume, is one of the marvels of modern technology, which is used at homes, offices, as well as in city fabric. It has gained vast acceptance as an effective security measure. (Agustina and Clavell, 2011; La Vigne et al., 2011) (Figure 1) shows the components of the CCTV system. Places are observed and are recorded, later is monitored by qualified security officers through the monitor.

Researches have been conducted on the success of surveillance cameras in reducing crime (Alexandrie, 2017). They reported changes in total crime, revealing crime reduction up to 24-28% in public spaces, concluding that video surveillance can reduce crime in numerous situations.

Although this system proves to be advantageous, the whole CCTV compounds from the cameras to the monitors are just the input units. CCTV surveillance demand 24/7 observation by humans for its effective usage which eventually is the consumption of a large number of human resources (Kong & Fu, 2018). Moreover, when problems like the privacy of the public arise, the users begin feeling uncomfortable and have a negative sensation that they are being scrutinized by another human being.

Figure 1. CCTV component diagram

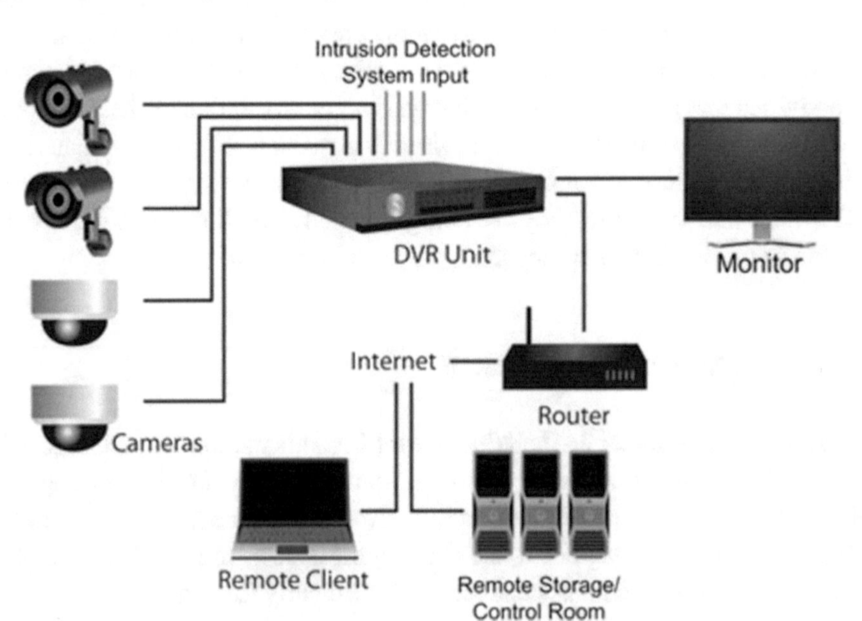

CCTV Surveillance System and Artificial Intelligence: Case Study

In (Dominguez et al., 2018), the proposal brings in the license plate recognition algorithms into the video processing units of the CCTV surveillance. The CCTV system here is able to recognize and register the license plate numbers of moving vehicles into its database. This automates the processes of monitoring and proves to prerequisite a lesser amount of human endeavor. (Figure 2) shows the flow chart of the algorithm implemented in the CCTV system.

Figure 2. License Plate Recognition Algorithm

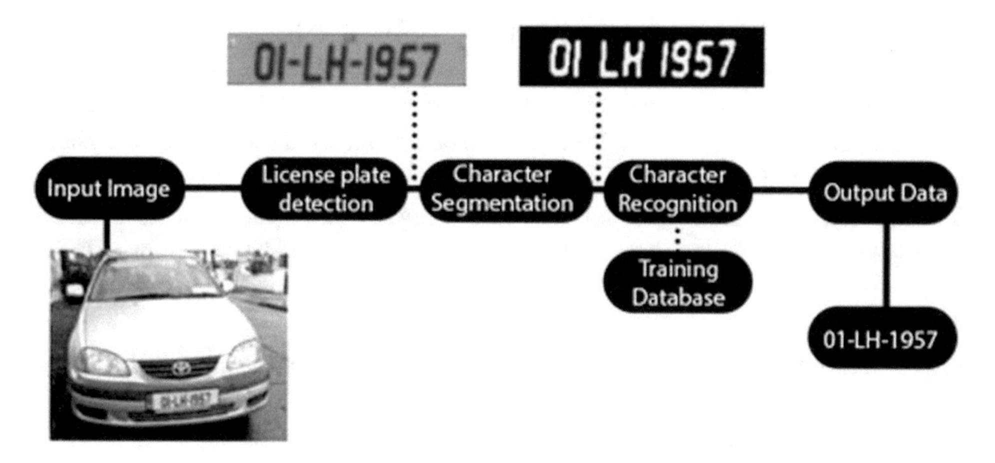

Each time an object is detected, the LPR algorithm is invoked. As shown in (Figure 2) the LPR has three main steps:

- License plate detection (LPD)
- Character segmentation
- Character recognition

LPD is the most expensive task among the rest as it requires a decent video processing algorithm demanding CPU or GPU hardware. As soon as LPD finds the License Plate in the vehicle, the other two steps are applied. If it is found a match, the algorithm continues, until trying to detect the whole plate. This algorithm could be running locally in the camera or in a centralized server that is receiving the objects detected (Dominguez et al., 2018).

Although the LPR implementation in CCTV makes the task easier, the algorithm tends to give out false positives, and this is due to insufficient resolution of the cameras. Thus, high-resolution camera sensors are to be used for surveillance purposes.

BACKGROUND

Machines with Artificial intelligence and Benefits of Automation Continuum

Artificial intelligence can be used to produce machines that perform tasks more efficiently than humans, letting them work in traumatic environments, with no pause throughout each hour of the year. AI has already directed to if not impossible deeds in fields like space travel and medicine (Kong & Fu, 2018).

Traditional automations provide a limited scope of reducing human work, but with both artificial intelligence and Automation, we cannot only just reduce human effort but also remove the need for intervention altogether. This Integrated System is called the Intelligent Robotic process automation or automation continuum. An ideal intelligent automation system functions using its main components of artificial intelligence. Based upon the need, different combinations, and proportions of these three or used separately can create a fully automated solution:

- **Machine Vision:** This refers to the ability of the system to understand visual inputs. The machine uses the inputted data (images or videos) as a base for a classification or identification mechanism.
- **Natural Language Processing:** Natural language processing is the ability to understand human voice and text inputs. This field has been worked out for a while now, making it efficient. Currently, machines can interpret the framework behind communication and take actions based on programmed data and related variables.
- **Machine Learning:** The machines can learn from the data inputted, outcomes of decisions and the environment variables to improve itself.

Computer Vision and Decision Making

Every human action can be imagined as a set of sequences of minor actions and each of these quanta of action is done for some reason (Kong & Fu, 2018). Example, a patient interacts with and responds to the environment using his hands, arms, legs, bodies, etc. in order to complete a physical exercise. Such actions can be observed, either with naked-eyes or graphic sensors. Through our human vision system, we can recognize the action and the purpose of the person. We effortlessly know that a person is exercising, and we could somewhat guess with confidence that the patient's action has obeyed with the prescribed instruction or not. However, monitoring human actions in real-world scenarios with human labor and resources are very expensive. Can a machine perform the same as a human?

One of the ultimate ambitions of AI research is to build a machine that can precisely comprehend humans' actions and purposes, with the intention of serving us better. Suppose a patient is undergoing a rehabilitation exercise at home, and the patient's robot assistant can recognize the patient's actions, analyzing the rightness of the exercise, and preventing the patient from further injuries. An intelligent machine like the patient's robot assistant would be significantly advantageous as it reduces the medical cost, trips to the therapist, etc.

Technological advances in computer science and engineering have been enabling machines to understand human actions in videos. (Ryoo, 2011) The two main topics in computer vision are:

- **Action recognition:** recognize a human action from video data. (Shotton et al., 2013) (Biswas and Basu, 2011)

For human action recognition, several methods could be applied using motion features, sensors, or deep learning. One of the key methods employed is through the use of motion energy image (MEI) and motion history image (MHI). (Figure 3) illustrates the outputs. The MEI depicts the motion characteristics while the MHI gives the sequence of the activity in the video (Bobick and Davis, 2011; Rodriguez-Moreno et al., 2019). The AI then uses a recognition method in order to match the action or activity with a stored library of actions based on matching the properties from both the images.

Figure 3. Examples of an input video frame, the corresponding motion energy image and motion history image used by computers to recognize human activities.

- **Action Prediction:** predict a human action from temporally partial video data. (Vrigkas et al., 2015)

The application of action prediction could be helpful in the early identification of activities that could be suspicious in nature or prediction of motion trajectory to prevent accidents. It can also be helpful in the interpretation of the videos that are incomplete to predict the missing or unobserved video or prediction of intentions based on an existing sequence of human activities for a particular process. (Kong & Fu, 2018) The limitation still exists in terms of catering logical reasoning in prediction and the importance is given to the appropriate information.

Human Activity Recognition (HAR) and Heinous Activity Sorting

Human Activity Recognition intent to understand human behavior which enables the computing systems to recognize and predict human activity (Bulling et al., 2014; Bishop, 2003; Duque Domingo et al., 2017), thereby facilitating AI-based Crime and Health management.

Let an object execute certain activities fitting a predefined activity set A:

$$A = \left\{A_i\right\}_{i=1}^{k} \tag{1}$$

where k denotes the number of activity types possible. There is a sequence of video data (camera) that captures the activity data

$$v = \{d_1, d_2, \ldots, d_t, \ldots, d_n\} \tag{2}$$

where d_t denotes video data at time *t*.

Our objective is to design a model $\mathfrak{F}$ to envisage the activity based on video data *v*

$$\widehat{A} = \left\{\widehat{A}_j\right\}_{j=1}^{m} = \mathfrak{F}\left(v\right), \widehat{A}_j \in A \text{ where m < k} \tag{3}$$

$$A^* = \left\{A^*_j\right\}_{j=1}^{n}, A^*_j \in A \text{ where n < m} \tag{4}$$

Where, A* is the set of all possible heinous activities including all "crime" and "health breakdown" activities. (Figure 5)

The goal of AI is to

- Identify heinous activities - i.e., working out a function $\phi\left(A^*, \widehat{A}\right)$ = *{true, false}*, where ϕ is a model verifying whether $\widehat{A} = A^*$ (Figure 4)
- Learn the model $\mathfrak{F}$ by minimising the variance between predicted activity $\widehat{A}$ and the predefined activity *A*.

Figure 4. Function Mapping F

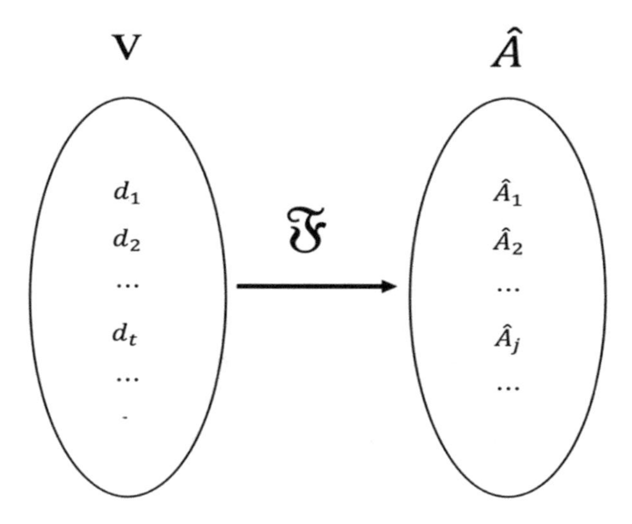

Figure 5. Venn Diagram of the Activity sets A, A and $\hat{A}$*

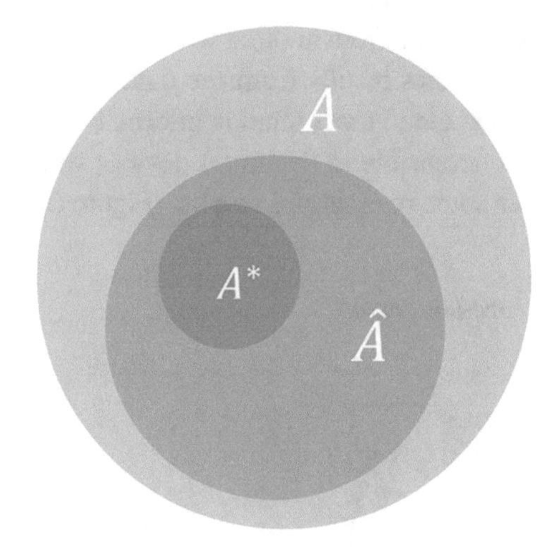

IoT and Big-Data

The Internet of Things, as the expression, it describes, is a concept of connecting man-made devices, using the sensor data, and interpreting some meaningful pieces of information required for efficient decision making. IoT has applications that are enormous in numbers – medical science, business, infrastructure, space science, manufacturing, architecture are few to be named. For instance, suppose for IoT technology is implemented in your house among all the smart devices, you wear a device like a bracelet, that could track your sleep cycles, and it wakes up on the right part of your sleep cycle, so you wake up with ease and comfort. Now the bracelet devices, communicates with the other smart devices, like- the geyser, coffee-machine, air-conditioners, lights, and schedules them according to your activity trends. IoT technology has proved to be efficiently managing tasks with precision and accuracy.

Another noticeable advantage of IoT is that it has proved to be energy efficient, and economical. For example, unless you are very careful about the energy savings, you are likely to leave fans, lights and air-conditions on while you are not present in the serviced spaces. This would be pocket draining and simply waste the available resource. However, an IoT implemented smart home would notice that you aren't present in the respective space- through the sensors of the smart devices- and effectively turn off the unnecessary devices. Now that could save a lot of energy and capital. Moreover, the IoT technology can specifically optimize the energy usage of the devices when the devices are functioning.

The IoT Architecture consists mainly of three-part: IoT smart devices, gateway and the IoT data (Jamali et al., 2020). The IoT smart devices are the closest devices to humans. They can be the source from which data can be extracted, they can be the device that mediates the data from another IoT smart device, and also, they can be the smart device that shows results from the data collected from other smart devices. IoT Data is a Data cloud that contains information from all the IoT smart devices, and this data is accessible to the smart devices so that they can interpret the required data to give some meaningful results. (Figure 6)

Figure 6. IoT Architecture Diagram

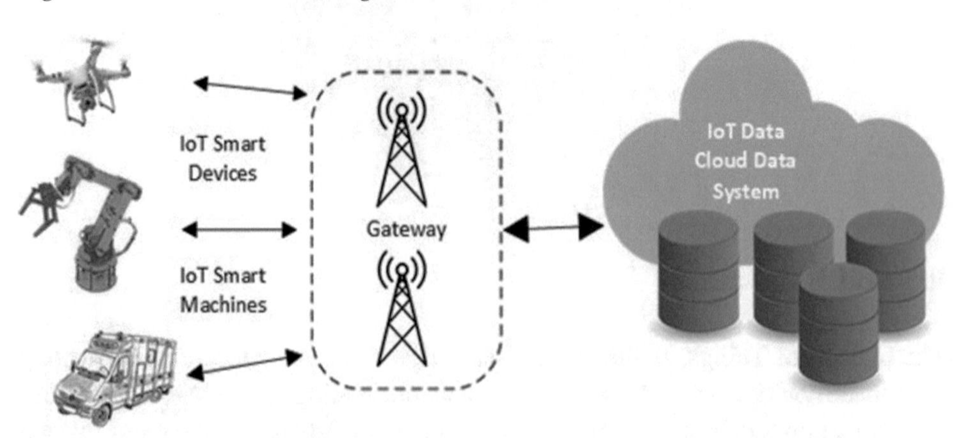

IoT has a vital role in building a perfect AI-Based Drone Surveillance, as the purpose required numerous parameters to be inputted to the system to get the way to reach the precise results. The Advancement in the cellular networks and the outbreak of the 5th Generation of cellular networks with up to 20 Gbps will facilitate this type of Innovations in City Management. (Eze et al., 2018) In this chapter, we will read in detail how such an AI-Based Drone Surveillance system would function and discuss various aspects of such a concept.

Thermal Maps and Human Behaviors

Heat is released as infrared radiations that could be detected by thermal sensor cameras. Heat energy is generated due to the vibration, wiggling, and bouncing off particles with each other. Infrared is a type of electromagnetic radiation or EM radiation. Like any other EM waves, they are also a bunch of photons carrying energy. How does it relate to temperature though? The explanation lies in a phenomenon called Black body Radiation, i.e., every object above absolute zero emits EM radiations. (Jain, 1991)

When heated molecules wiggles and vibrates around any charged particles inside them, like an electron, these particles are pushed or tugged. This mobility of charged particles is what produces EM radiations. So, we all are glowing from heat now. But we do not glow much in the visible spectrum, because the amount of radiation emitted at the wavelength depends on the temperature. All objects emit radiations of all wavelengths, but the hotter an object is the brighter and higher the frequency of the emission will be. At the temperature we normally encounter, we emit infrared radiations, and these can be detected by the thermal camera. Now the technology behind thermal imaging is advancing, enabling these cameras to produce detailed heat maps of the human body (Meola and Carlomagno, 2004).

Body Heat Maps and Human Emotion

A study was done to understand the human behavior or mood set by plotting the body heat map (Nummenmaa et al., 2014). The following plots (Figure 8) are obtained by experimenting with more than 700 people, the thermal map being mapped by a few experiments. (Figure 7) Participants marked areas in (A) whose activity they sensed was increasing (activation) and decreasing (deactivation) during various emotions. Subject wise activation and deactivation data (B) were inputted as integers, where the body was represented as 50,364 data points. Activation and deactivation maps were later merged (C) for statistical analysis.

Figure 7. Thermal Map generation with the emBODY tool.

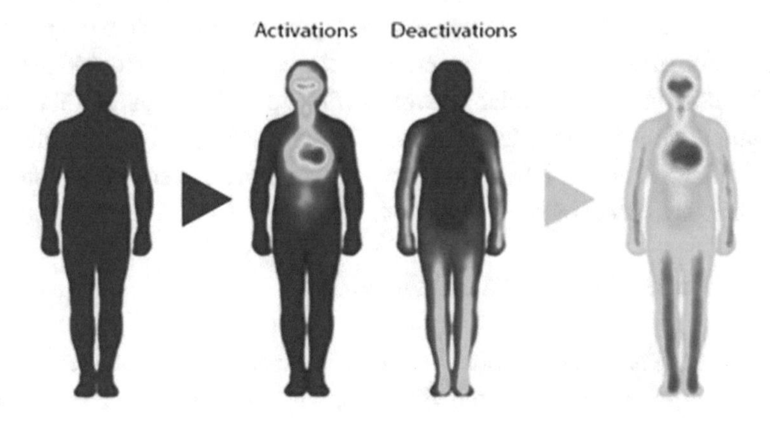

Night Vision and Predicting Crime Activity with Heat Maps

Since cameras are the devices of input, they are sensitive to the visible light spectrum. But that makes the surveying process inefficient in underexposed areas or simply darkness. The use of thermFigure 1 CCTV component diagram.al cameras as input devices is, therefore, more efficient.

Few studies have shown crime is involved when the committer is in the state of anger, similarly, the victim experiences fear. (De Haan & Loader, 2002) Such Emotions can be tracked and can be used to classify the activity executed by the person. Using modern thermal cameras, these maps can be prepared in real-time in higher resolutions and can be analyzed easily by AI and Deep-Learning. This technology can be used to distinguish the crime-victims and the culprit or can be used to determine the type of medical breakdown.

Figure 8. Thermal maps for various emotions related to heinous activities

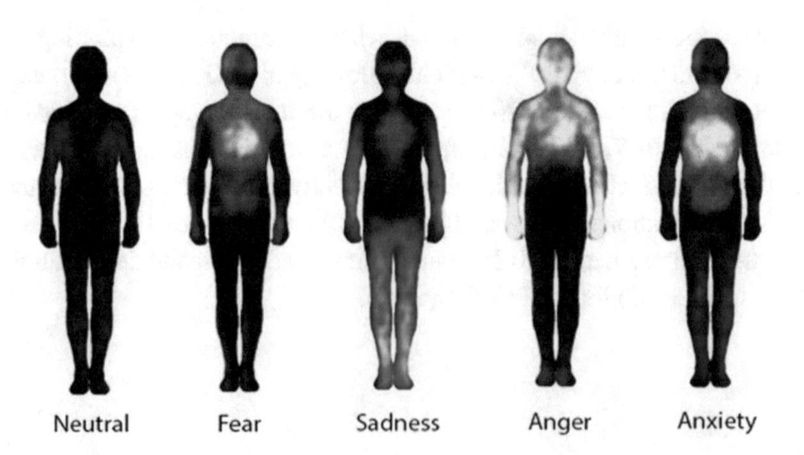

UAV APPLICATIONS: AN OVERVIEW

The global research and development budgets for UAV technology will reach 28642.8 cr rupees by 2020. (Statista Research Department, 2019) Thus, we could expect the increased application of drones, across all sectors (Figure 9). In the eon of IoT and 5G technology, drones can function and react to commands in real-time enabling instant feedback and response rates. Drones can increase the efficiency and productivity of various tasks while simultaneously alleviating the cost and work, this could be beneficial for various industries and sectors if utilized smartly.

Figure 9. Percentage increase in the growth of Drone Application in various sectors

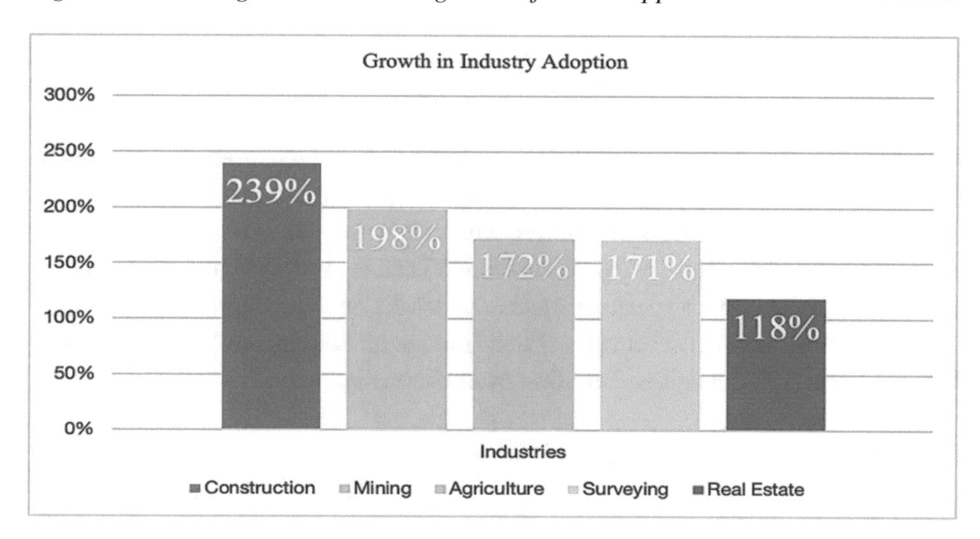

Drones in the Agricultural Sector

The global population is expected to strike a 9.6 billion checkpoint by 2050. (United Nations, 2017) So, to be able to feed this enormous pollution required more effective methods in the production of food, notifying the need for innovation needed in agricultural practices. For ages, our ancestors rely on intuition and experience, so the resultant harvest might not be highly profitable and sufficient, as the crops might fail if not cared well and checked frequently.

Drones can be the solution to this problem, as they could be used to monitor the irrigation cycles, nutrient levels, diseases and infection in the crop, collecting these data and meaningfully using them to produce a high yield and increase productivity. (Mogili and Deepak, 2018; Balaji et al., 2018)

Drones in the Industrial and Manufacturing Sector

The maintenance and Inspection activities in Manufacturing and industrial sector usually is challenging and has higher health risks. There are fields in these sectors that are hard to access and have higher risks and health hazards, putting the workers, and inspector's lives into risk. (Isah, 1996) But, with smart drone technology, such activities can be easily tackled with ease and in a safer manner. They can be functioning and performing tasks even if the machines are in working mode and they can be designed in such a way that they can fix minor errors and report major errors. Thereby, the process can be very cheaper and safer. (Omid and Torbjorn, 2019)

Drones in Infrastructure and Disaster Management Sectors

Smart drones can be used to manipulate the way services and deliveries are made to users. They are so not vulnerable that they could be used to deliver supplies to the needed in natural disasters and other emergencies. For example, the case study-based on Rwanda's medical supply drone system (Ackerman and Koziol, 2019).

Drones can be fasters, and smarter when implemented with the 3D printing technologies. (Mckinnon, 2016) Aside from Visual surveillance and reporting the parameters, the drones can perform mechanical tasks, such as 3D printing temporary shelters for the victims, and survivors of a flood or an earthquake. This 3D printing technology is called Additive Building Manufacturing.

Drones in the Construction Industry

The Utilization of Drone technology has been seen as rising day-by-day in the construction industry. Estimates have shown that by 2020, 78.71 Cr rupees will be invested on drone worldwide by the construction companies. (DroneDeploy, 2018) Thus, there can be a drastic advancement in the existing drone technology, due to market competitions.

Drones perform tasks like site supervision, Inspection of construction, Site views, etc. (Anwar et al., 2018). They are able to access areas, and spaces that are otherwise inaccessible. The process and transmit real-time data. With more advancement in technology and innovation, the scope of the utility of drones might expand in the construction industry.

Drones in Smart Cities and Intelligent Cities

Although, drones have uncountable applications in making our lives better, one of the most interesting applications is the utilization of the UAVs for automated city management and thereby creating an intelligent urban environment (Dung and Rohacs, 2018). The amalgamation of IoT, AI, and Machine Learning, and with adequate training, these mechanical flying robots can perform tasks that are nearly impossible for a group of humans. In the long run, they can prove to be less expensive and reduce the workload, also they can perform tasks faster than humans. (Udland, 2015)

First Aid System Based on UAV and Wireless Body Sensor Network: Case Study

In (Fakhrulddin et al., 2019), they have developed an advanced First aid delivery system, to work on elderly patients with heart conditions. An elderly person usually has fragile bones and hence, if they fall, there are higher chances of a fracture, hip break or a head injury, which might put their lives at risk. So, the author in (Yan et al., 2015) has developed the FDB-HRT Algorithm - a crossbred fall detection algorithm, which works on a combination of acceleration and a heart rate threshold. In this proposal, the author has designed a wearable device that could measure heart rate and the acceleration, and this wearable device is implemented with the FDB-HRT algorithm (Yan et al., 2015), and the wearable also has a geolocation circuit that could transmit the coordination and location of the patient. The Algorithm was 99.2% accurate for fall detections. Accordingly, the Wireless worn device would send signals to the drones and the drones are programmed to analyzes these signals to deliver advanced first aid kits to the patient by tracking the patient's location.

In this study (Fakhrulddin et al., 2019), it was found that the drones achieved a 105 s average time savings compared to the same tasked performed by a delivery vehicle in urban areas. Moreover, it was observed that there was a 31.81% timesaving, when the UAV was let to transport the first aid kit to the patients' location, then using an ambulance to do the same. Therefore, the study concludes that Drone Delivery is more efficient and faster as compared to traditional delivery methods.

Zip-Line's Blood Delivery Drones in Rwanda: Case Study

Zip-line is a California Based company developing technology and service medical supply delivery to remote and areas with poor infrastructure. Rwanda is one of the smallest countries in Central Africa. (Figure 10) Zip-line has been operating this drone-based delivery service to deliver blood toward another hospital elsewhere in Rwanda from the Kinazi's hospital. (Ackerman and Koziol, 2019)

It was being observed that a 50-kilometre trip to western Rwanda would take over an hour, while the drone makes its trip in over 14 minutes. (Ackerman and Koziol, 2019)

Figure 10. Zip-line's Blood Delivery Service to areas with poor infrastructure in Rwanda, Africa

DRONE SURVEILLANCE: IoT APPLICATIONS FOR HEALTH AND CRIME CONTROL

Drone surveillance is the application of Unmanned Aerial Vehicle or a matrix of them collectively surveying a specific subject or an environment (Burgard et al., 2005). Drone surveillance enables clandestinely conglomerate information about a subject from a distance or altitude. However, Drones integrated with Computer vision, Action recognition, Emergency Equipment, Artificial Intelligence & Machine Learning essentially are Flying Intelligent Robots. In fact, they are Ubiquitous Robots. Intelligent drones are equipped with a technology that uses artificial intelligence. These robots can optimize their path instantaneously and accomplish complex tasks due to their high mobility, which makes them a better replacement for crime and health management models in a smart city. The drones are the input devices or the sources of datum, according to IoT technology. They detect and analyze human activities in real-time. (Singh et al., 2018)The drones can be programmed to recognize human poses and activities (Singh et al., 2018). Thus, the drones can conclude whether an event in real-time is suspicious and if yes, the scene and the subject are analyzed with more sets of confirmation algorithms. The drone can now forward the data and conclusion to the Central Processing Units of the system. The CPU diverts the case along with the coordinates to the respective action taking units. The case would be then managed by the Emergency or the Security dispersing units, accordingly. Sometimes, in the case with certain strong suspicious confirmations, few drones can be programmed to handle the malicious, with features like shooting anesthesia or throwing a catch net. Few drones can be designed to carry a First Aid Kit, which delivers the kit to other humans nearby the scene so that they can help the victim.

UNMANNED CITY CRIME-HEALTH CONTROL (UCCHC) SYSTEM

The Unmanned City Crime-Health Control System is a proposed model of Ubiquitous Flying Robots working together to predict and report the occurrence of heinous activities like Human Committed crime activities, Health Breakdowns, Elder Falls, etc. By receiving these data, the system is supposed to take-actions according to the type of activity recognized. Although this technology is essential for all the spaces relating to human presence, this chapter focuses more on the public spaces, where the responsibility of management is not belonging to a private agency. The city fabric in this context refers to that space that is outdoors like parks, roads, streets, etc., excluding the interiors of buildings, and the periphery belonging to private agencies. However, the scope of UCCHC can be stretched to the building interiors by expanding the availability of the IoT smart devices inside the built environments to be accessed by the UCCHC system (Suriyarachchi et al., 2019).

Ideal Schematics of Drone Surveillance System (UCCHC)

The basic working system of the Unmanned City Crime-Health Control (UCCHC) system consists of (Figure 11)

- **Drone:** A hovering unmanned aerial vehicle, with 6 degrees of freedom, here, implemented with heinous activity prediction algorithm- IoT smart drones. These drones consist of:
  - **Flying mechanism**: Aerodynamic body, rotors, gyroscopes, accelerometer, proximity sensor, GPS.
  - **Sensor**: RGB/night vision camera, thermal camera.
  - **Processing:** CPU, data storage, transmitters and receivers.
  - **Power supply**: Battery, solar arrays (Kardasz et al., 2016; Liew et al., 2017)
- **Drone Matrix:** The drone matrix is the three-dimensional spatial network of actively functioning drones hovering above the city fabric, performing the task of surveillance, working together. They can be labeled by their spatial coordinates. The design of a default matrix layout is required for the efficient and effective performance of the UCCHC system.

- **Database:** The database containing catalogs of human activity classification, thermal-human behavioral maps, gesture datum, etc., that will be used by the algorithms to classify and predict heinous activities. This can be IoT cloud data.
- **Principal Server:** The CPU of the whole system, receiving data from drones and deciding whether to deploy emergency service or the crime control service to the affected area. This part of the system is prone to the trafficking of the data, such that the processing power must be able to handle intense data traffic. In fact, the server is responsible for repositioning and training the alignment of the drone matrix according to the trends of human activity in specific areas, weather patterns, population density, etc.
- **Deploys:** Deployment Stations for the specific service. Each deploys specialise in either crime control or health control, or the scopes could be expanded further. The number of deploys in the city fabric, effects how faster the problem it solves. The deploy location and numbers depends on numerous factors like the size of the city fabric to be surveillance, distance between two deploys, the population density, trends of heinous activities in the city fabric, rate of delivering service, radius of the area covered, etc. Similar to the layout of drones, these factors have to be further studied to design an efficient network of deploy stations.
- **Service Vehicles/Drones:** An Ambulance or a Crime control vehicle, a First Aid Drone. Again, depending on the Arrival rate, Population Density, Number of Deploys, Heinous Activity trends, etc. the number of service vehicles and drones per deploys can be determined.

Figure 11. IoT Architecture of the Unmanned City Crime-health control System

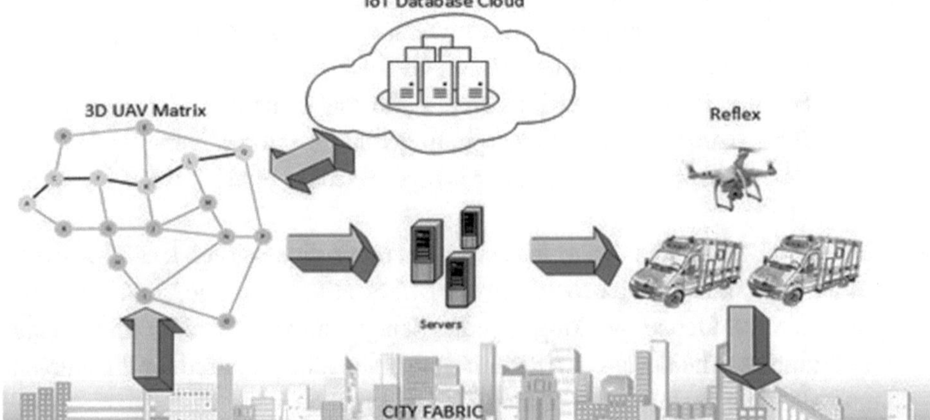

Grid Iron Layout, Radial Layout, Star Layout are a few examples of the arrangement of the system. The arrangement of the drone matrix is to be developed by factors like the topography of the city, road layout, population density, etc. One of the other important factors that the functioning of the drone depends, is the micro-weather, as this is required to be studied for an effective, safe and energy-efficient flight of the drones. Further studies and research can help to extract a fundamental efficient layout for city surveillance. The coordinates of each drone can be defined by the parameters x and y, where x-y is the plane of the drone matrix. (Figure 12) and (Figure 13) shows the illustration of the drone matrix over the city fabric and the airspace of the flight operations.

Figure 12. UAV matrix layout and Principal Server component diagram

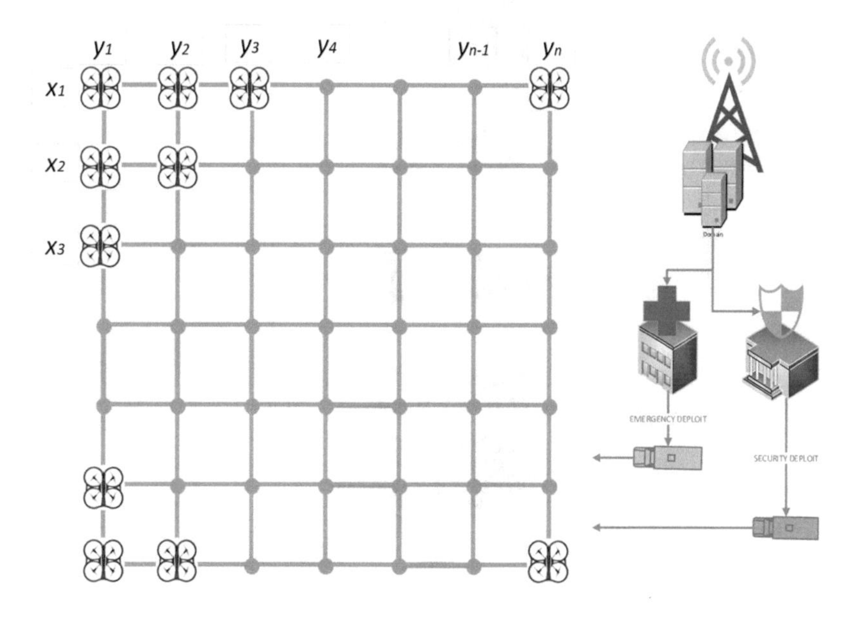

Figure 13. Illustration of the UAV Matrix and UAV air space

Figure 14. Survey Algorithm Flowchart for Drones

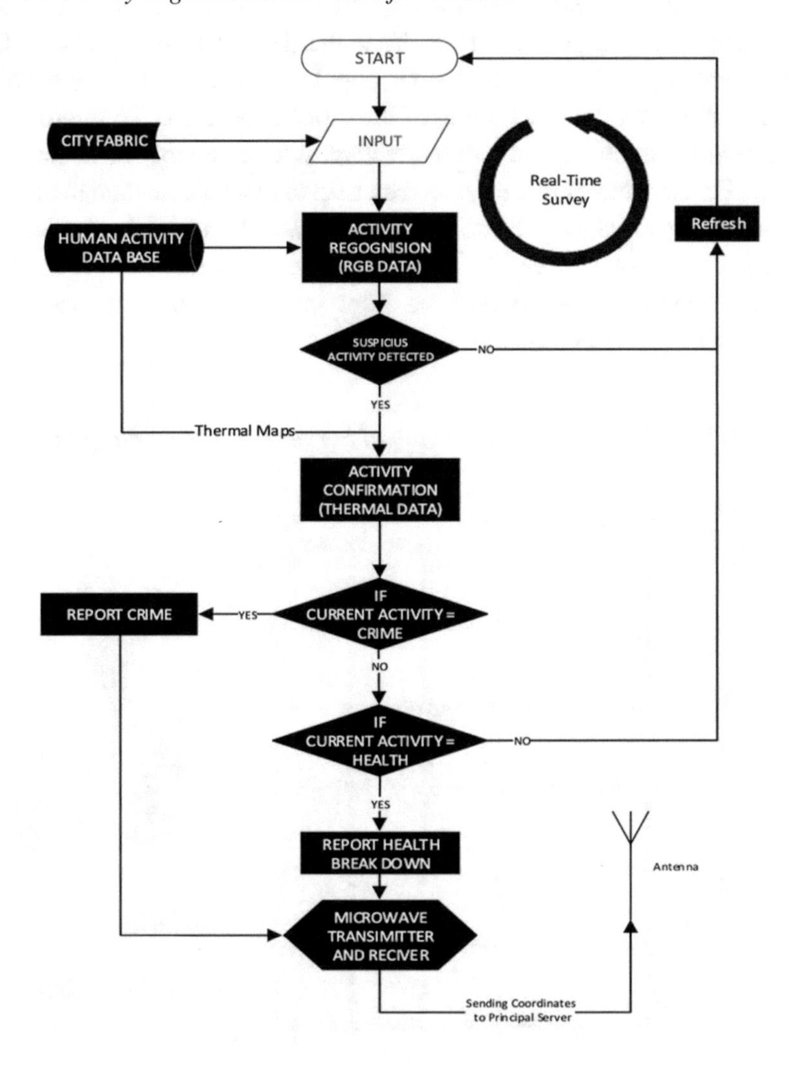

Human Actions to Digital Decisions

Drones are the building blocks of the surveillance system. The drones, in real-time have a computer vision over the city fabric. As soon as the drones recognize any malicious activity in its flight, it classifies whether the observed activity is a crime or health issue. The drone completes its task by transmitting the coordinates of the affected area along with the activity data to the server. (Figure 16) illustrates the survey algorithm flowchart for the drones.

The drone RGB sensors capture real-time video data from the city fabric and the activity recognition algorithm compare the data from the Human Activity Data Base in the IoT server with the real-time data. Usually, when no suspicious activity if detected the drones would simply refresh the data cache and repeat the cycle. The processing speed depends on the efficiency of the algorithm and the available data for comparison. But, in case, suspicious activity is detected, the drone analysis the thermal maps of the scene, for confirmation.

Through human emotion detection,

- The drone can predict whether the suspicious activity is - A case of crime or a medical emergency.
- The drones can predict the culprit and the victim, in a crime scene, with the human emotion classification using thermal imaging and heat maps.

Once, the drones process the required data, it then transmits the data to the IoT Server. By implementing AI-decision making algorithms, the processing is accomplished within the drones itself, and transmits only a few sets of data resulting in very less traffic of data across the surrounding environment, thus, encouraging an eco-friendly system.

It is to be noted that, the drones can be trained to recognized certain gestures, like rising the arms to call for help, waving arms up, and other gestures allotted understand the users. (Singh et al., 2018) For instance, suppose an elder walking in a street, feels some sort of dizziness, but no one is around him to help. In that case, all he has to do would be to raise his arms and wave it. As soon as the drone over the area notices this elder's gesture, the drone could analysis and notify the Central IoT server. This could be done, by gesture training the drones to recognize the waving arms are the signs of Medical Emergency. Also, if such a gesture is programmed to the UCCHC system, in a city, the citizens must be aware of such gestures and their meanings.

Low Light Surveillance: Night Vision

The RGB Sensor camera can be highly effective during the daytime (Raghuraman et al., 2015). But in low lights, these cameras become ineffective. It is, therefore, necessary to ensure that the camera has a night vision feature, to capture and survey during nighttime. But the use of thermal sensors and its data in the algorithm would simply tackle the issue of low exposure. But there is a point to emphasis from the study CCTV surveillance system and Artificial Intelligence: Case Study, that the resolution of the camera's sensors must be high, so the drone's heinous activity algorithms to function effectively and efficiently.

Figure 15. Survey Algorithm Flowchart for Drones

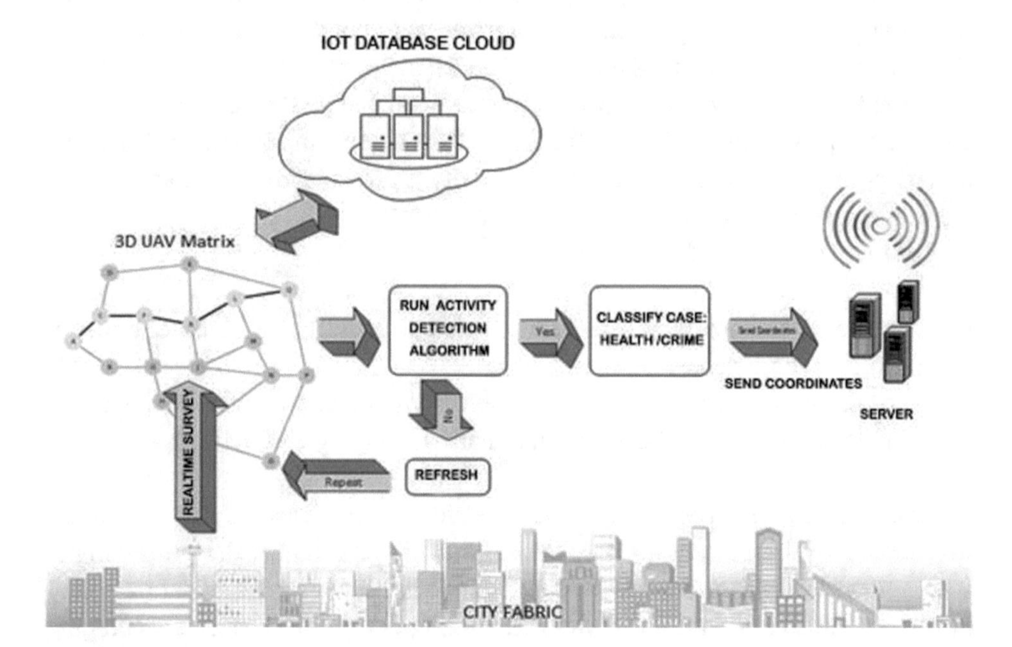

Digital Decisions to Reflex Actions

Possibilities of drone applications range from distributing food for hurricane victims to delivering medications for venomous animal bites to hikers in the wilderness. Drones are used to supply goods to various-disaster victims in remote locations. They can also help firefighters perceive the precise locations of fire and help emergency responders to locate victims.

However, in this system (Figure 17), drones in-case of recognizing heinous activities, the data is passed on to the main CPU. Where the CPU directs and triggers various service deploying units installed within the city fabric. These deploying units may include emergency healthcare vehicles, crime control vehicles, information kiosks, reflex drones that can deliver first aid kits, call for human help, or even warn the people around.

Although the services which include human assistance like, the ambulance, crime control vehicles-the overall process of surveillance, under an umbrella- is performing more efficient than traditionally it would do without UAVs, and AI, it is evident from the study that automated services, without the assistantships of humans, would perform faster.

Studies have shown that drones can function faster than normal vehicles in delivering services. This is because of its speed and traffic-free environment.

However, in the UCCHC system, the matrix of drones, depending on the density of drones, would be contributing to the air traffic and would alleviate the speeds of the drones. Moreover, problems like collision of drones might occur, but, through the application of AI, IoT along machine learning, these flying machines can be trained to tackle these drawbacks.

Figure 16. Working of UAV matrix layout and Principal Server

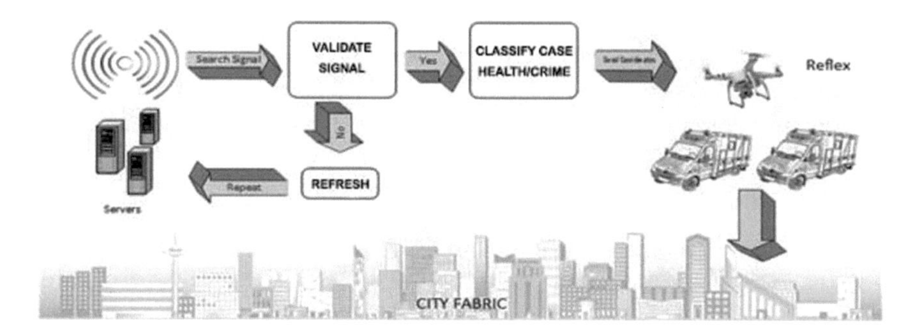

(Figure 17) shows two drones and sending data about the affected area to the principal servers and how the principal server classifies this received data and sends it to the respective service stations and further, the control station deploys its service to the affected area.

Figure 17. Working of UAV matrix layout and Principal Server

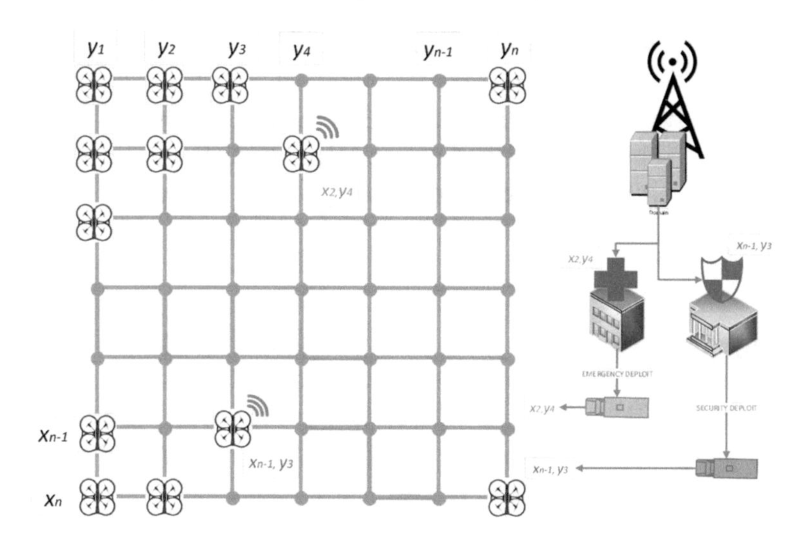

Figure 18. Algorithm Flowchart for the server

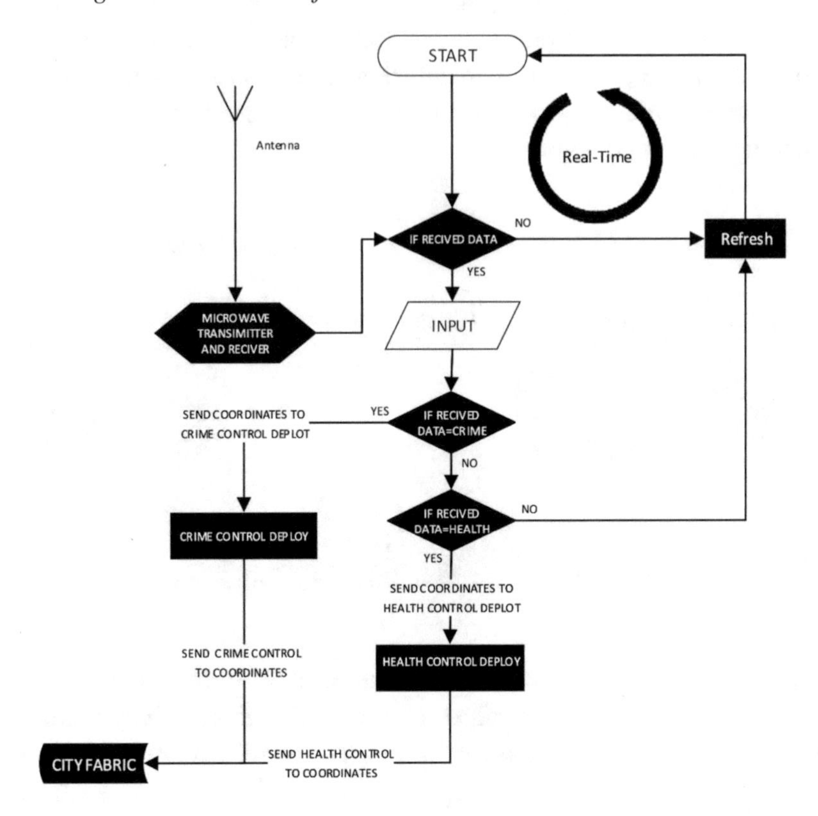

BENEFITS OF IMPLEMENTATION OF ARTIFICIAL INTELLIGENCE AND MACHINE LEARNING IN CITY SURVEILLANCE USING DRONE MATRIX

Technological advancement, liberal, and logical thinking are the key to breakdown and dissolve the problems that arise due to the UCCHC system like safety, security, privacy, emissions, energy, effects of microclimate, and weather.

Security

The security issue is becoming more significant in our day-to-day life. Places under surveillance typically allow certain human actions, and other actions are not allowed. With the input of UAV Matrix, a real-time visual surveillance system driven by action recognition and prediction algorithms may upsurge the likelihoods of apprehending a criminal on real-time video-footages and condense the menace caused by criminal actions (La Vigne et al., 2011; Alexandrie, 2017). The cameras also make some people feel more secure, knowing the criminals are being scrutinized.

Privacy

This system is part of the city fabric, and it doesn't include the spaces within the buildings. The City Fabric refers to the city volume excluding the premises of built structures and itself. So, the city fabric is the space that is communal. Also, such a space doesn't require the conduct of hidden and secretive activities. So, the issue of Privacy is not an effective cause to discard this idea of drone surveillance. Moreover, the fabric is supervised by non-Humans, and the data is just studied in detail unless the system recognizes any heinous activity. Also, the strict unmanned aerial vehicle permission and laws do ensure no third-party devices hovering above the covered space.

Control on the EMF exposure

In order to connect with the network of the UAV matrix, each drone must send out wireless or radio frequency signals, which is a form of Electromagnetic field radiations. So, won't this idea establish an environment that is hazardous to live and would be a sea of EMF radiations? All electronic devices which work wireless emit EMF radiations. EMF radiations consist of Extremely Low Frequency (ELF) Radiation and Radio Frequency (RF) Radiations. The ELF is emitted by every electronic device with a battery or electricity supply (Maleki et al., 2016). But, on top, RF radiations are emitted to connect with the surrounding system, similar to a wi-Fi router or cell phone connecting to a mobile signal tower.

Studies have estimated (GSMA Intelligence, 2017) about 77 percent of the world population uses devices that emit high levels of EMF, mobile phones. (National Cancer Institute, n.d.)When we are constantly exposed to EMFs, they can be very damaging. Researches and studies have concluded that the effects of EMF exposure vary from minor health concerns like headaches and skin rashes to very serious concerns like fertility problems, DNA fragmentation, cell damage, and cancerous tumors (National Cancer Institute, n.d.). These emissions are drastically reduced by limited transfer of data wirelessly. Since the drones are implemented with AI, most of the decisions are taken individually by the drones alone. This does not necessitate continuous interaction between the drone and the server.

Renewable Energy and Green Drones

Such a system uses renewable energy and even stored in extra batteries for emergencies these solar-powered drones. (Jain and Mueller, 2019) The Developer Aviation Industry Corporation of China (AVIC) had already test-flown the unmanned aerial vehicle in mid-2018, albeit with a smaller wingspan. The latest flight used a 20-meter wingspan prototype weighing 18.9kg intended to replicate the ultimate model. The drone reached and maintained a sailing altitude of over 20,000 meters (Nurkin et al., 2018). Airbus' Zephyr S, for example, the system managed to retain the drone buoyant for 2 weeks, and almost 26 days in mid-2018. This was because of the battery technology advancements and the usage of a light weighted solar array.

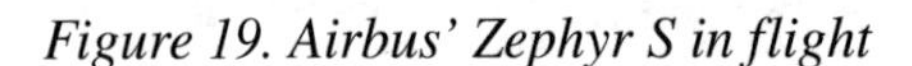

Figure 19. Airbus' Zephyr S in flight

Safety Measures

In some cases, due to the failure of the drone- flight failure due to winds, low power- can be problematic. There are chances of these machines falling from the sky with higher velocities and might have some sharp edge parts like the wing blades (Schenkelberg, 2016). They might harm people and animals if they fall over (Doward, 2019).

However, the power of machine learning has allowed previous machines to correct these errors due to failures (Wuest et al,. 2016; Attaran and Deb, 2018), and also, they can be programmed to fall in a population free area and reduce the impact on the city fabric. Frequent diagnostic self-checks of the drones would avoid such failure, and help the system fix the problems ahead.

Winds, Microclimate, and Weather

The weather patterns have been a very challenging aspect in the functioning of drones, since the invention of drones. Due to its lightweight and small body, drones are very sensitive to winds and breezes (Wang et al., 2019; Galway et al., 2011). New technologies have developed drones that can increase-decrease the rotation speeds of its rotors, to cope up with the effects of winds and to remain stable. However, this process can be very energy-consuming. Rains, Storms and Heavy winds can damage these tiny flying machines. But there are larger possibilities of advancement in drone technology, and these engineering problems will be classified in the coming years (Chatys and Koruba, 2005).

CONCLUSION

The very knowledge of this chapter is to encourage the scope of drone surveillance and implementing the surveillance with human activity prediction to serve as Unmanned City Crime-Health Control (UCCHC) System and sculpt an advanced productive infrastructure. Functioning of the UCCHC over city fabrics can ensure greater security of its users and has the potential to work effectively to respond to criminal activities and heath-breakdowns of its citizens. Emergency alerts and information are delivered instantaneously to the reacting bodies within the system. Such advancements in surveillance can promptly unravel and challenge the occurrence of monstrous actions. The implementation of such technology can be very advantageous for developing countries like India, where the population is very high and is problematic to manage a large number of users in its cities. Consequently, the number of casualties is relatively higher due to deferment in the identification of the problem and communications. So, the UCCHC system can drastically resolve the problems, ensuing more safe and secure cities for its users. Moreover, programmed city-surveillance in Indian cities can condense the wastage of human resources in city management and open up scopes for new research and employment. Along with safety and security, the technology would contribute to resource optimization as targeted resources and actions are to be devoted in a timely manner. The system could further provide an integrated opportunity to address various ICT led components for traffic, health, crime, security, safety, development, etc. in a smart city. The platform could act as a central force in devising plans and programs for urban planning and management. With numerous advantages and advent of technological innovations, the application of the UCCHC system is inescapable for the operation of the city mechanisms. As per the future application of UAVs, it goes beyond the speculation of the human mind. With the significant upsurge in the usage of UAVs, IoT and Cellular Networks likely in the coming years, this study could be an advantageous resource for implementers as well as future researchers.

REFERENCES

Ackerman, E., & Koziol, M. (2019). The blood is here: Zipline's medical delivery drones are changing the game in Rwanda. *IEEE Spectrum*, *56*(5), 24–31.

Agustina, J., & Clavell, G. G. (2011). The impact of CCTV on fundamental rights and crime prevention strategies: The case of the Control commission of video surveillance devices. *Computer Law & Security Review*, *27*(2), 168–174.

Alexandrie, G. (2017). Surveillance cameras and crime: A review of randomized and natural experiments. *Journal of Scandinavian Studies in Criminology and Crime Prevention*, *18*(2), 210–222.

Anwar, N., Izhar, M. A., & Najam, F. A. (2018, July). Construction monitoring and reporting using drones and unmanned aerial vehicles (UAVs). In *The Tenth International Conference on Construction in the 21st Century (CITC-10)* (pp. 2-4). Academic Press.

Attaran, M., & Deb, P. (2018). Machine learning: The new'big thing'for competitive advantage. *International Journal of Knowledge Engineering and Data Mining*, *5*(4), 277–305.

Balaji, B., Chennupati, S. K., Chilakalapudi, S. R. K., Katuri, R., & Mareedu, K. (2018). Design of UAV (drone) for crop, weather monitoring and for spraying fertilizers and pesticides. *Int J Res Trends Innov*, *3*(3), 42–47.

Bishop, C. M. (2003). *Neural Networks for Pattern Recognition* (Indian edition). Oxford University Press.

Biswas, K. K., & Basu, S. K. (2011, December). Gesture recognition using microsoft kinect®. In *The 5th international conference on automation, robotics and applications* (pp. 100-103). IEEE.

Bobick, A. F., & Davis, J. W. (2001). The recognition of human movement using temporal templates. *IEEE Transactions on Pattern Analysis and Machine Intelligence*, *23*(3), 257–267.

Bulling, A., Blanke, U., & Schiele, B. (2014). A tutorial on human activity recognition using body-worn inertial sensors. *ACM Computing Surveys*, *46*(3), 1–33.

Burgard, W., Moors, M., Stachniss, C., & Schneider, F. E. (2005). Coordinated multi-robot exploration. *IEEE Transactions on Robotics*, *21*(3), 376–386.

Chatys, R., & Koruba, Z. (2005). Gyroscope-based control and stabilization of unmanned aerial mini-vehicle (mini-UAV). *Aviation*, *9*(2), 10–16.

De Haan, W., & Loader, I. (2002). On the emotions of crime, punishment and social control. *Theoretical Criminology*, *6*(3), 243–253.

Dominguez, L., D'Amato, J. P., Perez, A., Rubiales, A., & Barbuzza, R. (2018, June). Running License Plate Recognition (LPR) algorithms on smart survillance cameras. A feasibility analysis. In *2018 13th Iberian Conference on Information Systems and Technologies (CISTI)* (pp. 1-5). IEEE.

Doward, J. (2019, June 9). *Military drone crashes raise fears for civilians*. https://www.theguardian.com/world/2019/jun/09/two-military-drones-crashing-every-month

DroneDeploy. (2018, June 7). https://www.dronedeploy.com/blog/rise-drones-construction/

Dung, N. D., & Rohacs, J. (2018, November). The drone-following models in smart cities. In *2018 IEEE 59th international scientific conference on power and electrical engineering of Riga Technical University (RTUCON)* (pp. 1-6). IEEE.

Duque Domingo, J., Cerrada, C., Valero, E., & Cerrada, J. A. (2017). An improved indoor positioning system using RGB-D cameras and wireless networks for use in complex environments. *Sensors (Basel)*, *17*(10), 2391.

Eze, K. G., Sadiku, M. N., & Musa, S. M. (2018). 5G wireless technology: A primer. *International Journal of Scientific Engineering and Technology*, *7*(7), 62–64.

Fakhrulddin, S. S., Gharghan, S. K., Al-Naji, A., & Chahl, J. (2019). An advanced first aid system based on an unmanned aerial vehicles and a wireless body area sensor network for elderly persons in outdoor environments. *Sensors (Basel)*, *19*(13), 2955.

Galway, D., Etele, J., & Fusina, G. (2011). Modeling of urban wind field effects on unmanned rotorcraft flight. *Journal of Aircraft*, *48*(5), 1613–1620.

GSMA Intelligence. (2017). *Global Mobile Trends 2017*. Retrieved from https://www.gsmaintelligence.com/research/?file=3df1b7d57b1e63a0cbc3d585feb82dc2&utm_source=Triggermail&utm_medium=email&utm_campaign=Post%20Blast%20%28bii-apps-and-platforms%29:%20Apple%20drops%20in-app%20tipping%20tax%20%E2%80%94%20Two-thirds%20of%20the

Isah, E. C., Asuzu, M. C., & Okojie, O. H. (1997). Occupational health hazards in manufacturing industries in Nigeria. *J Community Med Primary Health Care*, *9*, 26–34.

Jain, K. P., & Mueller, M. W. (2020, May). Flying batteries: In-flight battery switching to increase multirotor flight time. In *2020 IEEE International Conference on Robotics and Automation (ICRA)* (pp. 3510-3516). IEEE.

Jain, P. K. (1991). On blackbody radiation. *Physics Education*, *26*(3), 190.

Jamali, J., Bahrami, B., Heidari, A., & Allahverdizadeh, P. (2020). IoT Architecture. In J. Jamali, B. Bahrami, A. Heidari, P. Allahverdizadeh, & F. Norouzi (Eds.), *Towards the internet of things* (pp. 9–31). Springer International Publishing.

Kardasz, P., Doskocz, J., Hejduk, M., Wiejkut, P., & Zarzycki, H. (2016). Drones and possibilities of their using. *Journal of Civil and Environmental Engineering*, *6*(3), 1–7.

Kong, Y., & Fu, Y. (2022). Human action recognition and prediction: A survey. *International Journal of Computer Vision*, *130*(5), 1366–1401. doi:10.100711263-022-01594-9

La Vigne, N. G., Lowry, S. S., Markman, J. A., & Dwyer, A. M. (2011). *Evaluating the use of public surveillance cameras for crime control and prevention*. US Department of Justice, Office of Community Oriented Policing Services. Urban Institute, Justice Policy Center. doi:10.1037/e718202011-001

Liew, C. F., DeLatte, D., Takeishi, N., & Yairi, T. (2017). *Recent developments in aerial robotics: A survey and prototypes overview*. arXiv preprint arXiv:1711.10085.

Maleki, H., Zurek, R., Howard, J. N., & Hallmark, J. A. (2016). Lithium ion cell/batteries electromagnetic field reduction in phones for hearing aid compliance. *Batteries*, *2*(2), 19.

Mckinnon, A. C. (2016). The possible impact of 3D printing and drones on last-mile logistics: An exploratory study. *Built Environment*, *42*(4), 617–629.

Meola, C., & Carlomagno, G. M. (2004). Recent advances in the use of infrared thermography. *Measurement Science & Technology*, *15*(9), R27.

Mogili, U. R., & Deepak, B. B. V. L. (2018). Review on application of drone systems in precision agriculture. *Procedia Computer Science*, *133*, 502–509.

National Cancer Institute. (n.d.a). *About Cancer / EMF radiation and Cancer Risk*. https://www.cancer.gov/about-cancer/causes-prevention/risk/radiation/electromagnetic-fields-fact-sheet#r1

National Cancer Institute. (n.d.b). *About Cancer / Cellphones and Cancer Risk*. https://www.cancer.gov/about-cancer/causes-prevention/risk/radiation/cell-phones-fact-sheet

Nummenmaa, L., Glerean, E., Hari, R., & Hietanen, J. K. (2014). Bodily maps of emotions. *Proceedings of the National Academy of Sciences of the United States of America*, *111*(2), 646–651. doi:10.1073/pnas.1321664111 PMID:24379370

Nurkin, T., Bedard, K., Clad, J., Scott, C., & Grevatt, J. (2018). *China's Advanced Weapons Systems*. Jane's by IHS Markit.

Omid, M., & Torbjorn, N. (2019). Drones in manufacturing: Exploring opportunities for research and practice. *Journal of Manufacturing Technology Management*.

Raghuraman, S., Bahirat, K., & Prabhakaran, B. (2015, June). Evaluating the efficacy of RGB-D cameras for surveillance. In *2015 IEEE International Conference on Multimedia and Expo (ICME)* (pp. 1-6). IEEE.

Rodríguez-Moreno, I., Martínez-Otzeta, J. M., Sierra, B., Rodriguez, I., & Jauregi, E. (2019). Video activity recognition: State-of-the-art. *Sensors (Basel)*, *19*(14), 3160.

Ryoo, M. S. (2011, November). Human activity prediction: Early recognition of ongoing activities from streaming videos. In *2011 International Conference on Computer Vision* (pp. 1036-1043). IEEE.

Schenkelberg, F. (2016, January). How reliable does a delivery drone have to be? In *2016 annual reliability and maintainability symposium (RAMS)* (pp. 1-5). IEEE.

Shotton, J., Fitzgibbon, A., Cook, M., Sharp, T., Finocchio, M., Moore, R., & Blake, A. (2011, June). Real-time human pose recognition in parts from single depth images. In *CVPR 2011* (pp. 1297–1304). IEEE.

Singh, A., Patil, D., & Omkar, S. N. (2018). Eye in the sky: Real-time drone surveillance system (dss) for violent individuals identification using scatternet hybrid deep learning network. In *Proceedings of the IEEE conference on computer vision and pattern recognition workshops* (pp. 1629-1637). 10.1109/CVPRW.2018.00214

Statista Research Department. (2019). *Military drones (UAS/UAV): estimated U.S. and global R&D budget 2014-2023*. Statista Research Department.

Suriyarachchi, C., Waidyasekara, K. G. A. S., & Madhusanka, N. (2019, June). Integrating Internet of Things (IoT) and facilities manager in smart buildings: A conceptual framework. In *The 7th World Construction Symposium 2018: Built Asset Sustainability: Rethinking Design Construction and Operation* (Vol. 29, pp. 325-334). Academic Press.

Udland, M. (2015, October 12). *World Economic Forum: Why labour is becoming more expensive*. https://www.weforum.org/agenda/2015/10/why-labour-is-becoming-more-expensive/

United Nations. (2017). *World Population Prospects: The 2017 Revision*. Department of Economic and Social Affairs. Retrieved from https://www.un.org/development/desa/en/news/population/world-population-prospects-2017.html

United Nations. (2019). *World Urbanization Prospects: The 2018 Revision*. United Nations.

Vrigkas, M., Nikou, C., & Kakadiaris, I. A. (2015). A review of human activity recognition methods. *Frontiers in Robotics and AI*, *2*, 28.

Wang, L., Misra, G., & Bai, X. (2019). AK Nearest neighborhood-based wind estimation for rotary-wing VTOL UAVs. *Drones*, *3*(2), 31.

Wuest, T., Weimer, D., Irgens, C., & Thoben, K. D. (2016). Machine learning in manufacturing: Advantages, challenges, and applications. *Production & Manufacturing Research*, *4*(1), 23–45.

Yan, H., Xu, L. D., Bi, Z., Pang, Z., Zhang, J., & Chen, Y. (2015). An emerging technology–wearable wireless sensor networks with applications in human health condition monitoring. *Journal of Management Analytics*, *2*(2), 121–137.

ADDITIONAL READING

Cheng, J., Chen, X., & Shen, M. (2013). A Framework for Daily Activity Monitoring and Fall Detection Based on Surface Electromyography and Accelerometer Signals. *IEEE Journal of Biomedical and Health Informatics*, *17*(1), 38–45. doi:10.1109/TITB.2012.2226905 PMID:24234563

Vernon, D. (2019). Robotics and Artificial Intelligence in Africa. *IEEE Robotics & Automation Magazine*, *26*(4), 131–135. doi:10.1109/MRA.2019.2946107

KEY TERMS AND DEFINITIONS

Activity Dataset in Cloud Database: The database containing catalogs of human activity classification, thermal-human behavioral maps, gesture datum, etc., that will be used by the algorithms to classify and predict heinous activities.

City Fabric: City Fabric is the fabric of space in a city that excludes all the building envelopes. City fabric includes the Urban Infrastructures, Streets, Morphology etc. Usually this is the space that is maintained by the public governmental services.

Heinous Activities: Activities that are suspicious in nature, which can be used to predict the occurrence of unpleasant event. For example, in a crime scene, when a criminal is going to perform a crime, the all set of activities that happens just before he performs the crime, like lifting up weapons, moving towards victim in a suspicious manner, etc.

Human Activity Prediction: Ability to predict the future event that is going to happen, from a given sequence of image or video footage.

Human Activity Recognition: Ability to classify and recognize human activity with the help of image sequences or a video footage.

Motion Energy Image: It is a grayscale static image where; the frames of a video clip are collapse into single. This is used to predict and recognize the subjects in a video footage.

Motion History Image: It is a grayscale static image, which help in understanding the progress of subjects in a video footage. In this image, the frames of a video clip are collapse into single, with the brightness of the frames increasing the time stamps of the frame in the video. This is used to predict motion and motion flow.

UAV Matrix: The drone matrix is the three-dimensional spatial network of actively functioning drones hovering above the city fabric, performing the task of surveillance, working together. They can be labeled by their spatial coordinates.

Chapter 7
Mathematical Modeling of Unmanned Aerial Vehicles for Smart City Vehicular Surveillance Systems

Divya P. S.
Karunya Institute of Technology and Sciences, India

Manoj G.
Karunya Institute of Technology and Science, India

Jebasingh S.
Karunya Institute of Technology and Science, India

ABSTRACT

In the real world, smart city traffic management is a difficult phenomenon. Introducing the internet of things into traffic management systems in smart cities is a huge challenge. Smart city definitions differ from city to city and country to country, depending on the city's level of growth, willingness to change and reform, finances, and ambitions. Unmanned aerial vehicles (UAVs) have been used in a variety of applications for civil and defense infrastructure management. These uses include crowd surveillance, transportation, emergency management, and building design inspection. In smart cities, a variety of transport options exist with respect to public transport and private transport connectivity. The mathematical modelling-based vehicular network enables automobile manufacturers to incorporate smart features into vehicles at a low cost, boosting their market competitiveness. This proposal addresses the challenges concerning the surveillance system for smart city traffic management systems (TMSs).

DOI: 10.4018/978-1-7998-8763-8.ch007

INTRODUCTION

Unmanned Aerial Vehicles (UAV) are becoming more common in a wide range of scenarios and applications as a result of technical advancements in sensor, electronics, and telecommunication technologies. Furthermore, the UAVs can fly independently and/or remotely without the requirement for any human personnel on board. Their inclusion in the Urban Internet of Things (IoT) (Zanella et al., 2014) is a significant step forward. It has an impact in several civil and military applications that current UAV systems supports.

A Ground Control Station (GCS) and UAVs make up a conventional UAV system. UAVs are controlled by a GCS, which establishes a data control loop by wirelessly interconnecting with the UAV. The upstream flow in this data control loop consists of control messages, while the downstream flow provides telemetry and data. Downstream telemetry comprises GPS location, altitude, battery status, attitude, heading, and wind speed, as well as information about the present status of the UAV and the surrounding environment. The structure and frequency of control messages upstream depends on the application and goal. Upstream control and downstream data flows (Mulligan and Olsson, 2013) generally have strict Quality of Service (QoS) requirements due to the essential nature of many UAV applications.

BACKGROUND

The market for unmanned aerial vehicles (UAVs) has expanded in the last five years. Figure 1 income from commercial drones from 2016 to 2025 and expected revenue through 2025 (Chourabi et al., 2012), while UAV market values in various sectors. Infrastructure development, data mining from multi-dimensional maps, and precision agriculture can all benefit from UAVs and machine learning.

Figure 1. Commercial Reveue of UAV in USD

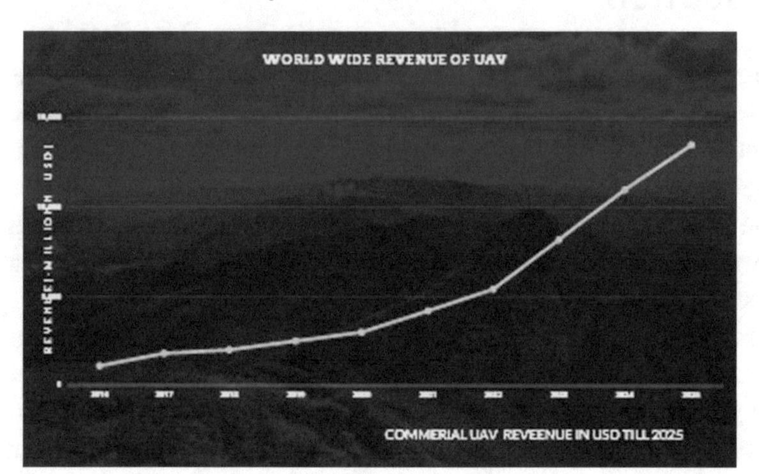

The integration of technologies necessary to deliver smart cities begins with the notion of smart cities, inhabitants with services that are more efficient and timelier. Drones have a lot of potential (Stöcker et al., 2017) in this competition for fast turning a future idea into reality and making cities a better place to live.

The Internet of Things (IoT) is envisioned as a near future communication paradigm that will allow everyday objects to be supplied with digital transceivers, communication, microcontrollers and other suitable protocols they will be able to communicate with each other and also add capability to interact with users via the Internet (Bellavista et al., 2013). The majority of Urban IoT or the foundation for smart city services is on a global or centralized architecture, in which a collection of dense and heterogeneous peripheral devices is placed across the city. These devices generate various forms of data, which is subsequently transmitted to a central control using appropriate communication technologies (Rodrigues et al., 2017), where it is stored and processed.

Figure 2. Smart City structure system for Urban Cities

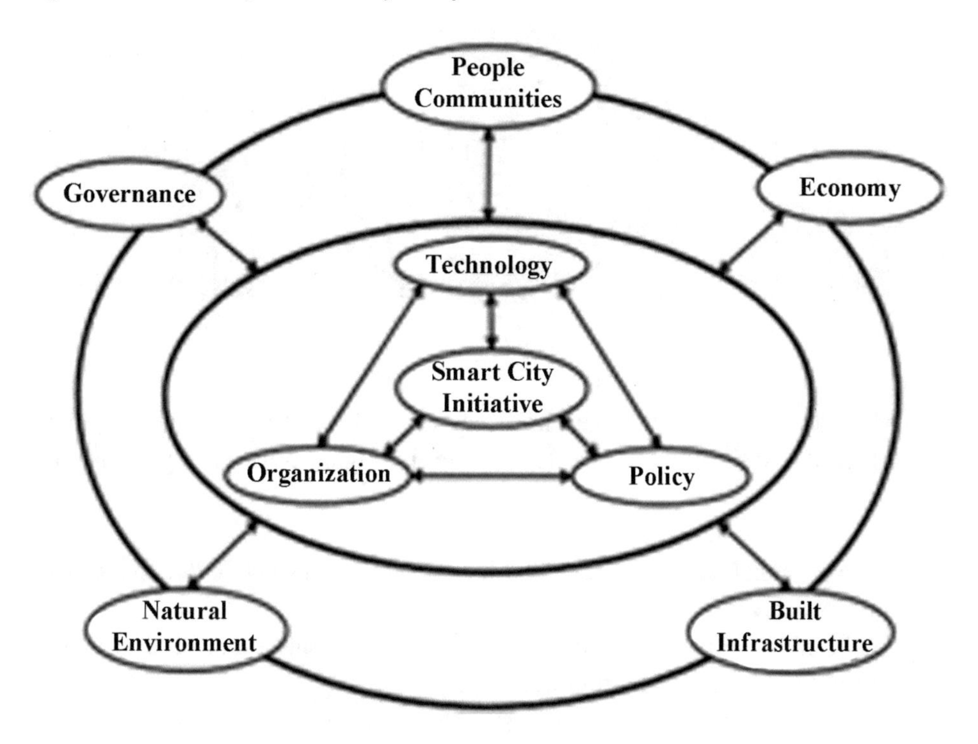

All of our life's economic, commercial, and social activities take place in cities. Cities are leaning on technology to help them cope with the situation the information as they become denser and more urban. The popularity of smart cities is increasing by the day, particularly in the aftermath of the global financial crisis. The global population (Mohammed et al., 2014) is increasing and by 2050, it is predicted to double. Furthermore, the world's population is increasingly concentrated in cities. Cities accounted for 54.5 percent of the world population in 2020 and this number is expected to rise to 60% by 2030 (Ha, 2020). In cities, we conduct all of our economic, commercial and social activities. Cities are relying on technology to help them accommodate more information as they get denser and more urban. The Figure 2 shows the smart city structure for the Urban Cities.

Figure 3. (UAVs) Unmanned Aerial Vehicles working in an urban setup

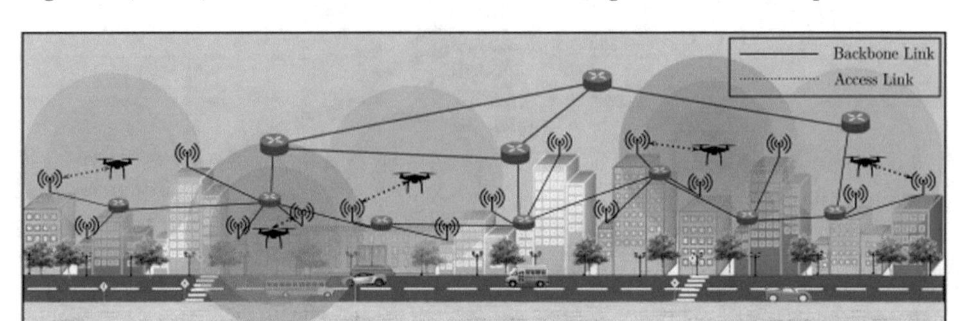

The major challenges involved in the urban environment (Gupta et al., 2016; Semsch et al., 2009) like the one depicted in the Figure 3 for ITS (Indian traffic System)

- Line of Sight (LoS) obstruction between the GCS and a UAV due to the urban infrastructure.
- Multi-path propagation because of the buildings and other civil structures present in the urban environment.
- Highly dynamic and crowded spectrum, since several technologies are operating in the same spectrum and each one is supporting many different users.
- Local or temporal congestion due to the multitude of other devices sharing the same network infrastructure.

Importantly, when the UAV is unplugged from the GCS, fail-safe measures are engaged or when the UAV doesn't get any message from GCS for a certain amount of time. The fail-safe mechanism causes a return-to-launch function and/or emergency landing using GPS. In all circumstances, however, the UAV fails to complete the task. Additionally, emergency landing may result in damage to the UAV (Kanistras et al., 2013), and pose a safety danger, especially in urban environments. Another solution would be to just hover the UAV until it gets a stable connection with the GCS, but this will be an energy inefficient way.

UNMANNED AERIAL VEHICLE (UAV) NETWORK

Unmanned aerial vehicles (UAVs), sometimes known as drones, are aircraft that can fly without the assistance of a human pilot. UAVs come in various sizes (Fu et al., 2016). Small UAVs can be employed in formations or swarms, whereas large UAVs can be utilised alone on missions. The later once are more useful in civil applications in urban environment. Before moving into the flight details, it is important to understand the controls available for UAV. There are four main controls for a quadcopter UAV (de Freitas et al., 2010), which are as listed below.

Mathematical Model of Dependability Taxonomy

The mathematical model for taxonomy is defined by Avizienis et al. defined three ideas are arranged in a tree. (Avizienis et al., 2004) is shown in the Figure 4. The dependability depends on the following attributes, means, and threads.

- **Attributes:** The attributes of the system quantifiable and evaluable properties characterizing system performance.
- **Means:** This strategy improves the value of qualities.
- **Threats:** Incidents that negatively impact system performance

Figure 4. Mathematical model for Dependability of the system

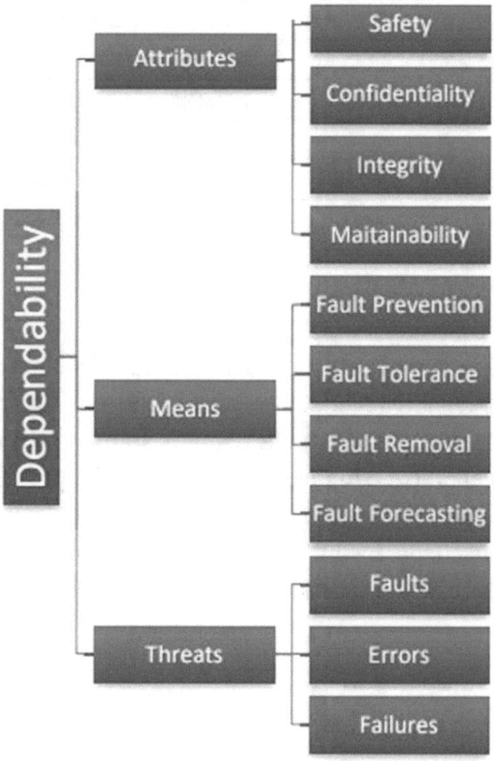

Mathematical Model Analysis

The purpose of the model analysis is to provide a probabilistic approach to the top event while taking fundamental events into account. The chance of the top event "communication breakdown" is predicted using an FTA (Cho et al., 2011) failing at a given time. Some reference data may be helpful in achieving the goal. However, finding failure rates for all basic occurrences is difficult. The Non electronic is a tool used to obtain data for fundamental events from the Parts Reliability Data Publication (NPRD-2016) database. This database lists each component, its environment, and other resources that differ from one component to the next. 3We select the minimum, maximum and mean failure rates, as well as the number of failures and hours from these sources. This database (George and Sousa, 2011) was used in our simulations using a real equipment failure rate. It offers the following reliability information for each one as shown in the table 1.

Table 1. Parameter of information

λmin	Minimum data failure rate
λmax	Maximum data failure rate
λmean	Mean failure data rate
Nmin	Minimum number failed
Nmax	Maximum number failed
tmin	Minimum number of operation hours
tmax	Maximum number of operation hours

Burr Distribution for UAV

The Burr distribution for UAV or the Burr Type XII a continuous probability distribution of a non-negative random variable is called a distribution. It is likewise known as the Singh-Maddala distribution and is solitary of a number of different distributions. The "generalized log-logistic distribution" appears on occasion. Table 2 shows the Burr Type XII distribution, which is part of a system of continuous distributions introduced by Irving W. Burr (1942), which comprises 12 major distributions. The Burr PDF expression is given as is given in the equation (1)

$$f_v(V) = \frac{ak\left(\frac{V}{\beta}\right)^{\alpha-1}}{\beta\left(1+\left(\frac{V}{\beta}\right)^{\alpha}\right)^{k+1}} \tag{1}$$

where k and α are continuous shape parameters (k>0,α >0), respectively, while β is a continuous scale parameter (β >0).
The cumulative distribution function (CDF) of Burr distribution is given as

$$F_v\left(V\right) = 1-\left(1+\left(\frac{V}{\beta}\right)^{\alpha}\right)^{-k} \tag{2}$$

Table 2. Parameter for Burr Distributions

Parameters	k>0, α>0 are shape parameters and β>0 scale parameter
Mean	$\mu_1 = \beta B\left(\beta - \frac{1}{\alpha}, 1 + \frac{1}{\alpha}\right)$ where B is the Beta function
Median	$(2^{1/\beta} - 1)1^{/\alpha}$
Mode	$\left(\frac{\alpha - 1}{\alpha\beta + 1}\right)^{1/\alpha}$
Variance	$-\mu_1^2 + \mu_2$
Skewness	$\frac{2\mu_1^3 - 3\mu_1\mu_2 + \mu_3}{\left(-\mu_1^2 + \mu_2\right)^{3/2}}$
Kurtosis	$\frac{-3\mu_1^4 + 6\mu_1^2\mu_2 - 4\mu_1\mu_3 + \mu_4}{\left(-\mu_1^2 + \mu_2\right)^2} - 3$, where moments $\mu_r = \beta B\left(\frac{\alpha\beta - r}{\alpha}, \frac{\alpha + r}{\alpha}\right)$

Dagum Distribution for UAV

Dagum distribution for UAV is an uninterrupted probability distribution defined over the positive real numbers. It is christened after Camilo Dagum. The Dagum distribution for UAV has three parameters k>0, α>0 are continuous shape parameters and β>0 is a scale parameter is shown in the table 3. The Dagum PDF is given by (Kleiber and Christian, 2008) is given in the equation 3:

$$f_v(V) = \frac{ak\left(\frac{V}{\beta}\right)^{\alpha k-1}}{\beta\left(1+\left(\frac{V}{\beta}\right)^{\alpha}\right)^{k+1}} \tag{3}$$

The (CDF) cumulative density function of Dagum distribution function are given as

$$f_v(V) = \left(1+\left(\frac{V}{\beta}\right)^{-\alpha}\right)^{-1} \tag{4}$$

Table 3. Parameter for Dagum Ditributions

Parameters	k>0, α>0 are shape parameters and β>0 scale parameter
Mean	$-\frac{\beta}{\alpha}\frac{\Gamma\left(-\frac{1}{\alpha}\right)\Gamma\left(\frac{1}{\alpha}+k\right)}{\Gamma(k)}$, if α>1 Indeterminate , otherwise
Median	$\beta\left(-1+2^{1/k}\right)^{-\frac{1}{\alpha}}$
Mode	$\beta\left(\frac{\alpha k-1}{\alpha+1}\right)^{1/\alpha}$
Variance	$-\frac{\beta^2}{\alpha^2}\left(2\alpha\frac{\Gamma\left(-\frac{2}{\alpha}\right)\Gamma\left(\frac{2}{\alpha}+k\right)}{\Gamma(k)}+\left(\frac{\Gamma\left(-\frac{1}{\alpha}\right)\Gamma\left(\frac{1}{\alpha}+k\right)}{\Gamma(k)}\right)^2\right)$, if α>2 Indeterminate, otherwise

Gamma Distribution for UAV

The Gamma for UAV is a continuous probability distribution family parameter is given in the table 4 for the UAV and the $\alpha>0$ is a continuous shape parameter, while $\beta>0$ is a continuous scale parameter. Its PDF is given as is given in equation 5

$$f_v(V) = \frac{1}{\Gamma(\alpha)\beta^{\alpha}} V^{\alpha-1} \exp\left(-\frac{v}{\beta}\right) \tag{5}$$

The cummulative density Function(CDF) for the Gamma Distibution is given as

$$f_v(V) = \frac{1}{\Gamma(\alpha)} \gamma\left(\alpha, \frac{v}{\beta}\right) \tag{6}$$

Where the $\gamma\left(\alpha, \frac{v}{\beta}\right)$ is the lower incomplete gamma function. The parameter of the Gamma Fucntion is given as

Table 4. Parameter for gamma distributions

Parameters	$\alpha>0$ shape parameter and $\beta>0$ scale parameter
Mean	$\alpha\beta$
Median	No simple closed form
Mode	$(\alpha-1)\beta$, for $\alpha\geq1$
Variance	$\alpha\beta2$
Skewness	$\frac{2}{\sqrt{\alpha}}$
Kurtosis	$\frac{6}{\alpha}$

Weibull Distribution for UAV

A Weibull distribution for UAV with a total length of 100,000 hours is taken into consideration since it is the most probable method for describing a random component lifetime distribution (Hallinan, 1993). Our objective is to enable successful communication between UAVs, drones, and GCC on the one hand, while also taking into account real-world limits like as large obstacles on the other. A Beta 1 distribution can be used to depict the failure rate (Pearson theory on probability distributions) because it is regarded a random variable that is specified between two limits [min, max] (Jiang and Swindlehurst, 2010). The Weibull Distribution for the UAV is calculated as follows:

$$f_v(V) = \frac{\alpha}{\beta}\left(\frac{V}{\beta}\right)^{\alpha-1} \exp\left[-\left(\frac{V}{\beta}\right)^{\alpha}\right] \tag{7}$$

Where a continuous shape parameter $\alpha>0$ and a continuous scale parameter $\beta>0$ are used.

The Cumulative Density Function (CDF) of the Weibull distribution function is given as

$$f_v(V) = 1 - \exp\left[-\left(\frac{V}{\beta}\right)^{\alpha}\right] \tag{8}$$

The parameter of the Weibull distribution is given as shown in the table 5 for the UAV,

Table 5. Parameters for Weibull Distribution

Parameters	α>0 shape parameter, β>0 scale parameter
Mean	$\beta\Gamma(1+1/\alpha)$
Median	$\beta(\ln 2)1^{/\alpha}$
Mode	$\beta\left(\frac{\alpha-1}{\alpha}\right)$ if a>1 0 if $\alpha \leq 1$
Variance	$\beta^2\left[\Gamma\left(1+\frac{2}{\alpha}\right)-\left(\Gamma\left(1+\frac{1}{\alpha}\right)\right)^2\right]$
Skewness	$\frac{\Gamma(1+3/\alpha)\beta^3-3\mu\sigma^2-\mu^3}{\sigma^3}$

The graphical representation of goodness of fit is represented in Figure 5. In station 1 and station 4 the curves whose peak is above the histogram may be neglected, since the peak above the histogram means that, the particular distribution is not fitting the UAV data.

Figure 5. Fitting of UAV speed of data with various distributions

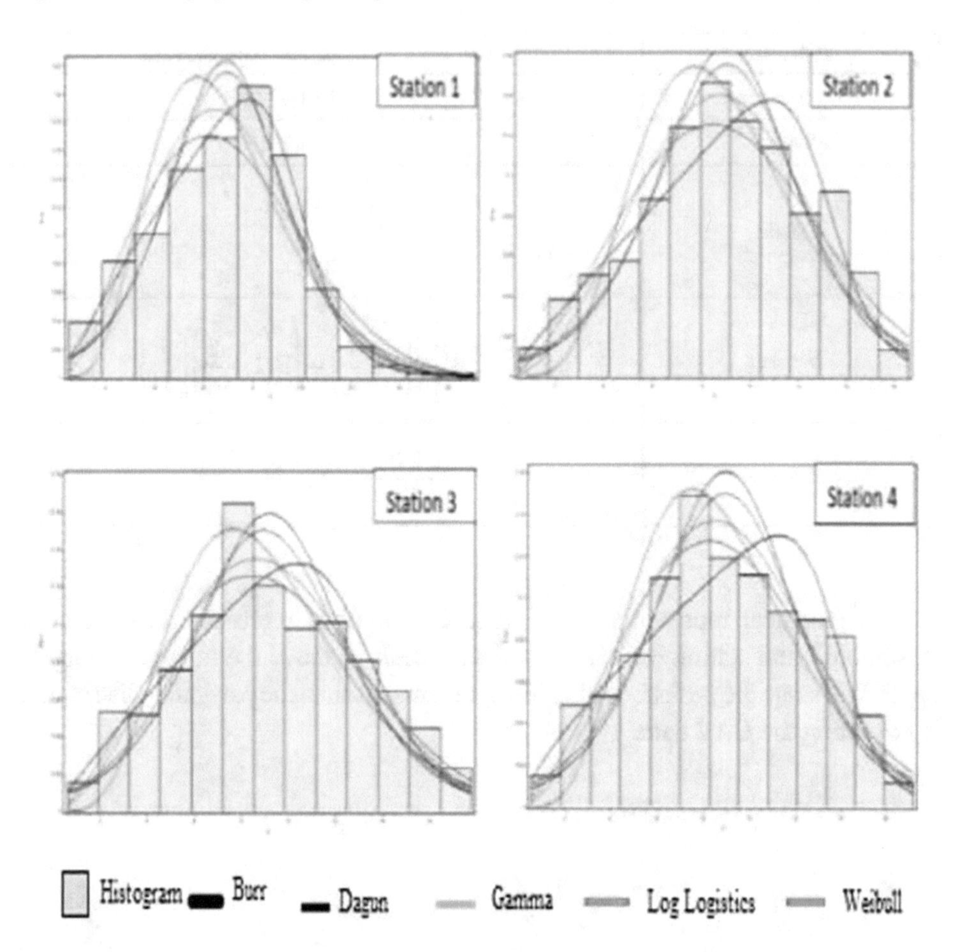

DISTRIBUTION EVALUATION APPROACH

Methods for calculating trip travel durations utilizing increasingly accessible data from mobile sources are still evolving and restricted (Weiger, 2007), particularly when it comes to probability distribution estimation. A Markov chain model was utilized in previous studies to estimate trip TTD (Total Time Delay).

Probability Distribution Estimation

The probability distribution estimate for a route (Hansen, 2016) using the Markov chain methodology (trip). Each circle represents a probable condition that a vehicle can encounter during a journey. The transition probability to the state a vehicle encounters on the current connection given the state on the prior connection is represented by the edge between close link states. The number of different connection states that can exist along a route.

GMC (Generalized Markov Chain) method (Rao et al., 2016) the product of the starting state probability and the probabilities of transitioning between neighboring states links determines the likelihood of each Markov path and the probability is given in the equation 9.

$$\text{Prob}\left(X_1 = x_1^{j1}, X_2 = x_2^{j2}, \ldots, X_N = x_N^{jN}\right) = \pi_1^{j1} \times p_{j_1,j_2} \times p_{j_2,j_3} \times \ldots \times p_{j_{N-1},j_N} \tag{9}$$

The distribution of Markov paths is calculated is given in the equation 10 using the MGF technique, as the total of connected conditional link TTDs.

$$Dist\left(X_1 = x_1^{j1}, X_2 = x_2^{j2}, \ldots, X_N = x_N^{jN}\right) = MGF\left\{Dist\left(x_1^{j1}\right), Dist\left(x_2^{j2}\right), \ldots, Dist\left(x_N^{jN}\right)\right\} \tag{10}$$

The task of estimating the probability distribution of trip travel times is defined as: Given a collection of link travel time probability distributions, estimate at every time, the probability distribution of trip travel durations for each OD pair.

The following are the primary assumptions used in prior research when employing a Markov chain methodology to estimate TTD:

1. The significance of conditional independence between link trip times or independent conditional on states.
2. The Constant transition probability for a given time (e.g., 1:15am)

Figure 6. Markov Distribution estimation for the UAV

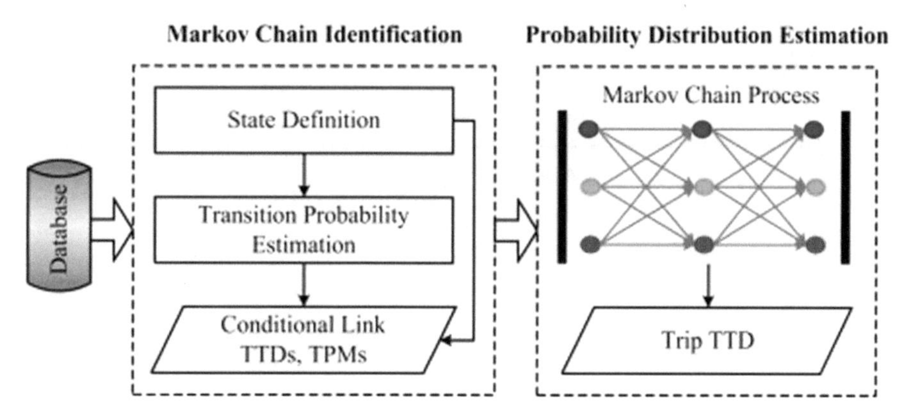

Markov Chain Identification

The goal of this stage is to specify traffic conditions and calculate the transition probability model. Link traffic state probabilities (Orfanus et al., 2016), link TTDs (dependent on states), and time-space dependent transition probabilities are the outputs. The states are defined using a Gaussian Mixture Model (GMM)-based clustering technique. The method ensures cluster homogeneity, differentiation in space and time, a big enough state to define the underlying traffic circumstances as well as computational efficiency. The utility functions are used to estimate utilizing a logit model to calculate transition probabilities.

Probability Distribution Estimation

Using a Markov chain technique, this calculates the trip TTD. Markov pathways are the permutations of link states throughout a route. The product of starting state probability and link transition probabilities is employed to calculate the Markov route probability. Using an MGF technique, the sum of correlated link TTDs conditional on states is used to calculate the Markov path TTDs.

Figure 7. Node point state distribution of the estimation framework

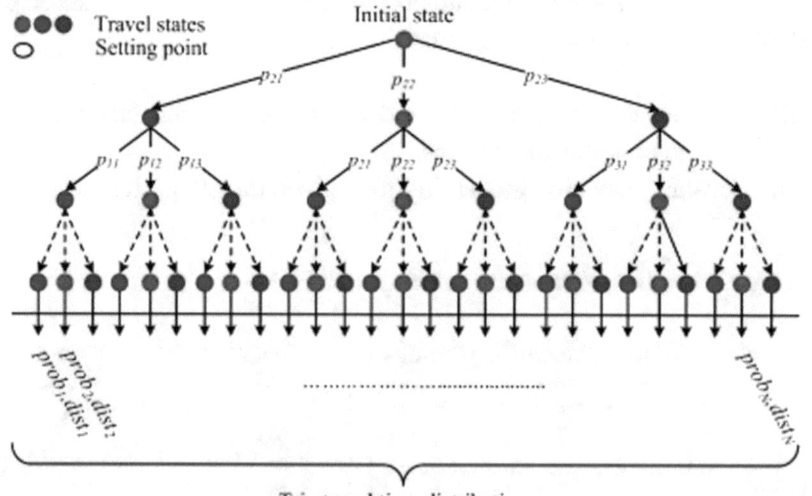

Sensitivity Analysis

It is also worth looking into how sensitive the results are to different travel distances and road conditions. The underlying properties of link TTDs were investigated using statistical tests such as unimodality and spatiotemporal correlation. The Hartigan dip test was used to determine whether a distribution was unimodal.

RESULTS AND DISCUSSION

The Mathematical modelling of the system is done with the TTR performance has an impact on service appeal, system efficiency and operating costs. AVL-AFC data that has been archived has the potential to improve all levels of transit management and performance (planning and policy, operations, control). The state-of-the-art in TTR and TTD analysis was re-evaluated in this study. TTR is commonly used in studies to determine the impact of strategic and operational initiatives. TTD offers all of the data needed for TTR analysis. While dedicated or mobile sensors can be used to derive or infer link-level TTD, methods for estimating trip TTD between an origin and a destination pair are currently being developed.

Figure 8. For all circumstances, the top three models are distributed at the route level.

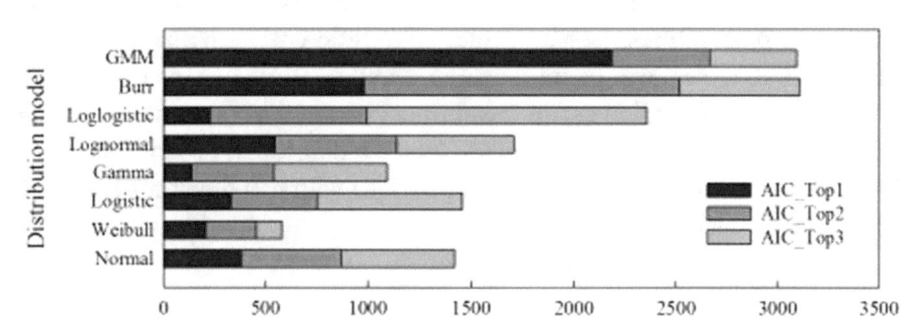

Table 6. Comparison of performance of candidate models at route and link level.

Model	Cov_sig[1]		Casess_pass ratio[2]	
	Route	Link	Route	Link
Normal	0.45	0.73	0.95	0.81
Weibull	0.78	1.03	0.83	0.63
Logistic	0.35	0.65	0.97	0.85
Gamma	0.40	0.77	0.95	0.71
Lognormal	0.43	0.76	0.95	0.72
Loglogistic	0.34	0.73	0.97	0.72
Burr	0.53	0.87	1.80	030
GMM	0.16	0.47	1.00	0.88

CONCLUSION

The mathematical modelling of the UAV is done and the smart city, traffic condition on the prior link, the traffic status on the current connection during the last time interval, and the current link's recurrent congestion index are all critical considerations for traffic state changes (link characteristics). The suggested link TTD prediction approach outperforms the alternatives predictions of deterministic and interval travel times. The model is quite generic since it uses explanatory factors' effects on transition probabilities. However, there is a little cost in terms of estimating precision. More factors will be investigated, and the efficacy of the model will be assessed in terms of predicting trip travel time distributions (a set of connections) and link travel time distributions. The TMS for the Indian traffic management system using UAV for the smart system modelling is done with real time traffic simulated mathematical modelling.

REFERENCES

Avizienis, A., Laprie, J. C., Randell, B., & Landwehr, C. (2004). Basic concepts and taxonomy of dependable and secure computing. *IEEE Transactions on Dependable and Secure Computing*, *1*(1), 11–33. doi:10.1109/TDSC.2004.2

Bellavista, P., Cardone, G., Corradi, A., & Foschini, L. (2013). Convergence of MANET and WSN in IoT urban scenarios. *IEEE Sensors Journal*, *13*(10), 3558–3567. doi:10.1109/JSEN.2013.2272099

Cho, A., Kim, J., Lee, S., & Kee, C. (2011). Wind estimation and airspeed calibration using a UAV with a single-antenna GPS receiver and pitot tube. *IEEE Transactions on Aerospace and Electronic Systems*, *47*(1), 109–117. doi:10.1109/TAES.2011.5705663

Chourabi, H., Nam, T., Walker, S., Gil-Garcia, J. R., Mellouli, S., Nahon, K., . . . Scholl, H. J. (2012, January). Understanding smart cities: An integrative framework. In *2012 45th Hawaii international conference on system sciences* (pp. 2289-2297). IEEE.

De Freitas, E. P., Heimfarth, T., Netto, I. F., Lino, C. E., Pereira, C. E., Ferreira, A. M., . . . Larsson, T. (2010, October). UAV relay network to support WSN connectivity. In International Congress on Ultra Modern Telecommunications and Control Systems (pp. 309-314). IEEE. doi:10.1109/ICUMT.2010.5676621

Fu, S., Saeidi, H., Sand, E., Sadrfaidpour, B., Rodriguez, J., Wang, Y., & Wagner, J. (2016, July). A haptic interface with adjustable feedback for unmanned aerial vehicles (UAVs)-model, control, and test. In *2016 American Control Conference (ACC)* (pp. 467-472). IEEE.

George, J., PB, S., & Sousa, J. B. (2011). Search strategies for multiple UAV search and destroy missions. *Journal of Intelligent & Robotic Systems*, *61*(1), 355–367. doi:10.100710846-010-9486-8

Gupta, L., Jain, R., & Vaszkun, G. (2015). Survey of important issues in UAV communication networks. *IEEE Communications Surveys and Tutorials*, *18*(2), 1123–1152. doi:10.1109/COMST.2015.2495297

Ha, T. (2010). The UAV Continuous Coverage Problem. *Theses and Dissertations*. 2094. https://scholar.afit.edu/etd/2094

Hallinan, A. J. Jr. (1993). A review of the Weibull distribution. *Journal of Quality Technology*, *25*(2), 85–93. doi:10.1080/00224065.1993.11979431

Hansen, R. L. (2016). *Traffic monitoring using UAV Technology*. The American Surveyor.

Jiang, F., & Swindlehurst, A. L. (2010, December). Dynamic UAV relay positioning for the ground-to-air uplink. In 2010 IEEE *Globecom Workshops*. IEEE.

Kanistras, K., Martins, G., Rutherford, M. J., & Valavanis, K. P. (2013, May). A survey of unmanned aerial vehicles (UAVs) for traffic monitoring. In *2013 International Conference on Unmanned Aircraft Systems (ICUAS)* (pp. 221-234). IEEE. 10.1109/ICUAS.2013.6564694

Mohammed, F., Idries, A., Mohamed, N., Al-Jaroodi, J., & Jawhar, I. (2014, May). UAVs for smart cities: Opportunities and challenges. In *2014 International Conference on Unmanned Aircraft Systems (ICUAS)* (pp. 267-273). IEEE. 10.1109/ICUAS.2014.6842265

Mulligan, C. E., & Olsson, M. (2013). Architectural implications of smart city business models: An evolutionary perspective. *IEEE Communications Magazine*, *51*(6), 80–85. doi:10.1109/MCOM.2013.6525599

Orfanus, D., De Freitas, E. P., & Eliassen, F. (2016). Self-organization as a supporting paradigm for military UAV relay networks. *IEEE Communications Letters*, *20*(4), 804–807. doi:10.1109/LCOMM.2016.2524405

Rao, B., Gopi, A. G., & Maione, R. (2016). The societal impact of commercial drones. *Technology in Society*, *45*, 83–90. doi:10.1016/j.techsoc.2016.02.009

Rodrigues, M., Pigatto, D. F., Fontes, J. V., Pinto, A. S., Diguet, J. P., & Branco, K. R. (2017, May). UAV integration into IoIT: opportunities and challenges. In ICAS 2017 (Vol. 95). Academic Press.

Semsch, E., Jakob, M., Pavlicek, D., & Pechoucek, M. (2009, September). Autonomous UAV surveillance in complex urban environments. In *2009 IEEE/WIC/ACM International Joint Conference on Web Intelligence and Intelligent Agent Technology* (Vol. 2, pp. 82-85). IEEE. 10.1109/WI-IAT.2009.132

Stöcker, C., Bennett, R., Nex, F., Gerke, M., & Zevenbergen, J. (2017). Review of the current state of UAV regulations. *Remote Sensing*, *9*(5), 459. doi:10.3390/rs9050459

Weiger, R. L. (2007). *Military unmanned aircraft systems in support of homeland security*. Army War Coll Carlisle Barracks PA.

Zanella, A., Bui, N., Castellani, A., Vangelista, L., & Zorzi, M. (2014). Internet of things for smart cities. *IEEE Internet of Things Journal*, *1*(1), 22–32. doi:10.1109/JIOT.2014.2306328

ADDITIONAL READING

Heinecke, H., Schnelle, K. P., Fennel, H., Bortolazzi, J., Lundh, L., Leflour, J., . . . Scharnhorst, T. (2004). *Automotive open system architecture-an industry-wide initiative to manage the complexity of emerging automotive e/e-architectures* (No. 2004-21-0042). SAE Technical Paper.

Hua, M. D., & Rifaï, H. (2010, December). Obstacle avoidance for teleoperated underactuated aerial vehicles using telemetric measurements. In *49th IEEE Conference on Decision and Control (CDC)* (pp. 262-267). IEEE. 10.1109/CDC.2010.5717431

Kanso, A., Elhajj, I. H., Shammas, E., & Asmar, D. (2015, July). Enhanced teleoperation of UAVs with haptic feedback. In *2015 IEEE International Conference on Advanced Intelligent Mechatronics (AIM)* (pp. 305-310). IEEE. 10.1109/AIM.2015.7222549

Chapter 8
Unmanned Aerial Vehicle Brands Fan Page Engagement Behavior Analytics

Senith S.
Karunya Institute of Science and Technology, India

Jegadeeswari M.
The Standard Fireworks Rajaratnam College for Women, Sivakasi, India

Alfred Kirubaraj
Karunya Institute of Science and Technology, India

Poornima Vijaykumar
St. Joseph's College of Commerce (Autonomous), India

Nisha Malini
Karunya Institute of Science and Technology, India

Praveen Kumar S.
Karunya Institute of Science and Technology, India

ABSTRACT

Over the last few years, the way people trade information and communicate with one another has changed tremendously. In business communication, social media channels such as Facebook, Twitter, and YouTube are becoming increasingly significant. Nevertheless, the study into online brand fan page is primarily focused on using website platforms rather than social media platforms. As a result, more research is needed to analyze UAV businesses' fan page engagement behavior in order to grow their fan base and further induce a fan's buying behavior using the honeycomb model's views. Consumers who have participated in an online brand fan page are the study's target group. A web-based survey was used to collect data. Identity, conversation, presence, sharing, reputation, relationships, and groups all had a significant beneficial effect on brand equity, according to the findings. This study confirms the impact of perceived value in improving various fan page behaviors, which aids in the identification and implementation of an online engagement plan for purchase.

DOI: 10.4018/978-1-7998-8763-8.ch008

INTRODUCTION

An unmanned aerial vehicle (UAV), sometimes known as a drone, is a plane that flies without a human pilot, troop, or customers. Unmanned aerial vehicles (UAVs) are portion of an unmanned aircraft system (UAS), which also contains a ground-based regulator and a transport network system with the UAV (Hu and Lanzon, 2018). UAV flight can be measured the slightest bit by a human machinist, as in a greatly channelled jet, or with varying degrees of independence, such as automatic pilot assistance, up to fully independent airplane with no human interaction (Sharma et al., 2020). UAVs were fashioned in the twentieth period for soldierly responsibilities that were "too dull, unclean, or unsafe for humans," and by the twenty-first period, they had developed essential possessions to most militaries (Cary and Coyne, 2020). Control technology industrialized and costs reduced, allowing them to be used in a diversity of non-military submissions. In-flight cinematography, merchandise distribution, farming, regulating and investigation, structure examinations, learning, rustling, and were competing are just a few examples (Tice, 1991) The first described use of an unmanned aerial vehicle for warfighting was in July 1849, when it served as an inflatable transporter (the predecessor to the aircraft carrier) in marine flight's first aggressive use of air control. During the blockade of Venice, Austrian forces tried to fire 200 combustible inflatables against the city (Hu et al., 2021) The balloons were mostly hurled from land, although some were launched from the Austrian cruiser SMS Volcano as well. At least one bomb landed in the city, but most of the balloons missed their aim because the wind changed after launch, and some floated back over Austrian lines and the launching ship Vulcano. Drone development began in earnest in the early 1900s, with the initial goal of supplying practise targets for military personnel (Alvarado, 2021) In 1916, A. M. Low's "Aerial Target" was the first effort at a powered UAV. Low acknowledged that Geoffrey de Havilland's monoplane was the single that floated lower than his wireless direction on March 21, 1917 (Koparan et al, 2020) Other British people unmanned growths surveyed during and after World War I, culminating in the 1935 deployment of a fleet of approximately 400 de Havilland 82 Queen Bee aerial targets (Koparan, 2018) In 1915, Nikola Tesla described an unmanned fleet of in-flight assault automobiles. These innovations also motivated Charles Kettering of Dayton, Ohio, to build the Kettering Bug and the Hewitt-Sperry Involuntary Aircraft. Originally intended to be an unmanned plane carrying a short-tempered consignment to a predefined target. Reginald Denny, a movie star and model aeroplane enthusiast, created the first scaled remote piloted vehicle in 1935 (Kaplan, 2013) Interest in UAVs rose in the highest echelons of the US military as applicable technologies matured and miniaturised in the 1980s and 1990s. The US Department of Défense awarded AAI Corporation and Israeli business Malat a contract in the 1990s. The AAI Pioneer

UAV, which AAI and Malat developed together, was purchased by the US Navy (Hallion, 2003) Many of these unmanned aerial vehicles (UAVs) saw action in the Gulf War in 1991. UAVs showed the promise of inexpensive, more accomplished aggressive machineries that could be deployed without endangering aircrews. The first generation of aircraft were mostly surveillance planes, but some had armaments, such the General Atomics MQ-1 Predator, which could fire AGM-114 Hellfire air-to-ground missiles. (Renner, 2016) The United States Air Force (USAF) employed 7494 unmanned aerial vehicles (UAVs) in 2012, accounting for about one-third of all aircraft in the service. UAVs were also used by the Central Intelligence Agency. UAVs were utilised in at least 50 nations by 2013. China, Iran, Israel, Pakistan, Turkey, created their own variants. Drone use has increased in recent years. There is no complete list of UAV systems because to their widespread use (Murphy, 2005). Drone use for shopper and universal flight doings has increased in tandem with the development of smart technology and improved electrical power systems. Quadcopter drones are expected to be the most popular hobby radio-controlled aircraft and toys by 2021, but their application in commercial and general aviation is limited due to a lack of autonomy and new regulatory settings that demand line-of-sight communication with the pilot. (Haydon, 2000) According to a report published in March 2021 by the UN Security Council's Panel of Experts on Libya, a Kargu 2 drone sought down and attacked a human target in Libya in 2020 (Mikesh, 1973). It's possible that this was the first occasion an independent slaughterer robot armed with lethal weapons assaulted people. This is the history of unmanned aerial vehicles (UAVs).

Applications of UAV

Self-directed drones have commenced to renovate numerous submission parts in current years due to their aptitude to fly yonder graphic line of vision (BVLOS) though maximising manufacture, lowering expenses and hazards, safeguarding spot protection, safety, and supervisory obedience, and defensive social workers in the event of a pandemic. (Nanalyze, 2019) They can also be utilised for consumer-related missions such as compendium distribution, as Amazon Prime Air has proved, and vital health-supply distributions (McNabb, 2020)

UAVs have a wide range of citizen, profitable, soldierly, and atmosphere requests. These are some of them:

Disaster relief, archaeology, biodiversity and environment conservation, law enforcement, crime, and terrorism are all examples of general recreation (Peck, 2020) Commercial aerial surveillance, cinematography, broadcasting, technical research, graphing, cargo transport, withdrawal, industrial, forestry, solar farming, updraft energy, ports, and farming are all examples of profitable aerial investigation

Warfare

Defense services throughout the world are progressively adopting UAVs for diverse requests such as scrutiny, logistics, announcement, outbreak, and contest, thanks to significant cost reductions and breakthroughs in UAV technology (Motwani, 2020) By 2020, seventeen countries will have armed UAVs, and over 100 countries will use UAVs for military purposes. Companies located in the United States, China, and Israel dominate the global military UAV market (Vergouw et al., 2016) In 2017, the United States had a market share of almost 60% of the military market. General Atomics, Lockheed Martin, Northrop Grumman, and Boeing are among the top five military UAV manufacturers in the world, followed by the Chinese business CASC. Since 2010 (Horowitz, 2020) China has built and increased its footprint in the military UAV market. The top 12 nations that are known to have purchased armed hums from China among 2010 and 2019 are all in the top 12.

Civil

Chinese businesses lead the citizen (commercial and general) drone industry. DJI, a Chinese drone maker, held 74% of the public marketplace share in 2018, with no other business secretarial for more than 5%, and $11 billion in global sales predicted for 2020 (Arnett, 2015) The US Central Subdivision suspended its fleet of DJI drones in 2020 as a result of greater monitoring of its operations, while the Righteousness Section restricted the use of public funding for the buying of DJI and other foreign-made UAVs. With a considerable market share gap, DJI is followed by Chinese firm Yuneec, US company 3D Robotics, and French company Parrot (Bateman, 2017), The US FAA has registered 873,576 UAVs as of May 2021, with 42 percent classified as commercial drones and 58 percent as recreational drones. Consumers are progressively buying drones with more advanced features, according to the 2018 NPD, with 33 out of a hundred evolution in both the $500+ and $1000+ market sectors (Friedman and McCabe, 2020) In comparison to the military UAV market, the civil UAV market is relatively new. At the same time, corporations are forming in both developed and developing countries (Miller, 2020) Many early-stage firms have gotten money and support from investors, as in the United States, and from government organisations, as in India. Some colleges and universities provide research and training programmes as well as degrees (Peterson, 2013) Operational and in-person training programmes for both recreational and marketable UAV use are also available from private businesses. Consumer drones are also commonly used by military organisations around the world due to their cost-effectiveness. Since civil drones are informal to use and have greater dependability, the Israeli military began using DJI Mavic and Matrices series of UAVs for light investigation missions in 2018 (Bloomberg.com, 2020) The US Army has also used DJI drones, which are the most extensively used commercial unmanned aerial systems. Chinese police have been using DJI surveillance drones in Xinjiang since 2017.

By 2021, the global UAV industry will be worth US $21.47 billion, with the Indian market around US $885.7 million. Drones with lights are increasingly being employed in night-time shows for artistic and commercial objectives.

Aerial Photography

Drones are well-suited to capture aerial pictures in photography and filmmaking, and they're extensively employed for it. Small drones eliminate the requirement for exact coordination between pilot and cameraman because both tasks are filled by the same person. In the case of large drones equipped with professional cine cameras, a drone pilot and a camera operator who controls camera angle and lens are normally present. The AERIGON cinema drone, for example, is flown by two persons and is utilised in the production of huge blockbuster movies. Drones allow access to dangerous, remote, and uncomfortable locations that are otherwise unreachable.

Agriculture and Forestry

There is a crucial need for more expedient and keener cultivated keys than old-style approaches as worldwide request for food manufacture grows exponentially, incomes are exhausted, woodland is concentrated, and farming labour is progressively in short supply, and the agricultural drone and robotics industry is expected to make progress. Agricultural drones have been utilised to aid in the development of sustainable agriculture in places like Africa (Drone Addicts, 2018)

Facebook Fanpage

The fan page is the brand's primary point of contact, serving as the hub for all network activities. Unlike profiles and groups, which require registration in order to participate, the fan page requires only a simple click on the "Like" icon to demonstrate interest in the subject and acquire the right to receive updates. You had to click on the symbol "Become a Fan" until April 2010, when Facebook changed it to "I like." This variation was created with the express intention of increasing the number of users who enjoy the material on a certain page. Because Facebook's developers found that clicking "Like" was perceived as less engaging by users, they devised this linguistic ruse to lighten the idea of participation in a community page, even if they were unfamiliar with its content. This is because in the real world, being a fan of something indicates you genuinely love a certain brand, whereas if the issue is simply interesting, we will most likely only say we "like" it. Let's take a look at how a fan page works and why it's such a useful tool for businesses. The page was created in 2007 to allow public personalities and individuals to add a "public" profile to their "private and personal" profile, where they could not market their activities, have more than 5,000 connections, and be acceptable to all users interested in seeing and sharing their content. Companies, consumer goods, small businesses, movies, television channels, actors, politicians, singers, newspapers, and non-profit organisations are among these. Facebook has designed the fan page to look and function similarly to a personal profile, allowing the owner to update his personal status with a message visible to all Facebook users. When a person clicks the "like" button, the news is shared with all of their friends on the homepage. This feature is critical since it has a viral impact and helps to distribute all of the content on a page. This is a significant benefit for the organisation because it is the user who, having been impressed by the material, promotes it spontaneously and becomes an active participant in the communication process. Another benefit is that the public profile is visible and available to both Facebook members and non-Facebook users, and that search engines simply index the page link, strengthening the company's Internet positioning strategy. Instead, a corporation that wishes to advertise and distribute its own home page can do so by acquiring ad space, which refers to the page's own contents. You have six possibilities when creating a public profile: local business, company, organisation, or institution, brand or product, artist, band, or public personality, entertainment, cause, or community. A fan page is similar to a corporate website within the social network. Companies have an "official" channel; therefore, it requires consent from the firm as well as documentation attesting to the brand owner's readiness to build one. Otherwise, because the registration terms have been broken, the corporation may request the cancellation or acquisition of the page. Information from Facebook (2009). Glynn (1981) is a man of many talents developing a loyal customer base may be the most crucial but difficult undertaking that marketers and managers face. The main issue is the difficulty of cultivating brand loyalty when new competitors constantly join the market, luring customers in with low prices and enhanced convenience. However, the fact that most people have a natural need to be a member of a group, thereby meeting an objective set of human wants, presents a huge potential McAlexander et al. (2002). Consumers are growing increasingly reliant on the consumption of various brands to express themselves. Customers who join a brand-based community have the option to be a part of a psychological and/or social group that is relevant to and hence supports a desired self-image. As a result, a brand-based community can serve as an aspirational group for existing non-users of the brand as well as a source of information for potential users. Marketers benefit from such communities because they have a consumer base that is both engaged and devoted to the brand.

METHODOLOGY

Research Questions

- Identifications of Brand fan page engagement behavior dimensions
- Exploring gender differences in brand Fan pages of UAV.
- How Brand Equity of a product is enhanced by the participation of customers in Fan pages?

These are the major research problems addressed in the study. The research problem can be summarized as a study

Sample and Sample Size

Members of three UAV brand Fan pages made up the main study's sample. For estimation and interpretation of results, a sufficient sample size is critical (Hair et al., 1998). However, there is no hard and fast rule for how big a sample size should be for structural modelling. To reduce the likelihood of model estimating problems, Kline (1998) proposed 100 to 150 cases. A sample size of 200 was proposed by Hoelter (1983) as a "critical sample size." For models with three or more indicators per factor, Anderson and Gerbing (1984) discovered that a sample size of 150 is sufficient to provide a correct solution.

The researcher conducted a pilot study with 47 fan page users to determine the viability and feasibility of the current study. The researcher had a talk with certain social media experts in order to pinpoint the source of the problem and ensure the proper completion of this study. Following the analysis, the necessary adjustments were made to make the study more accurate and relevant in the field of research.

RESULTS AND DISCUSSION

The content validity ratio for the complete Fan page brand equity dimensions is greater than 70, based on the above table 1. As a result, all elements of Fan page brand equity are seen positively. The variables are relevant and intertwined in a substantial way. As a result, the study and its contents are supported by the degree of validity. To study the significant difference in various dimensions of Engagement behaviour by the Male respondents

Table 1. Content validity ratio for brands fan page engagement dimensions

S.No	Dimensions	Content Validity Ratio
10	Identity	.80
11	Conversation	.80
12	Presence	.76
13	Sharing	.80
14	Reputation	.76
15	Relationship	.80
16	Groups	.76

The descriptive table 2 includes some extremely helpful descriptive statistics for the entire group, such as the mean and standard deviation for the dependent variables. The ANOVA test is used to compare the differences between the various dimensions of the Honeycomb model in terms of Gender group (Male) for Brands fan page. Where the compared groups reached statistical significance, the post-Hoc test of multiple comparison was used to determine which groups were the most influential.

Table 2. Engagement behaviour for brand's fan page by male respondents

Dimensions	Mean			Standard Deviation			F	Sig
	Carbon fiber drone	Phoenix drone	DJI inspire	Carbon fiber drone	Phoenix drone	DJI inspire		
Identity	2.64	3.62	3.80	0.70	0.74	0.84	76.6	0.00
Conversation	3.32	3.16	3.37	0.32	0.92	0.90	2.48	0.08
Presence	2.92	3.52	3.66	0.37	0.78	0.75	38.4	0.00
Sharing	2.09	3.15	3.27	0.34	0.97	0.95	73.9	0.00
Reputation	2.23	3.13	3.34	0.49	0.95	0.86	58.7	0.00
Relationship	2.48	3.25	3.55	0.53	0.84	0.84	54.9	0.00
Groups	3.03	3.57	3.70	0.32	0.73	0.79	34.0	0.00

Analysis of Variance

As shown in table 2, the F value for Identity is 76.6, and the significant value is 0.00, indicating that identity has significant effects at the 0.05 level, indicating that Identity has substantial differences with male respondents. We can see that the significance level of discussion is 0.08 (p=0.01), which is greater than 0.05, implying that there is no statistically significant difference between male and female respondents in terms of dialogue. The F value for Presence was determined to be 38.4 and the significant value is 0.00, indicating that Presence has significant effects at the 0.05 level, indicating that there are substantial differences between male and female respondents. Because the significance level for Sharing is 0.00 (p=0.00), which is less than 0.05, there is a statistically significant difference between male and female respondents in terms of sharing. Because the significance level for reputation is 0.001 (p=0.01), which is less than 0.05, there is a statistically significant difference in male respondents' reputation. Because the significance level for Relationship is 0.00 (p=0.00), which is less than 0.05, there is a statistically significant difference between male and female respondents' Relationship. Because the Groups significance threshold is 0.00(p=0.00), which is greater than 0.05, there is a statistically significant difference between groups among male respondents.

Post Hoc Test

We should run a Turkey's W multiple comparison to discover which means are different since we rejected the null hypothesis in the identity dimension (we observed differences in the means). Here's how such an analysis may look using the prior output.

Table 3. Multiple comparisons versus male for identity dimension

Dependent Variable: Identity Turkey HSD						
(I) Brands	(J) Brands	Mean Difference (I-J)	Std. Error	Sig.	95% Confidence Interval	
					Lower Bound	Upper Bound
Carbon fiber drone	PHOENIX DRONE	-.98010*	.08898	.000	-1.1896	-.7706
	DJI inspire	-1.16245*	.11903	.000	-1.4426	-.8823
PHOENIX DRONE	Carbon fiber drone	.98010*	.08898	.000	.7706	1.1896
	DJI inspire	-.18236	.11469	.251	-.4523	.0876
DJI inspire	Carbon fiber drone	1.16245*	.11903	.000	.8823	1.4426
	PHOENIX DRONE	.18236	.11469	.251	-.0876	.4523
*. The mean difference is significant at the 0.05 level.						

According to the table above, there were considerable discrepancies between male and female respondents when it came to the Identity dimension. Significant variations emerged across the groups of the "DJI inspire Fan page," according to the results of the Turkey's W multiple comparison study. The mean scores suggest that among the three groups, the group "DJI inspire Fan page" delivers superior identification through Fan page. We should run a Turkey's W multiple comparison to discover which means are different since we rejected the null hypothesis in the Presence dimension (we found differences in the means). Here's how such an analysis may look using the prior output.

Table 4. Multiple comparisons versus male for presence dimension

Dependent Variable: Presence *Turkey HSD*						
(I) Brands	**(J) Brands**	**Mean Difference (I-J)**	**Std. Error**	**Sig.**	**95% Confidence Interval**	
					Lower Bound	**Upper Bound**
Carbon fiber drone	PHOENIX DRONE	-.60084*	.07841	.000	-.7854	-.4163
	DJI inspire	-.74488*	.10489	.000	-.9918	-.4980
PHOENIX DRONE	Carbon fiber drone	.60084*	.07841	.000	.4163	.7854
	DJI inspire	-.14404	.10107	.329	-.3819	.0939
DJI inspire	Carbon fiber drone	.74488*	.10489	.000	.4980	.9918
	PHOENIX DRONE	.14404	.10107	.329	-.0939	.3819
*. The mean difference is significant at the 0.05 level.						

The table above shows that there were substantial variations between male and female respondents when it came to the Presence dimension. Significant variations emerged across the groups of the "DJI inspire Fan page," according to the results of the Turkey's W multiple comparison study. Among the three groups, the group "DJI inspire Fan page" delivers higher Presence to the members through Fan page, according to the mean scores.

We need run a Turkey's W multiple comparison to discover which means are different since we rejected the null hypothesis in Sharing dimension (we found differences in the means). Here's how such an analysis may look using the prior output.

The table 5 shows that there were substantial variations between Sharing dimension and Male respondents. Significant variations emerged across the groups of the "DJI inspire Fan page," according to the results of the Turkey's W multiple comparison study. Among the three groups, the group "DJI inspire Fan page" provides superior Sharing to the members through Fan page, according to the mean scores.

We should run a Turkey's W multiple comparison to discover which means are different since we rejected the null hypothesis in the Reputation dimension (we found differences in the means). Here's how such an analysis may look using the prior output.

Table 5. Multiple comparisons versus male for sharing dimension

Dependent Variable: Sharing Turkey HSD						
(I) Brands	(J) Brands	Mean Difference (I-J)	Std. Error	Sig.	95% Confidence Interval	
					Lower Bound	Upper Bound
Carbon fiber drone	PHOENIX DRONE	-1.05864*	.09540	.000	-1.2832	-.8341
	DJI inspire	-1.17400*	.12761	.000	-1.4744	-.8736
PHOENIX DRONE	Carbon fiber drone	1.05864*	.09540	.000	.8341	1.2832
	DJI inspire	-.11536	.12296	.617	-.4048	.1741
DJI inspire	Carbon fiber drone	1.17400*	.12761	.000	.8736	1.4744
	PHOENIX DRONE	.11536	.12296	.617	-.1741	.4048
*. The mean difference is significant at the 0.05 level.						

According to table 6, there were substantial discrepancies between the Reputation dimension and Male responders. Significant variations emerged across the groups of the "DJI inspire Fan page," according to the results of the Turkey's W multiple comparison study. The average scores show that the group "DJI inspire Fan page" has a higher Reputation among the members of the Fan page than the other three groups.

Table 6. Multiple comparisons versus male for reputation dimension

Dependent Variable: Reputation Turkey HSD						
(I) Brands	(J) Brands	Mean Difference (I-J)	Std. Error	Sig.	95% Confidence Interval	
					Lower Bound	Upper Bound
Carbon fiber drone	PHOENIX DRONE	-.90005*	.09474	.000	-1.1231	-.6770
	DJI inspire	-1.10646*	.12673	.000	-1.4048	-.8081
PHOENIX DRONE	Carbon fiber drone	.90005*	.09474	.000	.6770	1.1231
	DJI inspire	-.20641	.12211	.210	-.4939	.0810
DJI inspire	Carbon fiber drone	1.10646*	.12673	.000	.8081	1.4048
	PHOENIX DRONE	.20641	.12211	.210	-.0810	.4939
*. The mean difference is significant at the 0.05 level.						

We should run a Turkey's W multiple comparison to discover which means are different since we rejected the null hypothesis in the Identity dimension (we observed differences in the means). Here's how such an analysis may look using the prior output.

Table 7. Multiple comparisons versus male for relationship dimension

Dependent Variable: Relationship Turkey HSD						
(I) Brands	**(J) Brands**	**Mean Difference (I-J)**	**Std. Error**	**Sig.**	**95% Confidence Interval**	
					Lower Bound	**Upper Bound**
Carbon fiber drone	PHOENIX DRONE	-.76755*	.08869	.000	-.9763	-.5588
	DJI inspire	-1.06973*	.11864	.000	-1.3490	-.7905
PHOENIX DRONE	Carbon fiber drone	.76755*	.08869	.000	.5588	.9763
	DJI inspire	-.30218*	.11432	.023	-.5713	-.0331
DJI inspire	Carbon fiber drone	1.06973*	.11864	.000	.7905	1.3490
	PHOENIX DRONE	.30218*	.11432	.023	.0331	.5713
*. The mean difference is significant at the 0.05 level.						

According to the table above, there were substantial discrepancies between the Relationship dimension and male responders. Significant variations emerged across the groups of the "DJI inspire Fan page," according to the results of the Turkey's W multiple comparison study. The mean scores suggest that of the three groups, the group "DJI inspire Fan page" maintains a positive relationship with its members through its Fan page.

We should run a Turkey's W multiple comparison to discover which means are different since we rejected the null hypothesis in the Groups dimension (we found differences in the means). Here's how such an analysis may look using the prior output.

Table 8. Multiple comparisons versus male for groups dimension

Dependent Variable: Groups Turkey HSD						
(I) Brands	**(J) Brands**	**Mean Difference (I-J)**	**Std. Error**	**Sig.**	**95% Confidence Interval**	
					Lower Bound	**Upper Bound**
Carbon fiber drone	PHOENIX DRONE	-.54335*	.07503	.000	-.7200	-.3667
	DJI inspire	-.66667*	.10037	.000	-.9029	-.4304
PHOENIX DRONE	Carbon fiber drone	.54335*	.07503	.000	.3667	.7200
	DJI inspire	-.12331	.09671	.410	-.3510	.1043
DJI inspire	Carbon fiber drone	.66667*	.10037	.000	.4304	.9029
	PHOENIX DRONE	.12331	.09671	.410	-.1043	.3510
*. The mean difference is significant at the 0.05 level.						

Significant differences were found in the Groups dimension and among Male respondents, as shown in the table above. Significant variations emerged across the groups of the "DJI inspire Fan page," according to the results of the Turkey's W multiple comparison study. Among the three groups, the group "DJI inspire Fan page" delivers superior Groups to the members through Fan page, according to the mean scores.

TO STUDY THE SIGNIFICANT DIFFERENCE IN VARIOUS DIMENSIONS OF ENGAGEMENT BEHAVIOR BY FEMALE RESPONDENTS

The descriptive table includes some extremely helpful descriptive statistics for the entire group, such as the mean and standard deviation for the dependent variables. ANOVA is used to compare the differences in the various dimensions of the Honeycomb model in terms of Gender group (female) for Brands fan page. Where the compared groups reached statistical significance, the post-Hoc test of multiple comparison was used to determine which groups were the most influential.

Table 9. Honeycomb model for brands fan page by female respondents

Dimensions	Mean			Standard Deviation			F	Sig
	Carbon fiber drone	Phoenix drone	DJI inspire	Carbon fiber drone	Phoenix drone	DJI inspire		
Identity	2.85	3.11	3.07	0.75	0.67	0.68	7.75	0.00
Conversation	3.11	2.79	2.98	0.69	0.76	0.69	10.4	0.00
Presence	3.09	2.90	3.11	0.49	0.70	0.88	6.45	0.00
Sharing	2.40	2.62	2.77	0.80	0.78	0.87	6.29	0.00
Reputation	2.43	2.70	2.97	0.76	0.67	0.88	14.4	0.00
Relationship	2.56	2.80	2.83	0.63	0.60	0.71	8.32	0.00
Groups	3.09	3.07	3.20	0.46	0.50	0.66	2.06	0.12

Analysis of Variance

As shown in table 9, the F value for Identity is 7.75, and the significant value is 0.00, indicating that Identity has significant effects at the 0.05 level, indicating that Identity has substantial differences with Female respondents. We can observe that the significance level of discussion is 0.001(p=0.01), which is less than 0.05, indicating that there is a statistically significant difference between female and male respondents' conversations. The F value for presence was determined to be 6.45, and the significant value is 0.00, indicating that Presence has significant effects at the 0.05 level, indicating that there are substantial differences between male and female respondents. Because the significance level for reputation is 0.001 (p=0.01), which is less than 0.05, there is a statistically significant difference between Female and Male respondents' reputation. Because the significance level for relationship is 0.00 (p=0.00), which is less than 0.05, there is a statistically significant difference between Female and Male respondents' Relationship. Because the significance level of Groups is 0.12 (p=0.12), which is more than 0.05, no statistically significant difference between Groups by Female respondents exists.

POST HOC TEST

We should run a Turkey's W multiple comparison to discover which means are different since we rejected the null hypothesis in the Identity dimension (we observed differences in the means). Here's how such an analysis may look using the prior output.

Table 10. Multiple comparisons for female by identity

Dependent Variable: Identity Turkey HSD						
(I) Brands	**(J) Brands**	**Mean Difference (I-J)**	**Std. Error**	**Sig.**	**95% Confidence Interval**	
					Lower Bound	**Upper Bound**
Carbon fiber drone	PHOENIX DRONE	-.26166*	.06645	.000	-.4177	-.1056
	DJI inspire	-.22000	.10124	.077	-.4577	.0177
PHOENIX DRONE	Carbon fiber drone	.26166*	.06645	.000	.1056	.4177
	DJI inspire	.04166	.08680	.881	-.1622	.2455
DJI inspire	Carbon fiber drone	.22000	.10124	.077	-.0177	.4577
	PHOENIX DRONE	-.04166	.08680	.881	-.2455	.1622
*. The mean difference is significant at the 0.05 level.						

According to the table above, there were substantial discrepancies between Identity dimension and Female respondents. Significant disparities emerged across the groups of the "PHOENIX DRONE Fan page," according to the results of the Turkey's W multiple comparison study. Among the three groups, the group

"PHOENIX DRONE Fan page" provides greater Identity to the members through Fan page, according to the mean scores.

We should run a Turkey's W multiple comparison to discover which means are different since we rejected the null hypothesis in the Conversation dimension (we identified variations in the means). Here's how such an analysis may look using the prior output.

Table 11. Multiple comparisons for female by conversation

Dependent Variable: Conversation Turkey HSD						
(I) Brands	(J) Brands	Mean Difference (I-J)	Std. Error	Sig.	95% Confidence Interval	
					Lower Bound	Upper Bound
Carbon fiber drone	PHOENIX DRONE	.31208*	.07124	.000	.1448	.4794
	DJI inspire	.12847	.10854	.463	-.1264	.3833
PHOENIX DRONE	Carbon fiber drone	-.31208*	.07124	.000	-.4794	-.1448
	DJI inspire	-.18361	.09306	.120	-.4021	.0349
DJI inspire	Carbon fiber drone	-.12847	.10854	.463	-.3833	.1264
	PHOENIX DRONE	.18361	.09306	.120	-.0349	.4021
*. The mean difference is significant at the 0.05 level.						

According to the table above, there were substantial variations between Conversation dimension and Female respondents. Significant variances appeared across the groups of "Carbon fibre drone Fan page," according to the results of the Turkey's W multiple comparison study. Among the three groups, the group "Carbon fibre drone Fan page" gives greater Conversation to members through Fan page, according to the mean scores.

We should run a Turkey's W multiple comparison to discover which means are different since we rejected the null hypothesis in the Presence dimension (we found differences in the means). Here's how such an analysis may look using the prior output.

The table 12 shows that there were substantial variations between male and female respondents when it came to the Presence dimension. Significant variations emerged across the groups of the "DJI inspire Fan page," according to the results of the Turkey's W multiple comparison study. Among the three groups, the group "DJI inspire Fan page" delivers higher Presence to the members through Fan page, according to the mean scores.

Table 12. Multiple comparisons versus female by presence

Dependent Variable: Presence Turkey HSD						
(I) Brands	(J) Brands	Mean Difference (I-J)	Std. Error	Sig.	95% Confidence Interval	
					Lower Bound	Upper Bound
Carbon fiber drone	PHOENIX DRONE	.19446*	.06645	.010	.0384	.3505
	DJI inspire	-.02185	.10124	.975	-.2596	.2159
PHOENIX DRONE	Carbon fiber drone	-.19446*	.06645	.010	-.3505	-.0384
	DJI inspire	-.21631*	.08680	.034	-.4201	-.0125
DJI inspire	Carbon fiber drone	.02185	.10124	.975	-.2159	.2596
	PHOENIX DRONE	.21631*	.08680	.034	.0125	.4201
*. The mean difference is significant at the 0.05 level.						

We need run a Turkey's W multiple comparison to discover which means are different since we rejected the null hypothesis in Sharing dimension (we found differences in the means). Here's how such an analysis may look using the prior output.

Table 13. Multiple comparisons versus female by sharing

Dependent Variable: Sharing Turkey HSD						
(I) Brands	(J) Brands	Mean Difference (I-J)	Std. Error	Sig.	95% Confidence Interval	
					Lower Bound	Upper Bound
Carbon fiber drone	PHOENIX DRONE	-.22785*	.07604	.008	-.4064	-.0493
	DJI inspire	-.37074*	.11585	.004	-.6428	-.0987
PHOENIX DRONE	Carbon fiber drone	.22785*	.07604	.008	.0493	.4064
	DJI inspire	-.14289	.09933	.322	-.3762	.0904
DJI inspire	Carbon fiber drone	.37074*	.11585	.004	.0987	.6428
	PHOENIX DRONE	.14289	.09933	.322	-.0904	.3762
*. The mean difference is significant at the 0.05 level.						

The table above shows that there were substantial variations between Sharing dimension and Female respondents. Significant variations emerged across the groups of the "DJI inspire Fan page," according to the results of the Turkey's W multiple comparison study. Among the three groups, the group "DJI inspire Fan page" provides superior Sharing to the members through Fan page, according to the mean scores.

We should run a Turkey's W multiple comparison to discover which means are different since we rejected the null hypothesis in the Reputation dimension (we found differences in the means). Here's how such an analysis may look using the prior output.

Table 14. Multiple comparisons versus female for reputation

Dependent Variable: Reputation Turkey HSD						
(I) Brands	(J) Brands	Mean Difference (I-J)	Std. Error	Sig.	95% Confidence Interval	
					Lower Bound	Upper Bound
Carbon fiber drone	PHOENIX DRONE	-.27151*	.06843	.000	-.4322	-.1108
	DJI inspire	-.53944*	.10425	.000	-.7843	-.2946
PHOENIX DRONE	Carbon fiber drone	.27151*	.06843	.000	.1108	.4322
	DJI inspire	-.26793*	.08939	.008	-.4778	-.0580
DJI inspire	Carbon fiber drone	.53944*	.10425	.000	.2946	.7843
	PHOENIX DRONE	.26793*	.08939	.008	.0580	.4778
*. The mean difference is significant at the 0.05 level.						

According to the table above, there were substantial discrepancies between the Reputation dimension and Female respondents. Significant variations emerged across the groups of the "DJI inspire Fan page," according to the results of the Turkey's W multiple comparison study. According to the average scores, the group "DJI inspire Fan page" has an excellent reputation among three separate groups.

We should run a Turkey's W multiple comparison to discover which means are different since we rejected the null hypothesis in the Relationship dimension (we found differences in the means). Here's how such an analysis may look using the prior output.

According to the table 15, there were substantial discrepancies between the Relationship dimension and female respondents. Significant variations emerged across the groups of the "DJI inspire Fan page," according to the results of the Turkey's W multiple comparison study. The mean scores suggest that of the three groups, the group "DJI inspire Fan page" maintains a positive relationship with its members through its Fan page.

While analyzing the significant difference in various dimensions of honeycomb model by Male respondents it has been found that male respondents reveal significant difference in dimensions of Identity, Presence, Sharing, Reputation, Relationship and Groups and the result does not indicates significant difference in dimension of Conversation. Since we found differences in the means Posthoc test was carried out to know which Fan page is different from others. The result indicates that for the dimensions Identity, Presence Sharing, Reputation, Relationship and Groups "DJI inspire fan page" is different from others. While analysing the significant difference in various dimensions of honeycomb model by female respondents it has been found that female respondents reveal significant difference in dimensions of identity, Conversation, Presence, Sharing, Reputation and Relationship and the result does not indicates significant difference in dimension of Groups. Since we found differences in the means Posthoc test was carried out to know which Fan page is different from others. The result indicates that for the dimensions Presence, Sharing, Reputation and Relationship "DJI inspire fan page" is different from others and for Identity dimension "PHOENIX DRONE" fan page is different from others. Conversation dimension "Carbon fibre drone Fan page "is different from others.

Table 15. Multiple comparisons versus female for relationship

Dependent Variable: Relationship Turkey HSD						
(I) Brands	(J) Brands	Mean Difference (I-J)	Std. Error	Sig.	95% Confidence Interval	
					Lower Bound	Upper Bound
Carbon fiber drone	PHOENIX DRONE	-.23556*	.05962	.000	-.3756	-.0956
	DJI inspire	-.26889*	.09083	.009	-.4822	-.0556
PHOENIX DRONE	Carbon fiber drone	.23556*	.05962	.000	.0956	.3756
	DJI inspire	-.03333	.07788	.904	-.2162	.1495
DJI inspire	Carbon fiber drone	.26889*	.09083	.009	.0556	.4822
	PHOENIX DRONE	.03333	.07788	.904	-.1495	.2162
*. The mean difference is significant at the 0.05 level.						

CONCLUSION

Managerial Implications

Fan sites for brands could provide options for members to find like-minded consumers. Individuals get together in an online community because they share similar interests and goals. Individuals on the Brands Fan page can form a variety of subgroups based on similar or particular product service demands. Marketers must discover these prospective sub-groups and supply each one with more specialised and individualised services. In order to promote the hedonic character of their pages, marketers must use a range of strategies to categorise subgroups. For example, the Brands page may use a game platform for product announcements, such as a simple vote, online flash, or online puzzles. Including movies relating to new brand information could result in pleasurable encounters. RFID and other new technologies may be used to carry out community activities. Brand fan pages can improve member engagement by enhancing the hedonic experiences of being a member of a brand fan page, which directly influence the good impression of businesses. Marketers can use a variety of features to allow members to access all of the information on a brand's fan page. Pages may also run marketing efforts to encourage users to upload messages and photos to their brand's fan pages, thereby increasing member participation. At the same time, they may provide a component to enhance the members' hedonic experience, such as a free drink at check-in, as a reward for their involvement. Because they generate hedonic motivation, such as good experiences, pleasure, and positive feelings, these types of marketing may be more effective at engaging members in the activities of the brand fan page. Furthermore, marketing strategies that include free gifts or samples may be more effective than mere discounts and coupons in creating favourable client responses. Social media is a powerful tool for brands to retain relationships with their customers. By analysing user posts on brand pages, marketers may uncover the most significant features of their businesses. They can then take rapid action based on positive reviews and complaints about service. Marketers are capable of efficiently responding to consumer complaints. All of this may help firms' fan pages create positive brand images and develop strong relationships with their followers.

REFERENCES

Alvarado, E. (2021). *237 ways drone applications revolutionize business*. Drone Industry Insights.

Arnett, G. (2015). *The numbers behind the worldwide trade in drones*. Retrieved from https://www.theguardian.com/news/datablog/2015/mar/16/numbers-behind-worldwide-trade-in-drones-uk-israel

Bateman, J. (2017). China Drone Maker DJI: Alone atop the Unmanned Skies. *CNBC, 1*, 1.

Cary, L., & Coyne, J. (2011). ICAO unmanned aircraft systems (UAS), circular 328. *UVS International. Blyenburgh & Co, 2012*, 112–115.

Friedman, L., & McCabe, D. (2020). Interior Dept. Grounds Its Drones Over Chinese Spying Fears. *The New York Times*. Retrieved from https://www.nytimes.com/2020/01/29/technology/interior-chinese-drones.html

Glynn, T. J. (1981). Psychological sense of community: Measurement and application. *Human Relations*, *34*(9), 789–818.

Hallion, R. (2003). *Taking flight: Inventing the aerial age, from antiquity through the First World War*. Oxford University Press.

Haydon, F. S. (2000). *Military Ballooning during the Early Civil War*. JHU Press.

Horowitz, M. C. (2020). Do emerging military technologies matter for international politics? *Annual Review of Political Science*, *23*(1), 385–400. doi:10.1146/annurev-polisci-050718-032725

Hu, J., Bhowmick, P., Jang, I., Arvin, F., & Lanzon, A. (2021). A decentralized cluster formation containment framework for multirobot systems. *IEEE Transactions on Robotics*, *37*(6), 1936–1955. doi:10.1109/TRO.2021.3071615

Hu, J., & Lanzon, A. (2018). An innovative tri-rotor drone and associated distributed aerial drone swarm control. *Robotics and Autonomous Systems*, *103*, 162–174. doi:10.1016/j.robot.2018.02.019

Kaplan, P. (2013). *Naval Aviation in the Second World War*. Pen and Sword.

Koparan, C., Koc, A. B., Privette, C. V., & Sawyer, C. B. (2018). In situ water quality measurements using an unmanned aerial vehicle (UAV) system. *Water (Basel)*, *10*(3), 264.

Koparan, C., Koc, A. B., Privette, C. V., & Sawyer, C. B. (2020). Adaptive water sampling device for aerial robots. *Drones*, *4*(1), 5.

McAlexander, J. H., Schouten, J. W., & Koenig, H. F. (2002). Building brand community. *Journal of Marketing*, *66*(1), 38–54.

McNabb, M. (2020). *Drones Get the Lights Back on Faster for Florida Communities*. Retrieved from https://dronelife.com/2020/02/28/drones-get-the-lights-back-on-faster-for-florida-communities/

Mikesh, R. C. (1973). Japan's World War II balloon bomb attacks on North America. *Smithsonian Annals of Flight*. Advance online publication. doi:10.5479i.AnnalsFlight.9

Miller, M. (2020). *DOJ bans use of grant funds for certain foreign-made drones*. Retrieved from https://thehill.com/policy/cybersecurity/520269-justice-department-issues-policy-banning-use-of-grant-funds-for-certain/

Motwani, S. (2020, September). Tactical Drone for Point-to-Point data delivery using Laser-Visible Light Communication (L-VLC). In *2020 3rd International Conference on Advanced Communication Technologies and Networking (CommNet)* (pp. 1-8). IEEE.

Murphy, J. D. (2005). *Military aircraft, origins to 1918: An illustrated history of their impact*. ABC-CLIO.

Nanalyze. (2019). *How Autonomous Drone Flights Will Go Beyond Line of Sight*. Retrieved from https://www.nanalyze.com/2019/12/autonomous-drone-flights/

Peck, A. (2020). *Coronavirus Spurs Percepto's Drone-in-a-Box Surveillance Solution*. Retrieved from https://insideunmannedsystems.com/unintended-consequences-coronavirus-spurs-perceptos-drone-in-a-box-surveillance-solution/

Peterson, A. (2013). *States are competing to be the Silicon Valley of drones*. Retrieved from https://www.washingtonpost.com/news/the-switch/wp/2013/08/19/states-are-competing-to-be-the-silicon-valley-of-drones/

Renner, S. L. (2016). *Broken Wings: The Hungarian Air Force, 1918–45*. Indiana University Press. doi:10.2307/j.ctt2005t4h

Sharma, A., Vanjani, P., Paliwal, N., Basnayaka, C. M. W., Jayakody, D. N. K., Wang, H. C., & Muthuchidambaranathan, P. (2020). Communication and networking technologies for UAVs: A survey. *Journal of Network and Computer Applications, 168*, 102739. doi:10.1016/j.jnca.2020.102739

Tice, B. P. (1991). Unmanned aerial vehicles: The force multiplier of the 1990s. *Airpower Journal, 5*(1), 41–55.

Vergouw, B., Nagel, H., Bondt, G., & Custers, B. (2016). Drone technology: Types, payloads, applications, frequency spectrum issues and future developments. In *The future of drone use* (pp. 21–45). TMC Asser Press. doi:10.1007/978-94-6265-132-6_2

ADDITIONAL READING

Custers, B. (2016). *Future of Drone use*. TMC Asser Press. doi:10.1007/978-94-6265-132-6

Koparan, C., Koc, A. B., Privette, C. V., & Sawyer, C. B. (2019). Autonomous in situ measurements of noncontaminant water quality indicators and sample collection with a UAV. *Water (Basel)*, *11*(3), 604. doi:10.3390/w11030604

Koparan, C., Koc, A. B., Privette, C. V., Sawyer, C. B., & Sharp, J. L. (2018). Evaluation of a UAV-assisted autonomous water sampling. *Water (Basel)*, *10*(5), 655. doi:10.3390/w10050655

Layman, R. D. (1996). *Naval Aviation in the First World War: Its Impact and Influence*. Bloomsbury Academic.

Taylor, J. W. R., & Munson, K. (1977). *Jane's pocket book of remotely piloted vehicles: robot aircraft today*. Collier Books.

Chapter 9 Business Transformation and Enterprise Architecture Projects: Machine Learning Integration for Projects (MLI4P)

Antoine Trad
https://orcid.org/0000-0002-4199-6970
IBISTM, France

ABSTRACT

In this chapter, the author bases his research projects on his authentic mixed multidisciplinary applied mathematical model for transformation projects. His mathematical model, named the applied holistic mathematical model for project (AHMM4P), is supported by a tree-based heuristics structure. The AHMM4P is similar to the human empirical decision-making process and applicable to any type of project, aimed to support the evolution of organisational, national, or enterprise transformation initiatives. The AHMM4P can be used for the development of the enterprise information systems and their decision-making systems, based on artificial intelligence, data sciences, enterprise architecture, big data, and machine learning. The author tries to prove that an AHMM4P-based action research approach can unify the currently frequently used siloed machine learning trends.

DOI: 10.4018/978-1-7998-8763-8.ch009

INTRODUCTION

In this book chapter the author presents an Artificial Intelligence (AI) based generic concept for decision making that is based on Machine Learning; where the AHMM4P manages various types of algorithms. A transformation depends on the capacities of the decision-making system and the profile of the Business Transformation Manager (or simply the *Manager*) and his team; who are supported by a holistic framework (Trad & Kalpić, 2020a). The role of Machine Learning Integration for Projects (MLI4P) and the needed data and modules' modelling techniques are essential for managing various type of algorithms in an AI based transformation project. This chapter and the author's related research publications deal with Business Transformation Projects' (or simply *Project*) complexity as well as the support for the Decision-Making System for Projects (DMS4P) and Enterprise Architecture Integration for Projects (EAI4P). The proposed framework promotes the *Project's* technics to ensure success, by: 1) modelling artefacts; 2) implementing MLI4P components; 3) EAI4P support; 4) the use of a Generic Project Interface (GPI); and 5) complex algorithmics. The success of a *Project* depends on how an EAI4P and complex algorithmic modelling activities are synchronized (IMD, 2015).

Figure 1. EAI4P cycles synchronize with Project resources

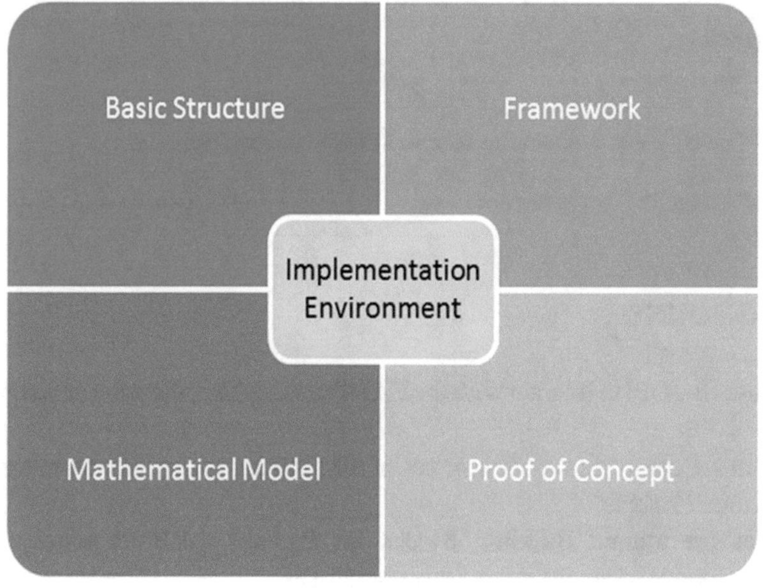

That is why the implementation of such *Projects* requires significant knowledge of EAI4P techniques. GPI handles MLI4P calls and its main mechanisms to support: 1) a generic data architecture; 2) implementation interfaces; and 3) data and modules modelling. GPI is a part of the Selection management, Architecture-modelling, Control-monitoring, Decision-making, Training management and Project management Framework (SmAmCmDmTmPmF, for simplification in further text the term Transformation, Research, Architecture, Development framework or *TRADf* will be used). As shown in Figure 1, *Project* resources interact with all the enterprise's (or simply an *Entity*) architecture phases, using the data Building Blocks for Projects (dBB4P) or the holistic brick (Trad & Kalpić, 2020a). GPI is MLI4P's main interface and the trends of using MLI4P for 2021, is tremendous, as shown in Figure 2 (Kapoor, 2021).

Figure 2. The growing role of MLI4P on Hyperautomation (Kapoor, 2021)

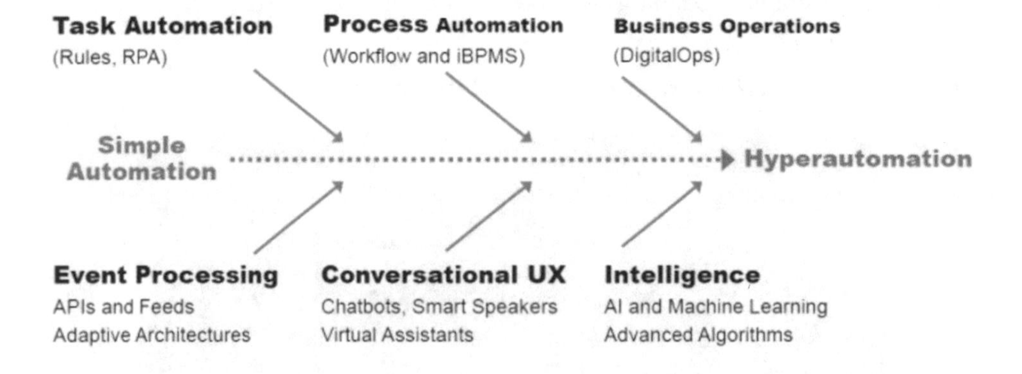

BACKGROUND

MLI4P uses the GPI to interact with the EAI4P and has the following characteristics:

- Is an AI composite model, or set of algorithms, which can be integrated in various *Projects*.
- Uses the atomic Building Blocks for Projects (aBB4P) concept; which corresponds to an autonomous set of classes.
- Uses a Natural Programming Language for Projects (NLP4P) for development of various types of interfaces.

The author's global research topic's and final Research Question (RQ) (hypothesis #1-1) is: "Which business transformation manager's characteristics and which type of support should be assured for the implementation phase of a business transformation project?" The targeted business domain is any business environment that uses: 1) complex technologies; and 2) frequent transformation iterations. For this phase of research, the sub-question (or hypothesis #2-3) is: "What is the impact of the MLI4P on *Projects*?"

MAIN FOCUS OF THE CHAPTER

In this chapter the focus is on MLI4P's usages, that are a part of the Architecture module (Am) and Decision making (Dm) modules, and it tries to prove that such a concept can be built on a loosely coupled EAI4P. It uses the Data Management Concepts for Artificial Intelligence (DMC4P) to interface various types of data sources. The MLI4PAI uses the AHMM4P, which manages algorithms that are used to analyse data and offer conclusions. *Projects* are increasingly complex, and data is global; these huge amounts of data are full of valuable operational information.

Artificial Intelligence Basics

AI is a concept that is older than Information and Communication Systems (ICS) and inspects if it is possible to create machines that contain *human like* cognitive abilities. This concept has influenced academicians, researchers and other scientific fields. It emerged as a practical domain in the middle of the 20th century. In 1950, Alan Turing (an English computer scientist, cryptanalyst, mathematician and theoretical biologist) developed a fundamental test for machine intelligence, which is known as the *Turing Test*. The term AI was coined in the proposal for a seminal AI conference that took place at Dartmouth in 1956 (Schmelzer, 2021).

Turing Test Basics

A *Turing Test* is a method used in AI to determine if a machine is capable of thinking like the human brain. Turing claimed that a machine can be estimated to be AI capable, if it can mimic human responses under certain constraints (Schmelzer, 2021).

Data Sciences Basics

Data Sciences Integration for Projects (DSI4P) basics are (Guru99, 2021):

- It involves extracting insights in the context of vast amounts of data, by using scientific methods, algorithms and processes. It helps in finding hidden patterns from raw data. DSI4P is the result of evolution of statistics, data analysis and Big Data for Projects (BGD4P).
- It is cross functional and tries to extract knowledge from structured or unstructured databases.
- It translates business problems into a RDP4P and then translates them into solutions.
- It includes, statistics, visualization, Deep Learning Integration for Projects (DLI4P) and ML4P concepts.
- Its process includes: discovery, data preparation, model planning, model building, operationalize and the communication of results.
- It predicts business solutions as it looks backward, where DSI4P looks forward.
- Possible applications are 1) Internet search; 2) Recommendation systems; 3) Image and Speech recognition; 4) Gaming world; 5) Online price comparison; and 6) many other…
- Its biggest challenges are: various information, data formats and sources.

MLI4P Basics

MLI4P's basics are (Schmelzer, 2021):

- It is a hallmark of intelligence and has the ability to learn from experience and helps machines to identify learning patterns, in order to make predictions; this is the essence of MLI4P.
- It uses algorithms that abstract learning from samples of *good* data and models.
- It involves different types of learning processes, with different levels of specialists' guidance and these processes are:
    - Supervised learning, which starts with human initiated training data that instructs algorithms on what to learn.
    - Unsupervised learning, is a method in which an algorithm, discovers information autonomous using unlabeled training data.
    - Reinforcement learning, uses algorithms to learn from trial and error; and is supported by initial instructions and ongoing oversight from specialists.

A Composite AI Model

AI, DSI4P and MLI4P, composite model's basics are (Schmelzer, 2021):

- DSI4P is powerful and when combining it with MLI4P gives the capability for generating insights from massive data.
- It leads to solving the challenge of complex *Project* problems; like:
  - Predictive analytics forecasts of customer behavior, business trends and events based on analysis of constantly changing datasets.
  - Intelligent conversational systems which support interactive communications with various parties.
  - Anomaly detection systems that can respond to continually evolving threats and enforce cybersecurity and fraud detection.
  - Hyperpersonalization systems enable targeted advertising, product recommendations, financial guidance and medical care and other services.
  - Major financial crimes.
- Even if they are separate concepts which individually offer capabilities, combing them is transforming the business ecosystems.
- The author's framework goes even further to combine them with EAI4P and other fields.

RESEARCH DEVELOPMENT PROCESS FOR AI

As shown in Figure 3, the Research and Development Process for Projects (RDP4P) focuses on the impacts of the mechanistic EAI4P integration and uses a mixed hyper-heuristics based methodology (Vella, Corne, & Murphy, 2009). The RDP4P is based on an extensive cross-functional Literature Research Process for Projects (LRP4P), a Qualitative Analysis for Artificial Analysis (QLA4P) methodology and on a POC for the proposed hypotheses.

Figure 3. The mixed method flow diagram (Trad & Kalpić, 2020a)

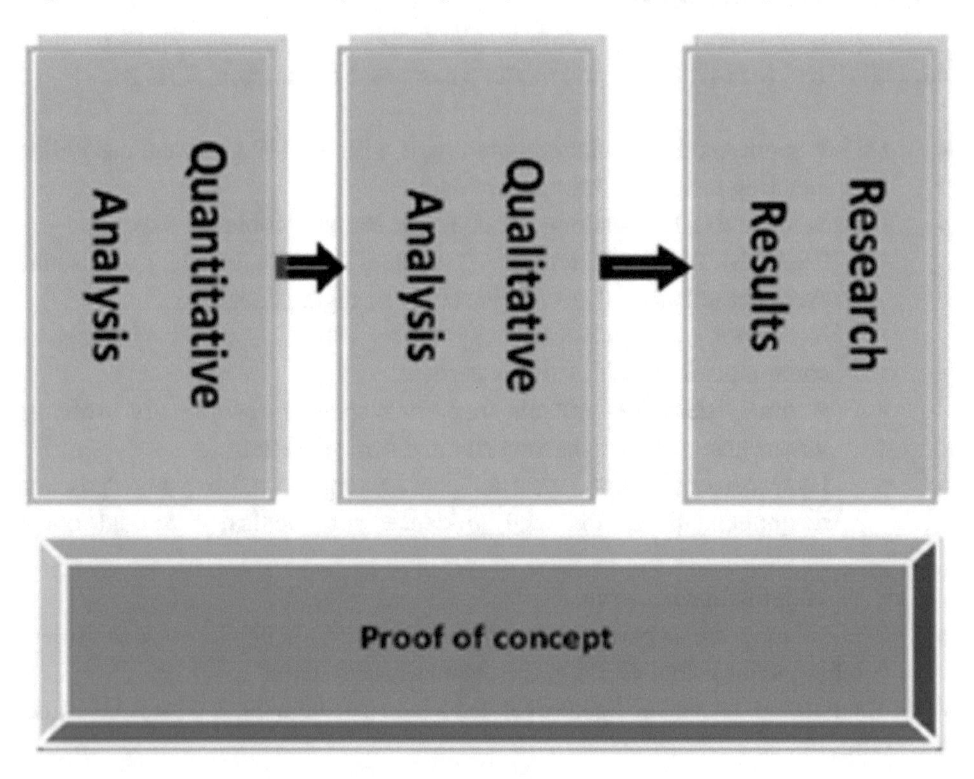

Projects and business engineering fields use a Proof of Concept (PoC) or prototyping to establish GPI's and MLI4P's (Camarinha-Matos, 2012): 1) feasibility; 2) viability; 2) major technical issues; and 3) offer recommendations.

Figure 4. The PoC's overall diagram of components

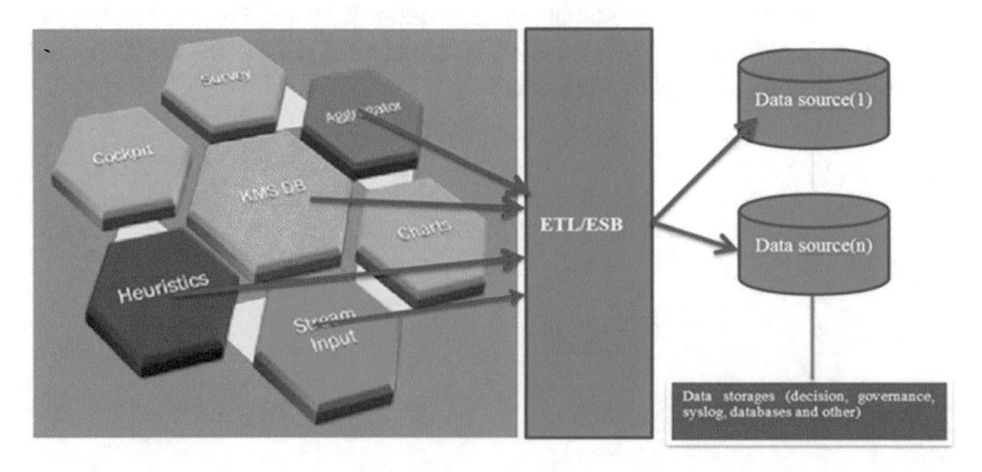

The PoC is used to prove the feasibility of the GPI and MLI4P, as shown in Figure 4. The PoC is based on a defined case that uses data class diagram for a data transaction, as shown in Figure 5; where an MLI4P module is called.

Figure 5. The PoC's class diagram package

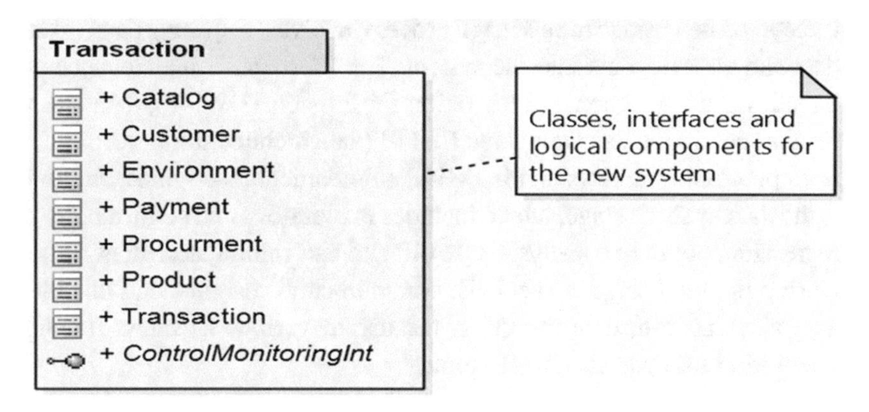

CRITICAL SUCCESS AREAS, FACTORS, GPI AND MLI4P

Critical Success Area (CSA) is a selected set of Critical Success Factors (CSF), where the CSF is a set of Key Performance Indicators (KPI), where each KPI corresponds/maps to a single *Project* requirement and/or problem type. For a given requirement (or problem), an *Entity* architect can identify the initial set of CSAs and their CSFs to be managed by the GPI and DMS4P. Hence the CSFs are important for the mapping between the problem types, knowledge constructs, organisational items. Therefore, CSFs reflect possible problem types that must meet strategic *Project* goals and predefined constraints. Measurements are used to evaluate performance in each of the CSA sets.

The Targeted Application Domain and Interaction

Main Stages

This chapter's targeted application domain characteristics are (OMNI-SCI, 2021):

- This exploration step: is the principal difference between DSI4P and Data Analytics for Projects (DA4P); where the DSI4P has a macro view, aiming to propose precise questions about data to extract and to have more insights.

- Modelling: fits the data into the model using ML4P algorithms and the model selection depends on the type of data and the business requirement.
- Definition: DSI4P includes the mining of large datasets of raw data, both structured and unstructured, in order to identify patterns and extract actionable insight from them. It includes, statistics, inference, computer science, predictive analytics, ML4P and BGD4P.
- Lifecycle, the 1st stage in an MLI4P process involves: acquiring data, extracting data and entering data into the system. The 2nd stage: is maintenance, which includes Data Warehouses Integration for Projects (DWI4P), data cleansing, data processing, data staging and EAI4P (data architecture).
- Data processing follows and is DSI4P's fundamental activities; and the next is the data analysis stage, which includes exploratory and confirmatory tasks, regression, predictive analysis, QLA4P and text mining activities.
- During the final stage, a specialist communicates insights; and this involves data visualization, data reporting, the use of various Business Intelligence Integration for Projects (BII4P) tools.

Combining Fields and a Holistic View

DSI4P and DA4P are different; DSI4P starts earlier, exploring a massive dataset, investigating its potential, identifying trends and insights, and visualizing them for others. DA4P comes in at a later stage. They report what they view, make prescriptions for improving performance based on their analysis, and optimize any data related tools. DA4P analyses a specific dataset of structured or numerical data using a given question. A DSI4P tackles larger masses of both structured and unstructured data; and he formulates, tests and assesses the performance of data questions in the context of a precise strategy. DA4P has more to do with placing historical data in context and less to do with predictive modelling and ML4P. DA4P needs precise questions and unlike DSI4P does not create statistical models or use ML4P tools. DA4P focuses on strategy for businesses, by comparing data assets to various *Entity's* hypotheses or plans. DA4P works with localized data that has already been processed.

RDP4P's CSF's

Based on the business case's (and its CSA) LRP4P process managed and weighted the most important CSFs that were used.

Table 1. The RDP4P's CSFs that have an average of 9.25.

Critical Success Factors	KPIs	Weightings
CSF_MLI4AI_LRP4AI	Proven	From 1 to 10. 10 Selected
CSF_MLI4AI_PoC	Feasible	From 1 to 10. 09 Selected
CSF_MLI4AI_References	Feasible	From 1 to 10. 09 Selected
CSF_MLI4AI_EA4AI	Proven	From 1 to 10. 10 Selected
CSF_MLI4AI_CSF_CSA	Feasible	From 1 to 10. 09 Selected
CSF_MLI4AI_Mixed_Methods	Feasible	From 1 to 10. 09 Selected
CSF_MLI4AI_Transformation_Setup	Feasible	From 1 to 10. 09 Selected
CSF_MLI4AI_DMS4AI_Interfacing	Feasible	From 1 to 10. 09 Selected

valuation

As shown in Table 1, the result's aim is to prove or justify the RDP4P's feasibility; and the result permits to move to the next CSA that is the AHMM4P.

AHMM4P'S SUPPORT FOR MLI4P

The *TRADf* is based on AHMM4P based GPI, which in turn supports the MLI4P, DSI4P, QNA4P and QLA4P based scenario(s), to interface the DMS4P, as shown in Figure 6.

Figure 6. The overview

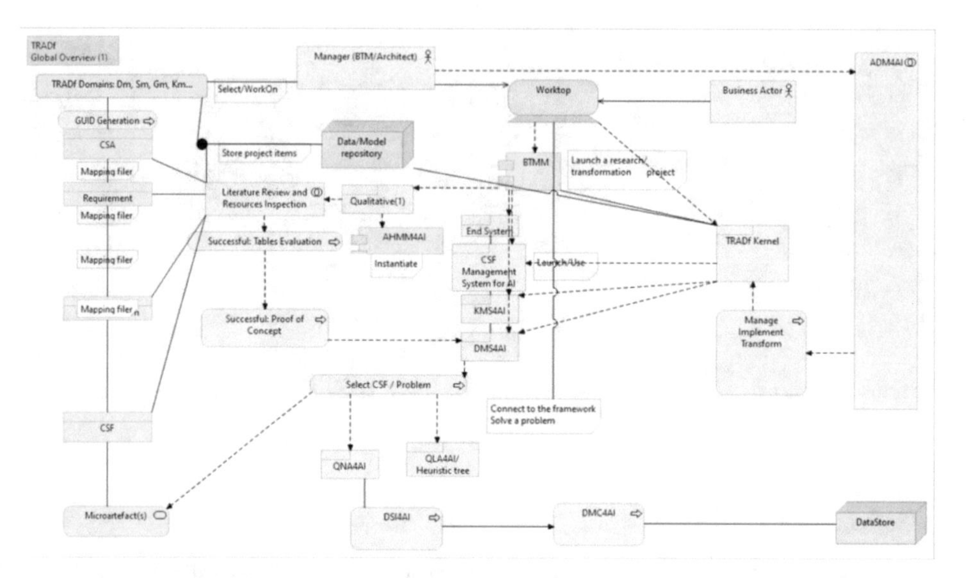

A Generic Holistic Approach

The *TRADf* proposes a holistic approach to analyse events and eventually manage possible risks to help Application and Problem Domains (APD) managers to avoid major *Project* pitfalls. Traditionally, complex risk concepts, were associated with a single origin or CSF; mainly personified to concretise a complex situation. Pitfalls may be defined as a violation of an internal risk's related CSF that can be due to various types of problems or constraints.

The Microartefacts' Distributed Architecture Model for the GPI

The AHMM4P has a dynamic defined nomenclature to facilitate GPI's integration with EAI4P model, and its Architecture Development Method for Projects (ADM4P). The AHMM4P is the *Entity's* holistic structural model that supports a set of multiple coordinated MLI4P processing to deliver solutions that correspond to various just in time processing schemes which use the same *Project*'s central pool of CSAs and CSFs. The basic AHMM4P nomenclature, is presented in Figure 7, to the reader in a simplified form, to be easily understood on the cost of a holistic formulation of the AHMM4P's basics for GPI, MLI4P, QNA4P and QLA4P. The DMS4P uses an AHMM4P's instance to solve a *Project* problem.

Figure 7. The applied AHMM4P's basics nomenclature (Trad & Kalpić, 2020a)

Basic MM's Nomenclature

Iteration	= An integer variable that denotes a *Project/ADM iteration*	
microRequirement	= KPI	(1)
CSF	= Σ KPI	(2)
CSA	= Σ CSF	(3)
Requirement	= U̲ microRequirement	(4)
microKnowledgeArtefact	= U̲ knowledgeItem(s)	(4)
neuron	= action ° data + microKnowledgeArtefact	(5)
microArtefact	= U̲ (e)neurons	(6)
microEntity or Enterprise	= U̲ microArtefact	(7)
Entity or Enterprise	= U̲ microEntity	(8)
microArtefactScenario	= U̲ microArtefactDecisionMaking	(9)
Decision Making/Intelligence	= U̲ microArtefactScenario	(10)
EnityIntelligence	= U̲ Decision Making/InteligenceComponent	(11)
MM(*Iteration*) as an instance	= EnityIntelligence(*Iteration*)	(12)

AHMM4P's instances supports the GPI; using CSFs weightings and ratings (in phase 1), and based on multicriteria evaluation (selected and defined constraints). The symbol å indicates summation of all the relevant named set members, while the indices and the set cardinality have been omitted. AHMM4P's role should be understood in a broader sense, more like set unions. As shown in Figure 17:

- The abbreviation "mc" can be used, and stands for micro, which depends on the granularity.
- The symbol å indicates summation of weightings/ratings, denoting the relative importance of the set members selected as relevant. Weightings as integers range in ascending importance from 1 to 10 (or another range defined by GPI based analysists).
- The symbol U̲ indicates sets union.
- The proposed AHMM4P supports GPI as an interface model; using CSFs weightings and ratings evaluation.
- The selected corresponding weightings to: CSF ϵ { 1 … 10 }; are integer values, that are presented in tables. The rules were presented in the RDP4P section.
- The selected corresponding ratings to: CSF ϵ { 0.00% … 100.00% } are floating point percentage values.

The AHMM4P's Structure for GPI and MLI4P based Solutions

The AHMM4P's has a composite structure that can be viewed as follows:

- The static view has a similar static structure like the relational model's structure that includes sets of CSAs/CSFs that map to tables and the ability to create them and apply actions on these tables; in the case of AHMM4P for GPI and MLI4P, is done by using QNA4P and QNA4P microartefacts and not tables (Lockwood, 1999).
- In the behavioural view, these actions are designed using a set of AHMM4P nomenclature, the implementation of the AHMM4P is in the internal scripting language, used also to tune the CSFs (Lazar, Motogna, & Parv, 2010).
- The skeleton of the *TRADf* uses microartefacts' scenarios to support just-in-time GPI requests.

Entity/Enterprise Architect as an Applied Mathematical Model

The EAI4P and its ADM4P are the kernel of this RDP4P and they are the basics of its *TRADf;* where the AHMM4P is GPI's skeleton. The LRP4P has shown that existing LRP4P's resources on MLI4P, are practically inexistent. This pioneering research work is cross-functional and links all the MLI4P or QLA4I/QNA4P based microartefacts to an *Entity*; where the main reasoning component is a MLI4P engine that is based on heuristics.

Heuristics, Empirics and Action Research

The MLI4P is based on a set of synchronized AHMM4P instances, where each AHMM4P can launch a QLA4P beam-search based heuristic processing (Kim and Kim, 1999; Della Croce and T'kindt, 2002). Weightings and ratings concept support the AHMM4P to process a GPI request for an optimal analysis or solution for a given *Entity's Project* problem. Actions Research Integration for Projects (ARI4P) (Berger and Rose, 2015) can be considered as a set of continuous beam-search heuristics processing phases and is similar to design, analysis and architecture processes, like the ADM4P (Järvinen, 2007). Fast changing *Entity's* change requests may provoke an important set of events and problems that can be hard to predict and solve; that makes the GPI various types of actions useless and complex to implement. The AHMM4P is responsible for the MLI4P that uses QLA4P heuristic process for *Entity*'s problem solving and synchronizes a set of AHMM4P instances which have also separate heuristics processes and are supported by a dynamic tree algorithm, as shown in Figure 8 (Nijboer et al., 2009) that manages tree nodes and their correlation with memorized patterns that are combinations of data states and heuristic goal functions. The AHMM4P capacities are measured by analysing the *TRADf's* AHMM4P tree.

Figure 8. The applied heuristics tree algorithm (Nijboer et al., 2009)

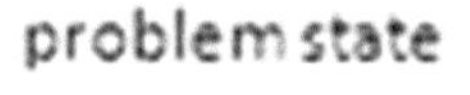

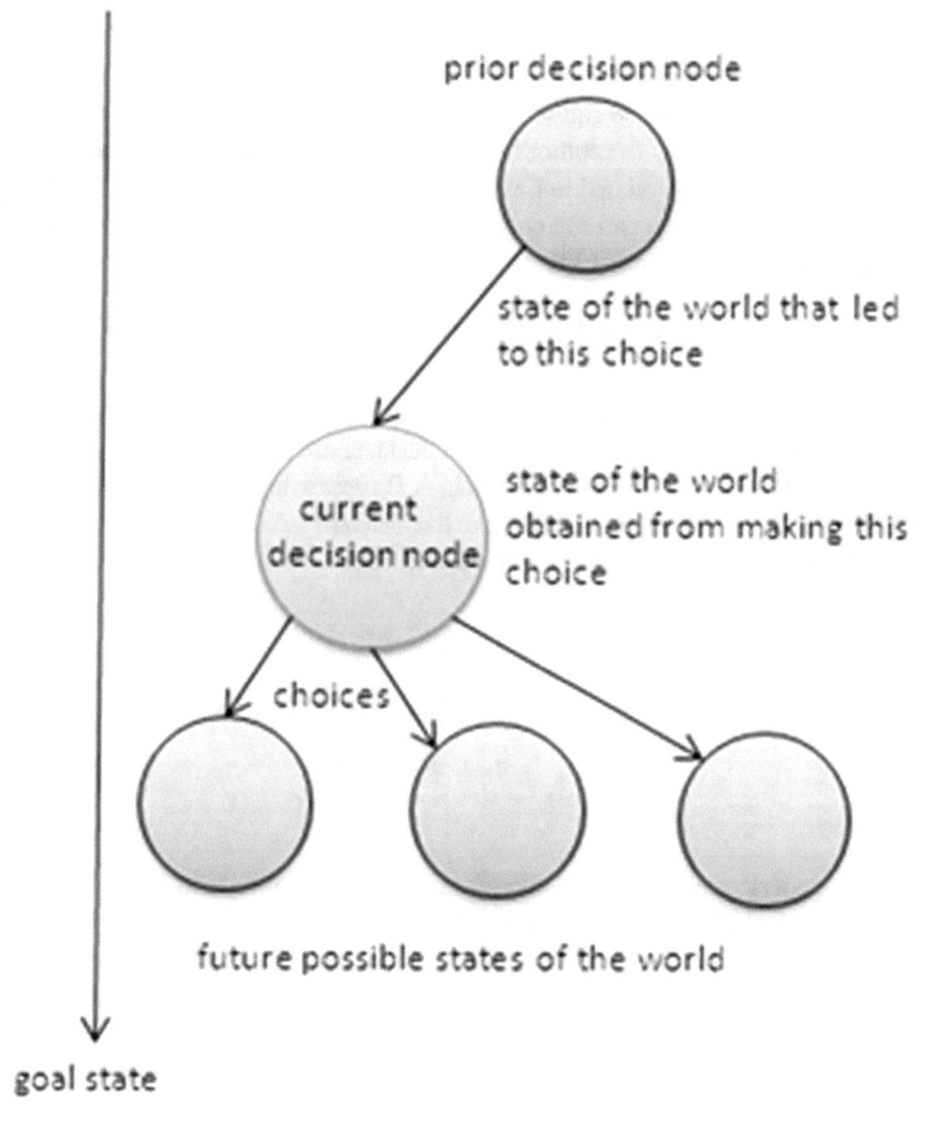

Holistic ARI4P based heuristics enables reflective practice that is the basis of a holistic approach to develop EAI4P based GPI solutions, where its kernel and skeleton are a dynamic DMS4P (Leitch, & Day, 2006). GPI is based on MLI4P, QLA4P and QNA4P methods (Loginovskiy et al., 2018).

Qualitative, Quantitative (or Mixed) related to the Notion of Time, Space and Scope

As already mentioned, the AHMM4P and its underlining set of created instances is mainly a QLA4P beam-search heuristic tree (Della Croce and T'kindt, 2002). In each of the tree's node a precise call to MLI4P functions (or other) can be executed, by precision or objectivity the author refers to inputted data, constraint (and/or rules) and above all: 1) Time related to timestamp tracing system; 2) Space, related to the analysed *Entity's* space; and 3) Scope of GPI's process. These facts, form the basis of an applicable GPI and MLI4P are based on AHMM4P instances.

The Applied GPI Transformation Mathematical Model

GPI is a part of the *TRADf* that uses microartefacts to support just-in-time DMS4P actions. The GPI based component and interface, are based on a light version of the ADM4P, having a systemic approach. A *Project* using GPI is the combination of GPI based, EAI4P methodology (like the TOGAF's ADM4P) and the proposed AHMM4P, that is presented in Figures 9 and 10.

Figure 9. The AHMM4P generic structure

The Generic AHMM's Formulation

$$\text{AHMM} = \underline{\bigcup}\ \text{ADMs} + \text{MMs} \qquad (13)$$

The generic AHMM can applied to any specific domain; in this chapter and RDP4P's phase, the *Domain,* is GPI based and the AHMM4P = AHMM(APD), as shown in Figure 10; where AHMM can applied to any domain and any concept.

The proposed combination can be modelled after the following formula for the GPI Transformation Mathematical Model (GPITMM) that abstracts the *Project* for a given *Entity*:

Figure 10. The AHMM4P structure

AHMM's Application and Instantiation for a Specific Domain

Domain	= Geopolitical Analysis (GA)	(14)
AHMM(*Domain*)	= $\biguplus$ ADMs + MMs(*Domain*)	(15)

(AHMM4P for an *Iteration*):

iAHMM4P = AHMM4P(*Iteration*);

$iAHMM4P = Weigthing_1 * iAHMM4P\_Qualitative + Weigthing_2 * iAHMM4P\_Quantitative$ (16)

The *Project's* AHMM4P (PAHMM4P) = $\sum$ iAHMM4P for an ADM4P's instance (17)

(GPITMM):

GPITMM = $\sum$ PAHMM4P instances (18)

The Main Objective Function for GPI based (MOFGPI) of the GPITMM's formula can be optimized by using constraints and with extra variables that need to be tuned using the AHMM4P. The variable for maximization or minimization can be, for example, the *Project* success, costs or other (Dantzig, 1949). For this chapter's PoC the success will be the main and only constraint and success is quantified as a binary 0 or 1. Where the MOF4P definition will be:

Minimize risk GPITMM Function (GPITMMf) (19)

The AHMM4P is based on a concurrent and synchronized *TRADf*, which uses concurrent threads that can make various AHMM4P instances run in parallel and manage information through the use of the AHMM4P's NLP4P. The GPITMM is the combination of the GPI, *Project* and EAI4P methodologies and a holistic AHMM4P that integrates the *Entity* or organisational concept, ICS that have to be formalized using a functional development environment like *TRADf's* NLP4P.

The AHMM4P's CSFs

Based on the LRP4P, the most important AHMM4P's CSFs that are used are evaluated to the following:

Table 2. The AHMM4P CSFS have an average of 10.0.

Critical Success Factors	KPIs	Weightings
CSF_MLI4AI_AHMMAI_TRADf_Integration	Proven	From 1 to 10. **10 Selected**
CSF_MLI4AI_AHMMAI_InitialPhase	Proven	From 1 to 10. **10 Selected**
CSF_MLI4AI_AHMMAI_PoC	Proven	From 1 to 10. **10 Selected**
CSF_MLI4AI_AHMMAI_Qualitative&Quantitative	Proven	From 1 to 10. **10 Selected**
CSF_MLI4AI_AHMMAI_Final_Instance	Proven	From 1 to 10. **10 Selected**
CSF_MLI4AI_AHMMAI_ADM4AI_Integration	Proven	From 1 to 10. **10 Selected**
CSF_MLI4AI_AHMMAI_APD_Interfacing	Complex	From 1 to 10. **08 Selected**

valuation

As shown in Table 2, the result's aim is to prove or justify that it is complex but possible to implement atheAHMM4P in the *Entity's* ICS. The next CSA to be analysed is DMC4P as an interface.

DMC4P AS AN INTERFACE

MLI4P's Integration Strategy

MLI4P's data strategy should be based on well-defined principles, like (IBM, 2015a):

- Data have to have a high level of quality.
- Data collections are derived from modelling concepts.
- Data is available to business personnel.
- Business processes for data management have to be automated.
- Data should not be redundant.
- Data have to be stored accurately and must be auditable.
- The cost of data collection procedures must to be optimized.
- MLI4P must adopt standards for common data models.

- Data events are managed by event handlers/algorithms.

Data Transformation Models

In MLI4P, the data transformation process, concerns the transforming data (and related services) from one format or class into another *unique* format or class. It is an essential process for data integration and data management tasks, like data wrangling, data warehousing, data integration and EAI4P interfaces development.

The Model's Unit of Work

A holistic alignment, identification and classification of all the RDP4P's resources must be done, so that the research process can start. A holistic alignment needs also, to define the Unit of Work (UoW) or the "1:1" mapping concept.

Data Solution Blocks

Upon a concrete *Project* requirement, the *Manager* issues a a contract to resolve this requirement by using EAI4P. EAI4P ensures that new requirements are managed accordingly to the *Project's* records and objectives. The requirement is linked to an instance of a newly created data Building Block for Projects (dBB4P) and its instance (a data Solution Block for Projects, dSB4P). The dBB4P is a part of GPI; and as shown in Figure 11, the *TRADf* uses the GPI and MLI4P that includes the pattern on how to integrate data solution blocks which are instances of the dBB4Ps.

DMC4P's CSF's

Based on the business case's (and its CSA) LRP4P process managed and weighted the most important CSFs that were used.

As shown in Table 3, the result's aim is to prove or justify the RDP4P's feasibility; and the result permits to move to the next CSA that is the data access, architecture and agility.

Figure 11. The PoC's dSB4P diagram

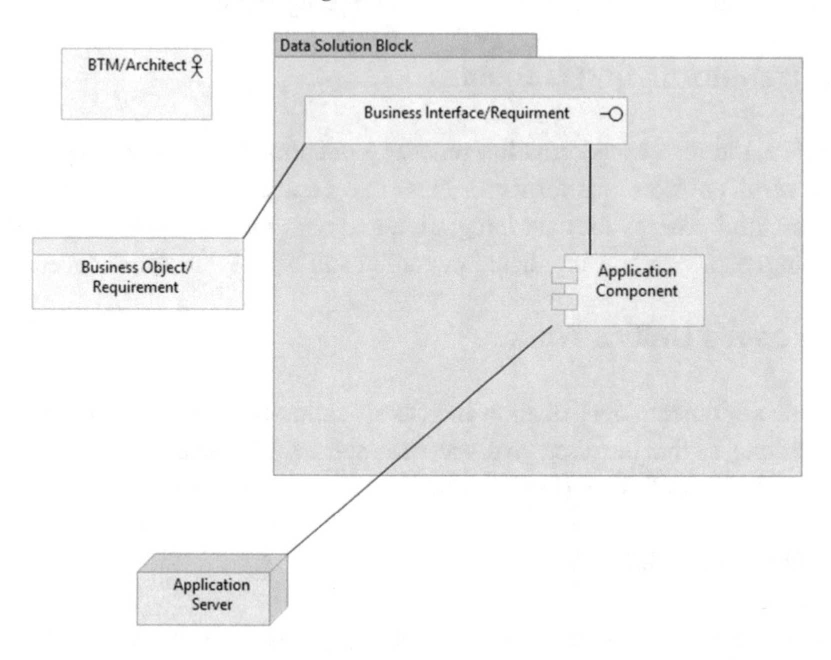

Table 3. The DMC4P's CSFs that have an average of 9.40

Critical Success Factors	KPIs	Weightings
CSF_DMC4AI_Interface_Strategy	Feasible	From 1 to 10. **09 Selected**
CSF_DMC4AI_Interface_Transformation	Proven	From 1 to 10. **10 Selected**
CSF_DMC4AI_Interface_Mapping	Proven	From 1 to 10. **10 Selected**
CSF_DMC4AI_Interface_EA4AI	Proven	From 1 to 10. **10 Selected**
CSF_DMC4AI_Interface_CSF_CSA	Feasible	From 1 to 10. **09 Selected**
CSF_DMC4AI_Interface_Modelling	Complex	From 1 to 10. **08 Selected**
CSF_DMC4AI_Interface_Replication	Proven	From 1 to 10. **10 Selected**
CSF_DMC4AI_Interface_Plaform	Feasible	From 1 to 10. **09 Selected**

valuation

THE MLI4P PLATFORM

Basic Roadmap

The major notions to implement a MLI4P oriented platform are (Veneberg, 2014);

- EAI4P supports the traceability and monitoring subsystem using standard patterns, but is not wise to out-source for GPI applications, because of the strategic goals. Combining DMS4P and MLI4P consolidates the *Entity's* support for business specialists, without the need to trace use data; EAI4P provides a meta-data model for operational and historic data that are needed for DMS4P routine operations.
- For massive digital data structures, Data Base Management System (DBMS) implemented to access, read and write datasets and d DBMS is a collection of inter-related data.
- DBMS' characteristics presents its capacities and actual DBMS technologies has three evolution eras of development: 1) navigational DBMS; 2) Relational DBMS (RDBMS); and 3) the Post-RDBMS (PRDBMS) types. RDBMS and PRDBMS are the most popular ones. Simple Query Language (SQL) is used with RDBMS, that manipulates tables.
- There are also non-relational database, labelled No Simple Query Language (NoSQL) Databases (NoSQLDB), that are the successors of RDBMS. RDBMS focuses on its relations between tables, where the NoSQLDB focuses on its objects' models. NoSQLDB include document-oriented databases and key-value stores. NoSQLDB are faster and more flexible than RDBMS. For GPI and MLI4P, NoSQLDBs can be used in combination with SQL-based environments.
- Data DWI4P used to store operational and historical data can be used as a reference source for MLI4P support for DMS4P operations.
- DWI4Ps lack data traceability and monitoring in respect for MLI4P processes, where *Project* teams need instructions on how to trace datasets and events.
- DWI4P is a set of decision support for the *Entity's* knowledge workers (executive, manager, analyst, architects).
- The Operational Data Store (ODS) is the data updated from an OnLine Transaction Processing (OLTP) response time; where the ODS is a hybrid environment in which *Project* data is transformed by an Extract Transform and Load (ETL) into a defined format. The ODS abstracts raw data DWI4P from *Entity*.

Enterprise Service Bus and Enterprise Application Integration

An Enterprise Service Bus (ESB) implements an enterprise-wide communication subsystem between mutually interacting applications in a SOA paradigm. It represents a communication architecture for global distributed computing and is a special variant of the more general client server concept, where *Entity* applications and modules may be a server or client. An ESB promotes modularity, agility and flexibility in relation to the ICS and *Project*. It is used in Enterprise Application Integration (EAI) approach for heterogeneous and complex services-based *Projects*. *Projects* must use ESB to glue the various data sources of the business environment, through the use of the technology stack and data connectors, which permit a holistic data services' management.

Extraction, Transform and Load

ETL processes are defined as accessing data stored in various locations and transforming them in order to enable their unification, quality or normalization. The MLI4P, proposes the separation of data processing activities; that enables data services to access data without bothering about various data sources' complexities. The MLI4P insures: 1) intra (or extra) data transparency; 2) managing accessibility; and 3) data quality control (Tamr, 2014). As shown in Figure 12, the ETL processes are responsible for the access of data from heterogeneous data sources. The challenge lays in the real-time data transformation and normalization processes (Trujillo & Luján-Mora, 2003).

BGD4P for MLI4P

BGD4P for Projects (BGD4P) has the following characteristics (SAS, 2021):

- BGD4P refers to data that is massive, fast or complex that is very complex to process using existing ICS' resources.
- Managing massive date for MLI4P can managed by BGD4P that is based on the now-mainstream definition of big data as the three V's: 1) Volume; 2) Velocity; and 3) Variety.
- **Volume:** *Entities* collect data from a variety of sources, including business transactions, smart (IoT) devices, industrial equipment, videos, social media and more. In the past, storing it would have been a major volume problem.
- **Velocity:** The evolution of ICS' and data streams with extreme speed, requests just-in-time processing. It is mainly a problem of performance and scalability.

- **Variety:** *Entity's* data have various types of formats, from structured, numeric data in traditional databases to unstructured text documents, emails, videos... That needs a unique transformation platform.

Figure 12. Various data sources (Tamr, 2014)

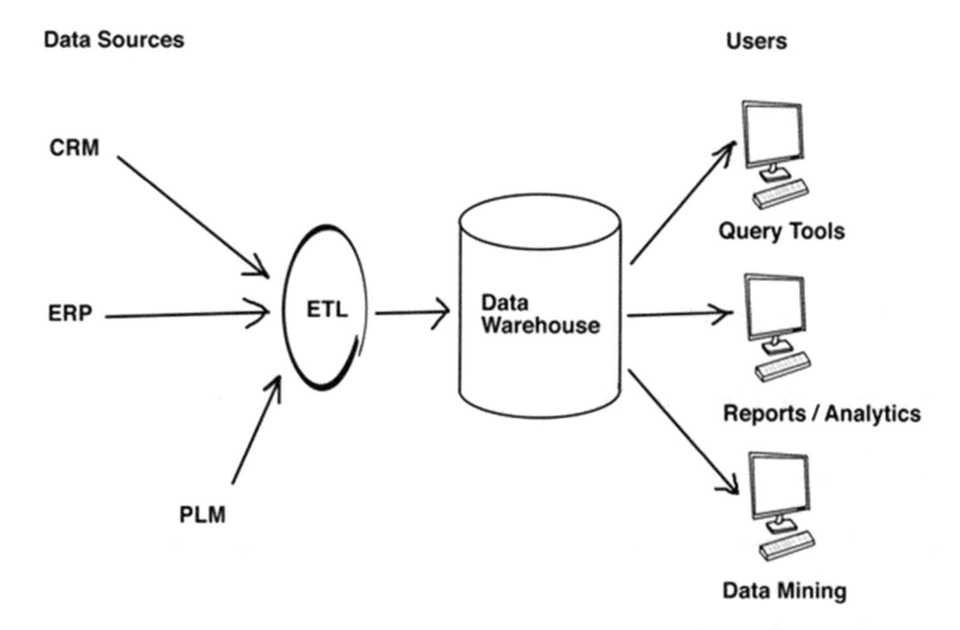

Platform's CSF's

Based on the business case's (and its CSA) LRP4P process managed and weighted the most important CSFs that were used.

As shown in Table 4, the result's aim is to prove or justify the RDP4P's feasibility; and the result permits to move to the next CSA that is EAI4P's integration.

Table 4. The Platform's CSFs that have an average of 9.0

Critical Success Factors	KPIs	Weightings
CSF_MLI4AI_Platform_Roadmap	Complex	From 1 to 10. **08 Selected**
CSF_MLI4AI_Platform_Interface	Complex	From 1 to 10. **08 Selected**
CSF_MLI4AI_Platform_Legacy_Mainframes	Proven	From 1 to 10. **10 Selected**
CSF_MLI4AI_Platform_EA4AI_ADM4AI	Proven	From 1 to 10. **10 Selected**
CSF_MLI4AI_Platform_CSF_CSA	Feasible	From 1 to 10. **09 Selected**
CSF_MLI4AI_Platform_Scalability	Feasible	From 1 to 10. **09 Selected**
CSF_MLI4AI_Platform_Performance	Feasible	From 1 to 10. **09 Selected**
CSF_MLI4AI_Platform_BigData	Complex	From 1 to 10. **09 Selected**

valuation

EAI4P'S INTEGRATION

EAI4P Principles and Basics

The main EAI4P's principles and basics are:

- MLI4P may be complex in large *Entities* dealing with types of risks and situations.
- DMS4P and DMC4P are implemented using existing BI solutions, combined with DWI4P for storing operational data.
- EAI4P is often used for strategy purposes and provides an overview of complex *Entity* architectures, showing business entities and relations.
- EAI4P accommodates DMC4P to enable data-driven EAI4P. Actually, there is no concept that enables BGD4P integrate in data-driven EAI4P.
- ADM4P supports a data-driven *Entity*; through a specific adaption of the ADM4P permits that DMC4P and BGD4P has on each phase within the ADM4P.

Figure 13. The layers of data-driven EAI4P.

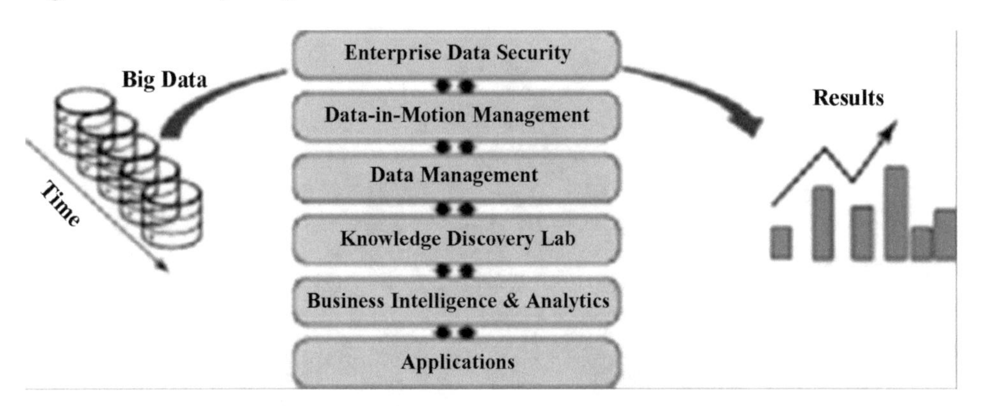

EAI4P Layers

EAI4P's main layers characteristics depend on (Sarkar, 2018):

- Lack of reproducibility and reusable artifacts: MLI4P analyses and modules are to be reproducible.
- Lack of collaboration: siloed should be removed, even though Specialists work in an isolated manner.
- Technical debt: there is a lack of standards and EAI4P concepts for MLI4P.
- Build reusable assets: it is important not just to focus on MLI4P but also on Non-Functional Requirements for Projects (NFR4P). Popular NFR4P includes, scalability, maintainability, availability and other.
- MLI4P needs to implement EAI4P based *Projects*, to define solutions, application and data models. For this goal there is a need to define layered EAI4P as shown in Figure 14.
- EAI4P based GPI is capable of: 1) envisioning end-to-end solutions: 2) to improve and transform the *Entity*.
- Structured evolution is essential to MLI4P that is an innovative field, that needs GPI as an interface.

Figure 14. Layered EAI4P Hierarchy (Sarkar, 2018)

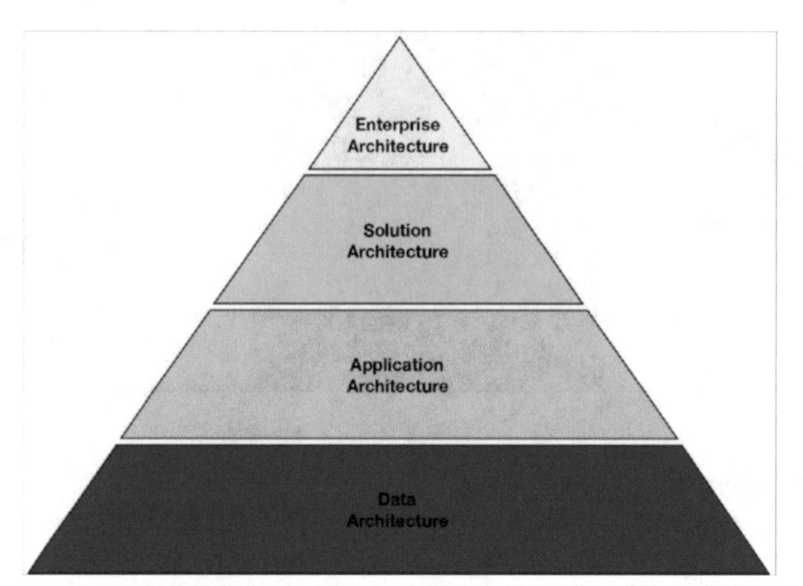

The Role of Resources and EAI4P Standards

The *TRADf* uses data technologies and existing data standards that include: 1) RDBMS standards; 2) vertical industry documents' standards; 3) business process standards; 4) data governance standards; 5) business services standards; 6) extensible mark-up language standards; 7) Object Oriented (OO) and ORM; 8) software development standards; 9) resources description standards. Standardization for inter-operability can be achieved by using an ESB and its ETL module (Tamr, 2014). The evolution of standards, like the Service Oriented Architecture (SOA) standards, have enabled MLI4Ps to become more receptive to development and integration with various standards. Standardized *Projects* have to be inter-operable and their focus must be on their: 1) data models; 2) data architectures; 3) software modelling and implementation concepts; and 4) data monitoring platforms. Regardless of the business domain, executive management understands the immense need for agility and the integration of MLI4P using GPI.

Application Reference Model

Application Reference Model (ARM), as shown in Figure 15, categorizes the system, applications' standards and ICS that support the delivery of service capabilities, allowing various *Entities* to share common solutions. EAI4P's data architect and a *Project's* data analyst, can use the standard ARM, which is not on for analytics, but is also used for EAI4P. ARM is a used, for mapping business APD to *Entity's* applications. ARM is a key for predictive analytics initiatives, where it defines the interface between the business and the ICS that supports it. GPI needs to create ARM artefacts to be used by the MLI4P (Doree, 2015).

Figure 15. An ARM implementation

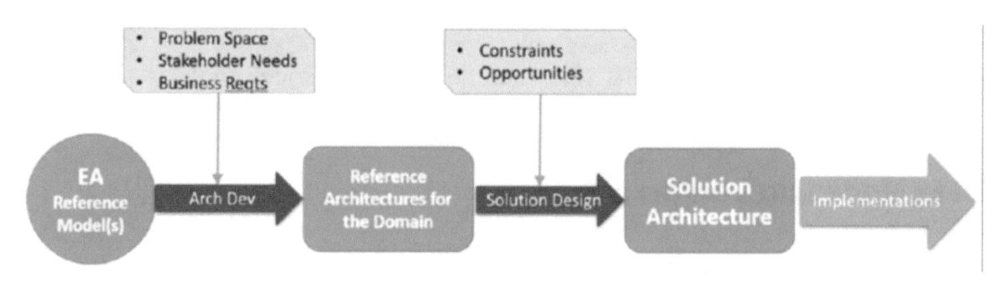

EAI4P's CSF's

Based on the business case's (and its CSA) LRP4P process managed and weighted the most important CSFs that were used.

Table 5. The EAI4P's CSFs that have an average of 9.0

Critical Success Factors	KPIs	Weightings
CSF_EA4AI_Integration_Principles	Feasible	From 1 to 10. **09 Selected**
CSF_EA4AI_Integration_MLI4AI_Strategy	Feasible	From 1 to 10. **09 Selected**
CSF_EA4AI_Integration_Standards	Feasible	From 1 to 10. **09 Selected**
CSF_EA4AI_Integration_EA4AI	Proven	From 1 to 10. **10 Selected**
CSF_EA4AI_Integration_CSF_CSA	Feasible	From 1 to 10. **09 Selected**
CSF_EA4AI_Integration_Layers	Complex	From 1 to 10. **08 Selected**
CSF_EA4AI_Integration_Modelling_Languages	Proven	From 1 to 10. **10 Selected**
CSF_EA4AI_Integration_ARM	Complex	From 1 to 10. **08 Selected**

valuation

As shown in Table 5, the result's aim is to prove or justify the RDP4P's feasibility; and the result permits to move to the next CSA that is modelling and implementation environments.

MODELLING AND IMPLEMENTATION ENVURONMENTS

The Modelling Basic Approach

The modelling basic approach's characteristics are (OMNI-SCI, 2021):

- MLI4P needs capable and autonomous specialists because it is a complex environment.
- MLI4P needs many skills that include from statistical analysis and other fields.

Data Modelling Languages

EAI4P Modelling Language

EAI4P modelling languages like ArchiMate, which is an open and independent modelling language for *Entity* functional, enterprise and data architectures. ArchiMate is supported by many vendors and consulting companies; and it provides instruments to enable *Entity* architects to describe, analyse and visualize the relationships among business (including data) domains in an clear manner (The Open Group, 2013b).

Unified Modelling Language

Unified Modelling Language (UML) can be used with MLI4P in the following contexts (Sikander, & Khiyal, 2018):

- To model data concepts and implementations; like the Data Flow Diagram (DFD) is an artefact that represents a flow of data through a process or a system (usually an ICS).
- The DFD also provides information on the outputs and inputs of each object (like a table) and the process itself. The DFD has no control flow, there are no Dynamic Rules for Projects (DR4P) and no loops. Specific operations based on the data can be represented by a flowchart.
- A UML Diagram, is essential to illustrate a conceptual model to a precise *Project* problem with the component or class diagrams. And to present the used algorithm, using the sequence or activity diagrams.
- In the context of the process of data analysis, integrates three models as shown in Figure 16, it is the process of inspecting, cleaning, converting and modelling of data, in order to obtain particular results. Raw data is collected from various sources and transformed into usable information streams for

DMS4P and MLI4P processing. There are different phases in the process of DMS4P and MLI4P processing; these processes are iterative. The UML state diagram for MLI4P process is shown in Figure 17.

Figure 16. The MLI4P process flowchart (Sikander & Khiyal, 2018)

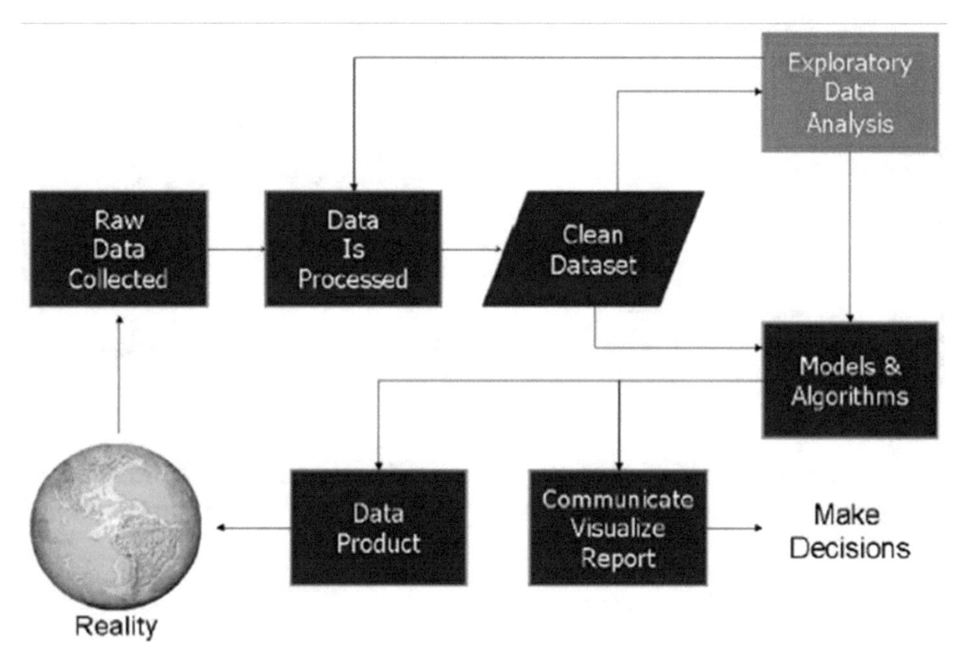

Figure 17. An MLI4P process flowchart, using the UML state diagram (Sikander & Khiyal, 2018)

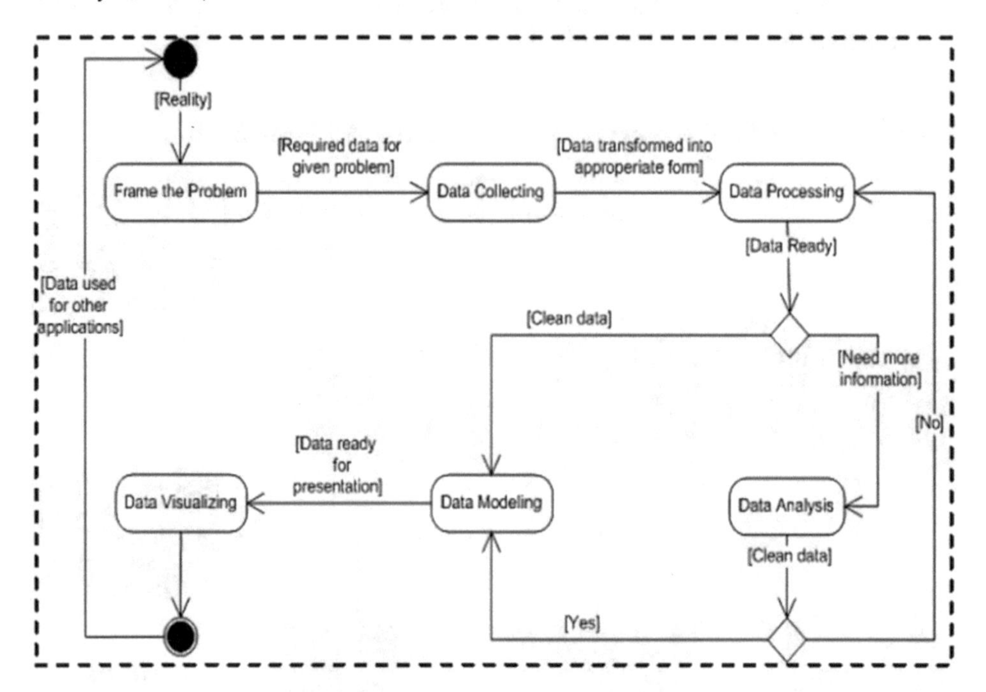

Data Modelling Diagrams

A data model is needed to describe the structure and relationships among a set of data managed in the *Project* and stored in a data management system. The most common is the relational model that is often represented in entity-relationship diagrams or class diagrams; these diagrams basically show data entities and their relationships. GPI for a *Project* can be seen as a data architectural view (Merson, 2009); where the modelling process incorporates various conceptual views.

Data Dissemination View

The purposed data dissemination diagram shows the relationship within the MLI4P: 1) data entities; 2) business data services; and 3) application components/services. This diagram as shown in Figure 20, presents the logical data entities to be physically managed/accesses by application components or services. That permits a flexible architecture and modelling of the data sources. This diagram includes data services that abstract various data sources (TOGAF Modelling, 2015a).

Figure 18. The PoC's data dissemination view

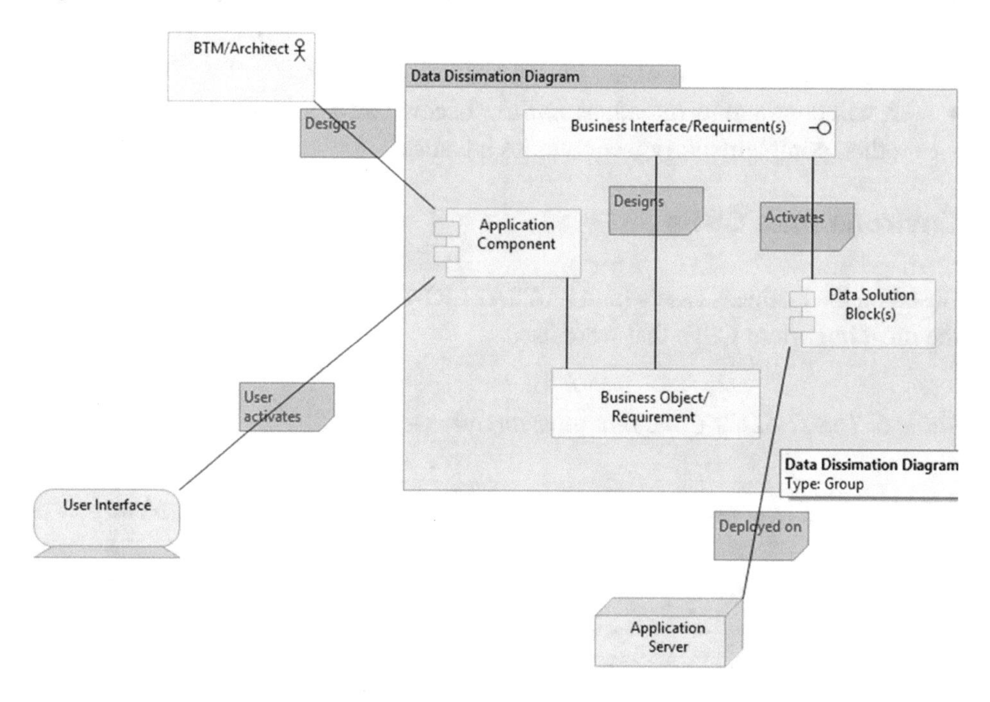

Functional Development

TRADf's internal NLP4P tool can be used for various application domains and in general for hard systems' thinking. The author recommends the use of an interpretable scripting for building a GPI (Moore, 2014). The AHMM4P based process is domain-driven and is founded on *TRADf* that in turn is based on a NLP4P to manage heuristics/ rules, *Entity*/EAI4P, QLA4P and QNA4P microartefacts (Simonin, Bertin, Traon, Jezequel & Crespi, 2010). The complexity lies in how to integrate the AHMM4P and its programming NLP4P in GPI. The main characteristics and facts related to functional languages are (Clancy, 2019):

- Functional programming reaches for stardom in finance.
- Financial institutions are adopting functional programming or NLP4P as an alternative to the dominant imperative approach.
- NLP4P results with less bugs as it is stricter, and easier to check and to test.
- It has in-built concurrency constructs, and are optimal for distributed ledgers. Distributed ledgers may find a wide application in the derivatives industry.
- It lacks skilled developers, as well as its memory and speed drawbacks.

- They will become more relevant with the adoption of the Distributed Ledger Technology (DLT) in finance.
- NLP4P make testing easier because of immutability.
- It will continue to spread, as *Entities* become aware of its advantages, but as other, complementary technologies advance.

Environments CSFs

Based on the business case's (and its CSA) LRP4P process managed and weighted the most important CSFs that were used.

Table 6. The TRADf's CSFs that have an average of 9.40

Critical Success Factors	KPIs	Weightings
CSF_Modelling_Implementation_MLI4AI_Interface	Proven	From 1 to 10. **10 Selected**
CSF_Modelling_Implementation_Modelling_Languages	Proven	From 1 to 10. **10 Selected**
CSF_Modelling_Implementation_Diagrams	Feasible	From 1 to 10. **09 Selected**
CSF_Modelling_Implementation_Functional_Dev	Feasible	From 1 to 10. **09 Selected**
CSF_Modelling_Implementation_CSF_CSA	Proven	From 1 to 10. **10 Selected**
CSF_Modelling_Implementation_ADM4AI_Integration	Proven	From 1 to 10. **10 Selected**
CSF_Modelling_Implementation_APD_Interfacing	Complex	From 1 to 10. **08 Selected**

valuation

As shown in Table 6, the result's aim is to prove or justify the RDP4P's feasibility; and the result permits to move to the next CSA that is the GPI's components integration

GPI'S MAIN COMPONENTS INTEGRATION

GPI's Construct

The GPI's construct has the following characteristics (OMNI-SCI, 2021):

- It should simulate the human brain's functioning using ICS; that includes learning, logical reasoning and auto-correction.
- It supports the *Entity's* system to learn, auto-correct and reasoning and to draw inferences in an independent manner.

- It is generic and has a holistic approach because it can handle a many types of activities, like the humans, all of which demand reasoning, judgment, and thought.
- It uses specialized methods to handle specific tasks.
- It uses neural networks and it needs also to teach its DMS4P to think like a human brain and that needs an extraordinary amount of data. This is the intersection of DSI4P (=the field), GPI (=the goal) and ML4P (=the process).

ML4P's Integration

The ML4P's integration using GPI, has the following characteristics (OMNI-SCI, 2021):

- GPI, DSI4P and ML4P work together. DS4P is the field of MLI4P that feeds the ICS with huge amounts of data in order to learn to support DMS4P's processes.
- A GPI based DMS4P can do a lot with human interaction. As humans feed the DMS4P massive quantities of data it can become capable to solve *Project* problems.
- Humans feed various types of data to the DMS4P, so it can learn all of the data's associated knowledge items/features. The learning process is iterative and when the DMS4P is ready to be used.
- MLI4P is a natural extension of statistics and evolved with ICS technologies to handle massive amounts of data. In contrast, DS4P is part of MLI4P, but it is a complex process which allows DMS4P to learn without adding ICS modules.
- In MLI4P s use algorithms to evolve the DMS4P, but those algorithms rely on source data. The DMS4P uses that data as a learning sets, so it can improve its algorithm, tuning and testing it, optimizing itself as it goes. It fine-tunes the various parameters of its MLI4P algorithms by using various statistical techniques, including naive Bayes regression, and supervised clustering.
- Many used techniques require human inputs which are also a part of MLI4P. A DMS4P can train another DMS4P to detect data structures using unsupervised clustering to optimize a classification algorithm; but to complete the process, a human must classify the structures that the DMS4P identifies.
- The scope of DSI4P also goes far beyond MLI4P, encompassing data that is generated not by any mechanical process, computer or machine; like in the case, where MLI4P also includes survey data, data from clinical trials, or really any other kind of data that exists.

- MLI4P involves deploying data not just to train DMS4Ais; and is not limited to statistical data processing, MLI4P includes automating ML4P and data-driven decisions. It also encompasses data integration, data engineering, and data visualization, along with distributed EAI4P, and the creation of dashboards and other BI tools. In fact, any deployment of data in production mode is also within the scope of MLI4P.
- When a Specialist creates the insights created from data, a DMS4P learns from those insights that were already perceived by the Specialist. A DMS4P can build its own insights on the existing algorithmic structure, the starting point relies on some kind of used structured data.
- A DSI4P specialist must have MLI4P skills; where data for a DSI4P specialist may or may not involve data from DMS4P's processing.

BGD4P's Integration

The integration of BGD4P has the following characteristics (OMNI-SCI, 2021):

- Unifying data streams from various sources, like, online purchases, multimedia forms, instruments, financial logs, sensors, text files and others. Data can be: 1) unstructured; 2) semi-structured; and 3) structured.
- Unstructured, includes data from: blogs, digital audio/video feeds, digital images, emails, mobile devices, sensors, social networks and tweets, web pages, and online sources.
- Semi-structured data comes from: system log storages, XML files and text files.
- Structured data, has been already processed by OLTP, RDBMS, transaction data processes and other formats' processing.
- BGD4P is optimal for the transformed system and it supports the processing of tremendous amounts of data from various sources; which is impossible to process with BI or DA4P tools. BGD4P support for MLI4P offers *Entities* with advanced, complex algorithms and other tools for analysing, cleansing, processing and extracting meaningful insights from data storages.
- MLI4P is not a tool, skill or method; it is a scientifical RDP4P based approach that uses AHMM4P and the ICS to process BGD4P.
- The foundations of MLI4P, combines cross-functional activities, like, data cleansing, intelligent data capture techniques and data mining and NLP4P programming. MLI4P is capable of capturing, maintaining and preparing BGD4P for complex intelligent analysis. This is the main difference between MLI4P from classical data engineering; where data engineering prepares datasets delivered from MLI4P processing which does the intelligent analysis.

- BGD4P is the raw material used in the context of MLI4P to deliver velocity, variety and volume (or the 3Vs); that supports the techniques for analysing huge volumes of data.

Use of Statistics

Statistics Methods' Integration

The Integration of statistics methods have the following characteristics (OMNI-SCI, 2021):

- MLI4P is a broad, cross functional domain that is used in applied business management, ICS, economics, mathematics and software engineering along with statistics.
- MLI4P main challenges require the collection, processing, management, analysis and visualization of large quantities of data. Specialists use tools from various APDs, including statistics, to achieve these goals.
- There is a close relation between MLI4P and BGD4P. BGD4P exists for unstructured formats and includes some non-numeric data. Therefore, the task of processing data by MLI4P involves removing noise and extracting useful datasets.
- Statistics is also a broad field demanding APD expertise and it focuses on the study of numerical and categorical data; where statistics is an APD that sees use in numerous other verticals.
- MLI4P employs statistical protocols to design the RDP4P and to ensure that its results are valid.
- Statistical methods support the MLI4P to explore and describe data while summarizing them. Finally, statistical protocols are essential to accurate prediction.

Deep Learning Integration Using GPI

The DLI4P has the following characteristics (OMNI-SCI, 2021):

- It is a function of GPI processes data and generates patterns to be used by the DMS4P.
- It is a type of ML4P, that is focused on deep neural networks (or called deep neural learning.) that can master unstructured or unlabelled data, without any type of human assistance.

- It uses hierarchical artificial neural networks to be used in ML4P. These artificial neural networks are like the human brain.
- Traditional DA4P uses data in a linear fashion, whereas the DLI4P uses a hierarchy of functions that enables a nonlinear approach to problems' solving.
- BGD4P is mainly unstructured, so the DLI4P is important subset of MLI4P.

MLI4P for Business

The MLI4P for business has the following characteristics (OMNI-SCI, 2021):

- MLI4P and analytics are used together, when DSI4P is applied in a business activity.
- MLI4P supports businesses to understand better their specific needs of customers by using the existing data.
- MLI4P can train models for search and to deliver business recommendations.

Business Intelligence Integration Using GPI

The BII4P has the following characteristics (OMNI-SCI, 2021):

- BII4P is a subset of data analysis, and it analyses existing data for insights into business trends.
- BII4P gathers data from internal and external sources, prepares and processes it for a specific use, and then creates dashboards with the data to resolve business problems.
- MLI4P is a more exploratory, future-facing approach and it analyses all relevant data, current or past, structured or unstructured. Having the goal of smarter, more informed DMS4P processing.

GPI's CSF's

Based on the business case's (and its CSA) LRP4P process managed and weighted the most important CSFs that were used.

Table 7. GPI's CSFs that have an average of 8.50

Critical Success Factors	KPIs	Weightings
CSF_MLI4AI_GAIP_Construct	Feasible	From 1 to 10. **09 Selected**
CSF_MLI4AI_GAIP_DSI4AI	Complex	From 1 to 10. **08 Selected**
CSF_MLI4AI_GAIP_Statistics	Feasible	From 1 to 10. **09 Selected**
CSF_MLI4AI_GAIP_BGD4AI	Complex	From 1 to 10. **08 Selected**
CSF_MLI4AI_GAIP_DLI4AI	Feasible	From 1 to 10. **09 Selected**
CSF_MLI4AI_GAIP_BII4AI	Complex	From 1 to 10. **08 Selected**
CSF_MLI4AI_GAIP_Finance	Feasible	From 1 to 10. **09 Selected**
CSF_MLI4AI_GAIP_Transformation	Complex	From 1 to 10. **08 Selected**

valuation

As shown in Table 7, the result's aim is to prove or justify GPI's feasibility; and the result permits to move to the next CSA that is the assembling the MLI4P component.

ASSEMBLING MLI4P COMPONENTS

The Analysis Processes

The Process

MLI4P's Application Programming Interface (API) can be used to avoid common pitfalls, because of expensive, complex and monolithic nature, the barrier to integrate MLI4P using GPI, has been simplified with the introduction of simpler entry-level and distributed platform becomes a viable *start small and grow* approach (Gartner, 2020).

People

To motivate stakeholders, it is imperative that there is a strong relationship between the MLI4P and the *Entity's* business results. This is complex because of ICS' silo structure; when integrating data silos, reducing duplicates, improving data quality and creating a semantically consistent view of master data; are all very important (Gartner, 2020).

Technology

To assess impacts related to *Projects*, firstly is to assess MLI4P's solutions to a given problem and whether the *Entity* is ready to be applied. It is recommended to start with a PoC and assess an end solution. Secondly, as it is imperative not to start with ICS/technology integration, and where EAI4P needs to inspect the existing *Entity's* processes significantly (Gartner, 2020).

Data-Driven Business Transformation Projects

Data-driven *Projects* (or simply *DDProject*) has the following characteristics (IBM, 2019):

- DMS4P based on data is daily business and there is a need to adopt avantgarde technologies and methodologies that facilitate data-centric DMS4P that is the nucleus of a *Project*; where GPI is the skeleton that supports ML4P and DLI4P.
- GPI supports MLI4P's integration, it has the potential to improve data querying accuracy and performance, and to optimize *Project* resources.
- Data platforms and analytics references from the LRP4P, reveal the extent to which *Entities* use GPI for ML4P as a critical goal of their data analytics initiatives. Two-thirds of *Entities* confirm that GPI like approach for ML4P is important for their data platform and DA4P initiatives, the increase to 88% among the most data-driven *Entities* (in which all strategic decisions are data-driven).

Integration Objectives

The integration's main objectives are (IBM, 2019):

- That operational data queries are overloaded on the ICS, consume excessive platform resources and also requires manual resources to be transformed.
- To improve data query performance and accuracy, where MLI4P-enabled DBMS querying can have a dramatic impact on increasing the overall accuracy of the obtained results.
- To support business analysts in analytics tasks and that the technical environment is simplified to enable the use DMS4P to all team members.
- To improve Specialists productivity, by using MLI4P to prepare data, that is one of the three most significant barriers to ML4P adoption. MLI4P-enabled DBMS can accelerate data development.

- To automate DBMS' administration, by using automation of administration tasks and scripts.

Main Components

DSI4P main components (Guru99, 2021):

- Statistics is the most critical unit of DSI4P basics responsible for collecting and analysing numerical data in large quantities to get useful insights.
- Visualization technique helps you to access huge amounts of data in easy to understand and digestible visuals.
- It explores the building and study of algorithms which learn to make predictions about unforeseen/future data.
- DLI4P method is new DSI4P research where the algorithm selects the analysis model to follow.

MLI4P Algorithms

MLI4P supports *Entities* by efficiently modelling large datasets, where the *Project* must choose the right algorithm(s), which depends on the expected outcome. Models' development is not an easy task, because there are various algorithms for different goals and dataset; like the linear regression algorithm, which is easier to train and implement than other algorithms. As shown in Figure 19, the main MLI4P algorithms are: 1) supervised learning; 2) unsupervised learning; and 3) semi-supervised learning or reinforcement learning (Kelley, 2020).

Figure 19. The main algorithms

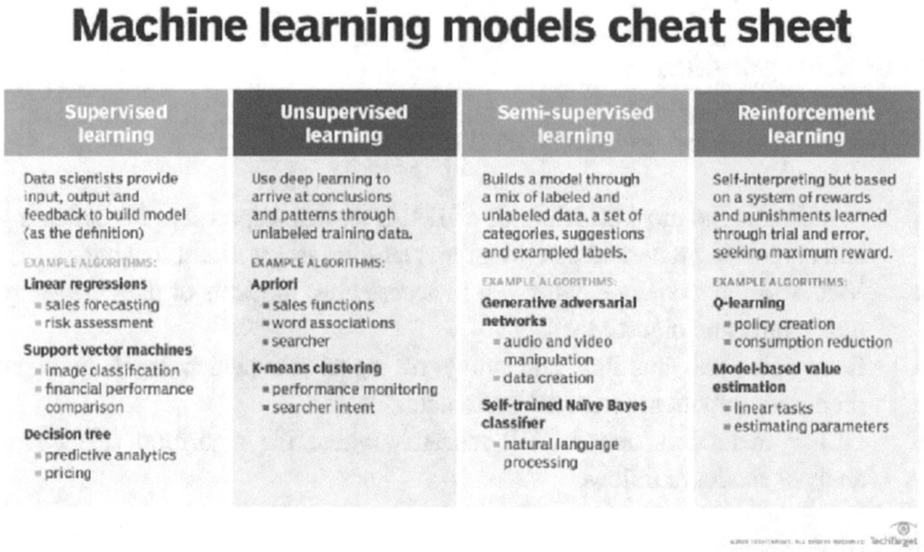

Machine learning models cheat sheet

Supervised learning	Unsupervised learning	Semi-supervised learning	Reinforcement learning
Data scientists provide input, output and feedback to build model (as the definition)	Use deep learning to arrive at conclusions and patterns through unlabeled training data.	Builds a model through a mix of labeled and unlabeled data, a set of categories, suggestions and exampled labels.	Self-interpreting but based on a system of rewards and punishments learned through trial and error, seeking maximum reward.
EXAMPLE ALGORITHMS:	EXAMPLE ALGORITHMS:	EXAMPLE ALGORITHMS:	EXAMPLE ALGORITHMS:
Linear regressions ▪ sales forecasting ▪ risk assessment	**Apriori** ▪ sales functions ▪ word associations ▪ searcher	**Generative adversarial networks** ▪ audio and video manipulation ▪ data creation	**Q-learning** ▪ policy creation ▪ consumption reduction
Support vector machines ▪ image classification ▪ financial performance comparison	**K-means clustering** ▪ performance monitoring ▪ searcher intent	**Self-trained Naïve Bayes classifier** ▪ natural language processing	**Model-based value estimation** ▪ linear tasks ▪ estimating parameters
Decision tree ▪ predictive analytics ▪ pricing			

Supervised Models

Supervised models need specialists to feed the algorithm with datasets; using input and parameters for output, as well as information on accuracy during the training phase; and the main algorithms are (Kelley, 2020):

- Linear Regression (LR), is the most popular type of MLI4P algorithm and it maps simple correlations between two variables in a set of data. Inputs and outputs are quantified to present a relationship, including how a change in one variable affects the other. LRs are then plotted via a line on a graph. LR is frequently used in sales forecasting and risk assessment for *Entities* that seek to make long-term business decisions.
- Support Vector Machines (SVM), is an algorithm that separates data into classes. During model's training, SVM finds a line that separates data in a given set into classes and maximizes the margins of each class; afterwards the model can apply them to future data. This algorithm is optimal for training data that can be separated by a line (a hyperplane). Nonlinear data can be fed into a facet of SVM (nonlinear SVMs). SVMs are used in financial sectors, as they offer high accuracy on both current and future datasets. The algorithms can be used to compare relative financial performance, value and investment gains virtually.

Decision Trees

This algorithm characteristics are (Kelley, 2020):

- It takes data and graphs it, to branches to present solutions.
- Classifies response variables and predict response variables based on previous decisions.
- Are a visual method for mapping decisions and their results; delivers also explanations.
- Maps to various decisions and their impact on an end results.
- Because of their long-tail visuals, they work best for small datasets, decisions and concrete variables.
- Their use cases involve augmenting option pricing, from mortgage lenders classifying borrowers to product management teams.
- Remain popular because they outline multiple outcomes and tests without needing specialists to deploy multiple algorithms.

Unsupervised Models

These algorithms are not trained by specialists, in which it tries to identify patterns in data by combing sets of unlabelled training data and locating correlations. The used models have no information on what to search for and the main algorithms are (Kelley, 2020):

- The Apriori algorithm (based on the Apriori principle) is most used in market basket analysis to mine item sets and to generate association rules. It checks for a correlation between two items in a dataset to determine if there's are positive or negative correlation between them. It is used by sales teams to notice which products customers are likely to purchase in combination with other products. Besides sales functions, it can be used for e-commerce.
- K-means clustering algorithm (K-mean), is an iterative method for sorting data points into groups, which are based on similar characteristics. It is for accurate, streamlined groupings processed in a relatively short period of time, compared to other algorithms. It is popular among search engines to produce relevant information and *Entities* looking to group user behaviours.

Semi-supervised Models

Semi-supervised learning teaches an algorithm by using labelled and unlabelled datasets; using a set of labelled categories, suggestions and examples. They create their own labels by exploring the dataset; and the main algorithms are (Kelley, 2020):

- Generative Adversarial Networks (GAN), are deep generative models that gained popularity and have the ability to imitate data in order to model and predict. They work by pitting two models against each other in a competition to develop the best solution to a problem. One neural network, a generator, creates new data while another, the discriminator, works to improve on the generator's data. After many iterations of this, datasets become more and more lifelike and realistic. Popular media uses GANs, to do things like face creation and audio manipulation. GANs are also impactful for creating large datasets using limited training points, optimizing models and improving manufacturing processes.
- Self-trained algorithms, in which developers can modify models using the Naïve Bayes classifier, which allows self-trained algorithms to perform classification. In a self-trained model, researchers train the algorithm to recognize object classes using a labelled training dataset. Then they use the model classify the unlabelled data. Once that cycle is terminated, they persist the correct self-categorized labels to the training datasets.
- Reinforcement learning algorithms are based on the concepts of rewards and punishments, learned from trial and error. This model has a defined goal and tries to maximize the reward for getting closer to that goal based on limited data and learns from its previous actions. It can be model-free, where it creates interpretations of data through constant trial and error.
- Q-learning algorithms are model-free, and they seek to find the best method for satisfying the defined goals, by seeking the maximum reward by trying to optimize the set of actions. It is frequently combined with DLI4P, like Google's DeepMind. Q-learning breaks down in various algorithms: 1) Deep deterministic policy gradient; and 2) Hindsight experience replay.
- Model-based value estimation can quickly arrive at near-optimal control with learned models under fairly restricted dynamics classes. They are designed for specific use cases.

MLI4P's CSF's

Based on the business case's (and its CSA) LRP4P process managed and weighted the most important CSFs that were used.

Table 8. The MLI4P's CSFs that have an average of 8.40.

Critical Success Factors	KPIs	Weightings
CSF_MLI4AI_Assembling_Analysis_Processes	Complex	From 1 to 10. **08 Selected**
CSF_MLI4AI_Assembling_DDProject	Complex	From 1 to 10. **08 Selected**
CSF_MLI4AI_Assembling_Integration_Components	Complex	From 1 to 10. **08 Selected**
CSF_MLI4AI_Assembling_EA4AI_ADM4AI	Proven	From 1 to 10. **10 Selected**
CSF_MLI4AI_Assembling_CSF_CSA	Feasible	From 1 to 10. **09 Selected**
CSF_MLI4AI_Assembling_SupervisedLearning	Complex	From 1 to 10. **08 Selected**
CSF_MLI4AI_Assembling_UnSupervisedLearning	Complex	From 1 to 10. **08 Selected**
CSF_MLI4AI_Assembling_SemiSupervisedLearning	Complex	From 1 to 10. **08 Selected**

valuation

As shown in Table 8, the result's aim is to prove or justify the RDP4P's feasibility.

THE PROOF OF CONCEPT

Author's RDP4P, is based on AHMM4P and ARI4P that can adapt to any type of MLI4P problem and used algorithm.

Basics

This RDP4P's hypothesis by using a PoC, which has been developed using the Microsoft Visual Studio 2019. The PoC contains *TRADf*'s major components for the MLI4P's processing, and primarily will be tested using the ARI4P's mixed reasoning engine, which is based on the heuristics model. This PoC serves to confirm the research's RQ. The used goal function calculates the best solution for the encountered MLI4P problems. The PoC's results are presented in the form of a set of recommendations.

CSFs, Rules and Constraints Setup

The ARI4P's process execution starts with the use of the imputed data collection in the *TRADf's* data storage and then these data are filtered using the selected set of CSFs. The execution of the QLA4P part follows. The inputted data collection is considered to be the root or initial node that helps in the establishment of the basic state that is enhanced with the adopted MLI4P's solution(s). The ARI4P's tree reasoning goal is to select the optimal solution(s).

The Tree and Resources

The DMS4P's decision tree's collection of nodes contains the following resources, as shown in Figure 20, which are based on the unbundling of the system: 1) the executed actions; 2) the constraints; 3) the MLI4P related problems; and 4) the solutions. Each tree node is a CSF suggestion and it is linked to a concrete data state, which in turn contains an aggregate of a resource linked with a 1:1 mapping link.

Figure 20. A view on the DMS4P's decision tree's solution nodes

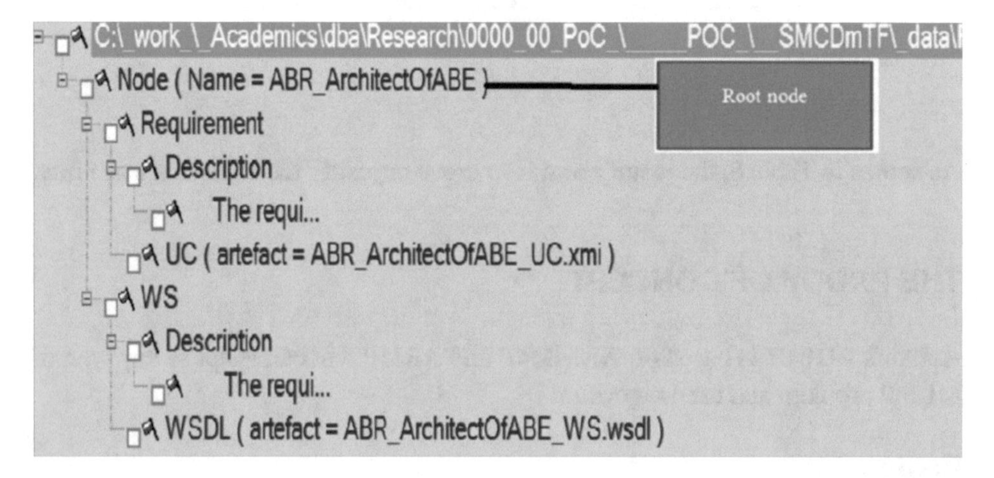

Possible Solutions

The selected CSFs were fed in the DMS4P's heuristics engine in order to reveal the optimal MLI4P prerequisites for a selected *Project* problem or request. The CSFs were configured and weighted; afterwards they were processes in order to deliver a set of possible solutions. The DMS4P starts with the initial set of selected CSFs that correspond to a specific problem or request; then the grounded hyper-heuristics processing is launched to find a set of possible solution(s) in the form of possible improvements or suggestions of needed actions (Jaszkiewicz, & Sowiñski, 1999). The author's aim is to convert their relevant research outcomes into a set of managerially useful recommendations for MLI4P's integration; and the *TRADf* a hyper-heuristics tree processing model template that is suitable for a wide class of problem instances (Vella, Corne, & Murphy, 2009).

The PoC Processing

The PoC uses an internal set of CSFs' that are presented in Tables 1 to 8. These CSFs have bindings to specific RDP4P resources, where the AHMM4P formalism was designed using an NLP4P microartefacts. In this chapter's tables and the result of the processing of the DMS4P, as illustrated in Table 9, shows clearly that the MLI4P is feasible.

Table 9. The MLI4P based RDP4P's outcome

	Critical Success Factors	KPIs	Weightings Ranges	Values
1	RDP5AI	HighlyFeasible	From 1 to 10.	9,25
2	AHMM4AI	PossibleClassification	From 1 to 10.	10
3	DMC4AI	AutomatedExists	From 1 to 10.	9,4
4	PLATFORM	IntegrationPossible	From 1 to 10.	9
5	EA4AI	AdvancedStage	From 1 to 10.	9
6	ENVIRONMENT	Advanced	From 1 to 10.	9,4
7	GAIP	IntegrationPossible	From 1 to 10.	8,5
8	ASSEMBLING	IntegrationPossible	From 1 to 10.	8,4

EvaPA

RESULT: 9,11875

The AHMM4P's main constraint is that CSAs for simple research components, having an average result below 8.5 will be ignored. In the case of the MLI4P's implementation an average result below 6.5 will be ignored. As shown in Table 9 the average is 9.10. The AHMM4P based MLI4P processing model represents the relationships between this research's requirements, NLP4P generic and microartefacts, unique identifiers and the CSAs. The PoC was achieved using *TRADf* client's interface. From the *TRADf* client's interface, the NLP4P development setup and editing interface can be launched. Once the development setup interface is activated the NLP4P interface can be launched to implement the needed microartefact scripts to process the defined three CSAs. These scripts make up the kernel knowledge system and the DMS4P set of actions that are processed in the background. The MLI4P uses the DMS4P that automatically generates actions which make calls to QLA4P and QNA4P modules, that manages the edited NLP4P script and flow, as shown in Figure 21.

Figure 21. The edited NLP4P script and flow

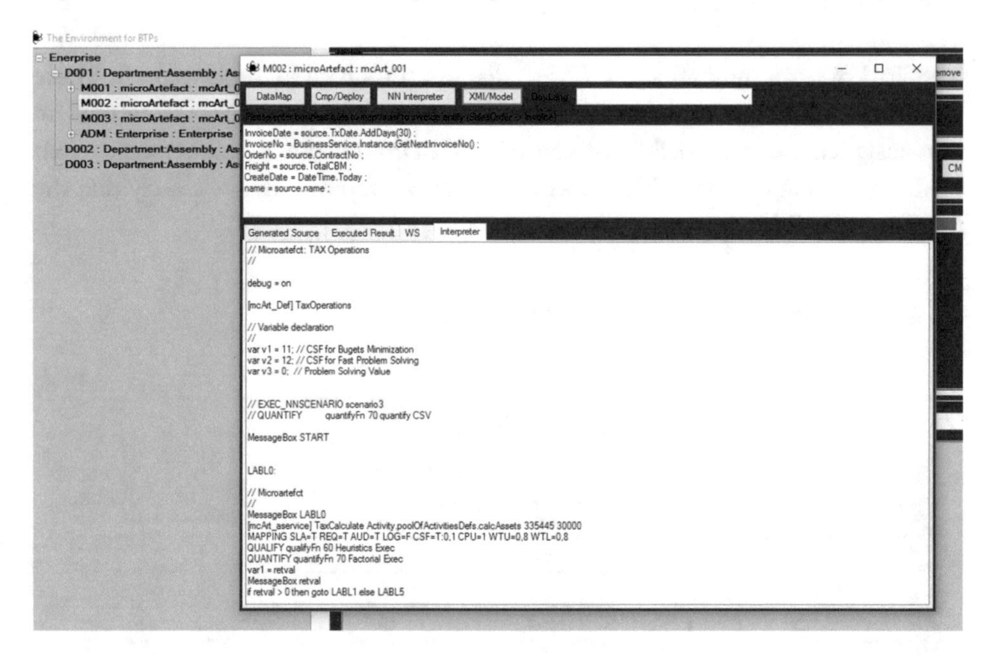

MLI4P structures, serve various APDs and in this PoC the functional domain are information analysis and decision making of a business transaction's log; where the data that results from business transactions are logged and used in the data analysis process, as shown in Figure 22.

Figure 22. The PoC's data interface view

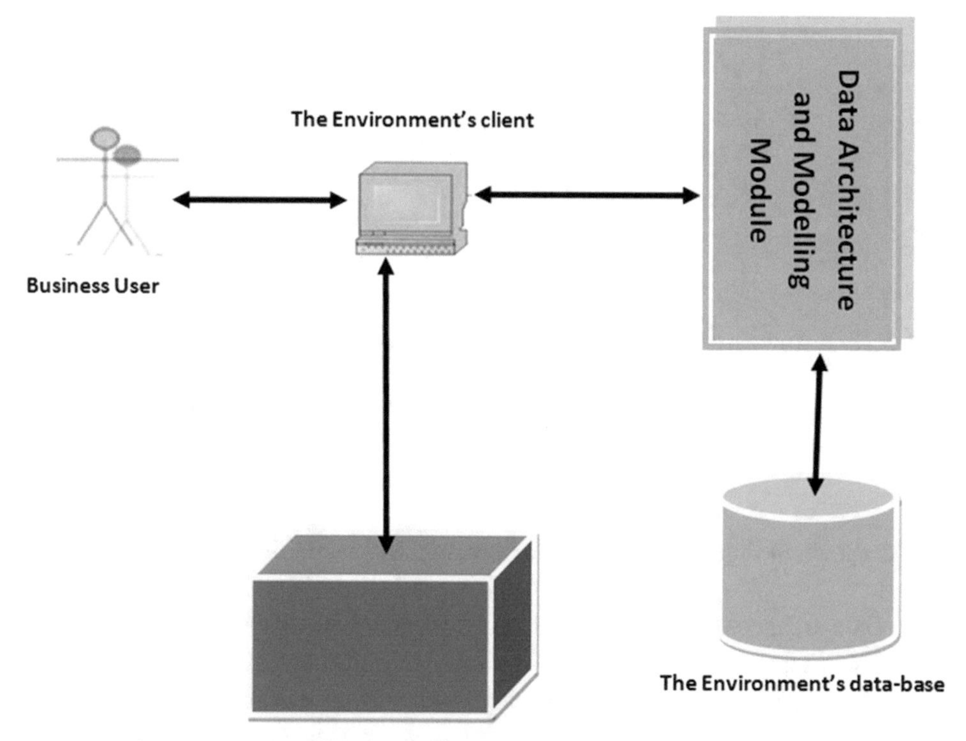

Model's Analysis

MLI4P's PoC took into account the integration of information analysis processes that are used by a virtual user who simulates a business analyst. In this PoC a spreadsheet was built to use various data sources that fed the prototyped business transactions system, as shown in Figure 22.

Layers

The prototyped business transaction is based on the PoC's class model, as shown in Figure 23.

Figure 23. The PoC's transaction view

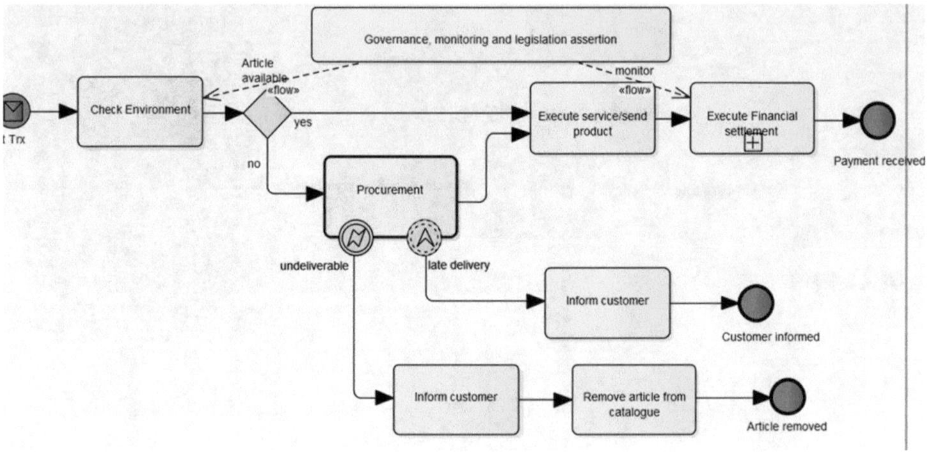

As shown in Figure 24, the data architecture and modelling PoC's layers are:

- Data architecture using TOGAF and GPI to link MLI4P to various artefacts.
- The client layer that contains the following packages based on: RDBMS, Microsoft Excel, Flat Files Interfaces and a Client web interface.
- The data services layer that contains a data services hub.
- The data management layer that contains the following packages: an entity relational model, an extensible mark-up language transformer and a NoSQL database interface.
- The control-monitoring layer contains a generic logger interface.
- The platform layer that contains the following packages: an *Entity* service bus, an object database connector, and a java database connector.

Figure 24. The PoC's layer's view

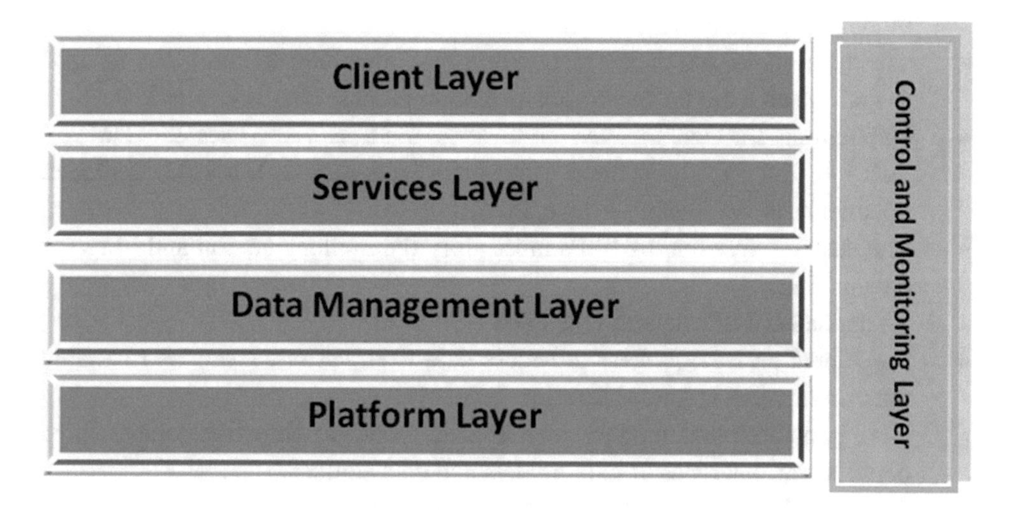

Figure 25. TOGAF data architecture

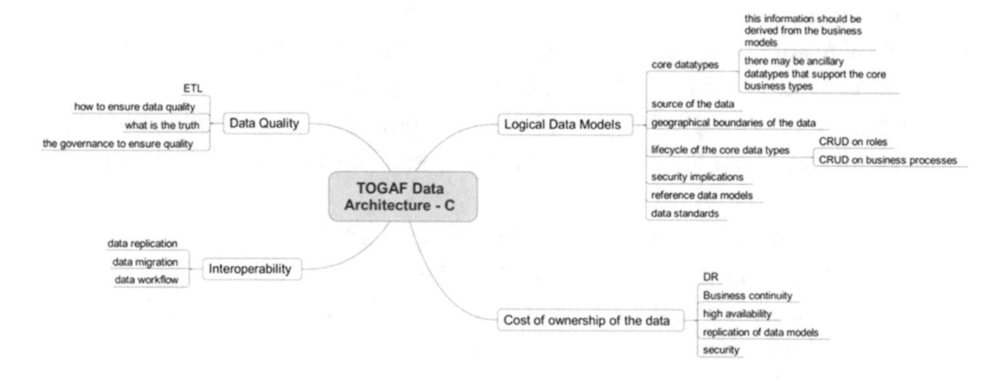

SOLUTIONS AND RECOMMENDATIONS

This chapter's and its PoC's list of the most important technical and managerial recommendations:

- The *Manager*s should be supported with a MLI4Ps tools that can integrate with the EAI4P principles (IBM, 2015a).
- The role of data standards is important; today there are many standards concerning the MLI4P. The *Manager* must propose a solution on how to integrate these various data standards in the *Project*.

- GPI services enable MLI4P's integration concept that glues the various models to the business environment.
- The meta-model and the corresponding "1:1" mapping approach can be used to manage an agile business transformation process (OASIS, 2014).
- Unbundling through the use of business data services uses the DMS4P well integrated in the context of a system that applies a holistic approach (Daellenbach, McNickle, & Dye, 2012).
- The data services and dBB4Ps must unify the implementation and usage of data models.
- To define MLI4P's default CSFs and CSAs.
- Implement an ETL process to access data from various locations to enable their unification and normalization for MLI4P usage.
- Data transformation models need a meta-model to show the relationships between various types of data models and data sources.
- Define a data normalization and inter-operability paradigm that must be based on various standards.
- Define a business process and a data architecture and modelling integration paradigm to enhance the KMS4P.
- Develop an agile MVC4P pattern that is dynamically created for each business transaction. This MVC4P uses a dBB4P and its instance
- Define the ADM4P's integration with the proposed concept where the data building and solution blocks are the basic artefacts that circulate through the ADM4P.
- Define business architecture integration paradigm that is needed to manage the implementation of the dBB4Ps (The Open Group, 2014a).
- Define conceptual views that can be built to simplify the application of the MLI4P. Such views can also be the used to simplify the *Entity*'s data-model that links various data sources.
- Create basic class diagrams for the MLI4P(s) to be the central artefact that is defined in the initial phase and then calibrated in all other phases.
- Define the usage of business transactions' data models in various class diagrams.
- An in-house ARI4P can replace commercial MLI4P tools.

CONCLUSION

The RDP4P is based on MLI4P and is mainly motivated by high failure rates in *Projects*. The MLI4P feasibility is presented in the PoC and is to present the needed GPI and MLI4P concepts for a *Project*'s. There are severa trends shaping the future of MLI4P; first, MLI4P tasks in the life-cycle are automated that increases the *Entity's* ROI. Another important element is that MLI4P resources become accessible to more people. A third important trend is the tension between the right to privacy, the need to regulate and the state of transparency. MLI4P has the ability to make ML4P algorithms transparent, that makes regulatory oversight possible.

FUTURE RESEARCH DIRECTIONS

The *TRADf* future research will focus on DLI4P.

REFERENCES

Camarinha-Matos, L. M. (2012). *Scientific research methodologies and techniques-Unit 2: Scientific method. Unit 2: Scientific methodology. PhD program in electrical and computer engineering*. Uninova.

Clancy, L. (2019). *Functional programming reaches for stardom in finance*. CME Group. https://www.risk.net/risk-management/6395366/functional-programming-reaches-for-stardom-in-finance

Daellenbach, H., McNickle, D., & Dye, Sh. (2012). *Management Science. Decision-making through systems thinking* (2nd ed.). Plagrave Macmillian.

Dantzig, G. (1949). *Programming of Interdependent Activities: II Mathematical Model*. Academic Press.

Della Croce, F., & T'kindt, V. (2002). A Recovering Beam Search algorithm for the one-machine dynamic total completion time scheduling problem. *The Journal of the Operational Research Society*, *53*(11), 1275–1280. doi:10.1057/palgrave.jors.2601389

Doree, J. (2015). *Data Science, Data Architecture-Predictive Analytics Methodology: Building a Data Map for a Greenfield Data Science Initiative*. Datasciencearchitect. https://datasciencearchitect.wordpress.com/tag/togaf/

E-business definition. (2014). Accessed and reviewed in April 2019. http://dictionary.reference.com/browse/e-business

Gartner. (2020). *Three Essentials for Starting and Supporting Master Data Management*. ID G00730039. Gartner Inc. https://www.gartner.com/doc/reprints?id=1-24MIGFGU&ct=201119&st=sb

Guru99. (2021). *Data Science Tutorial for Beginners: What is, Basics & Process.* https://www.guru99.com/data-science-tutorial.html

IBM. (2015a). *Modelling the Entity data architecture*. IBM.

IBM. (2019). *AI and Data Management-Delivering Data-Driven Business Transformation and Operational Efficiencies*. IBM USA.

IMD. (2015). *IMD business school and Cisco join forces on digital business transformation*. IMD.

Järvinen, P. (2007). Action Research is Similar to Design Science. *Quality & Quantity, 41*(1), 37–54. Retrieved August 10, 2018, from https://link.springer.com/article/10.1007/s11135-005-5427-1

Jaszkiewicz, J., & Sowiñski, R. (1999). The 'Light Beam Search' approach - An overview of methodology and applications. *European Journal of Operational Research, 113*(2), 300–314. doi:10.1016/S0377-2217(98)00218-5

Kapoor, A. (2021). *Artificial Intelligence and Machine Learning: 5 Trends to Watch Out for in 2021.* AI Zone. DZone. https://dzone.com/articles/artificial-intelligence-amp-machine-learning-5-dev

Kelley, K. (2020). *9 types of machine learning algorithms with a cheat sheet.* Techtarget. https://searchenterpriseai.techtarget.com/

Kim, K., & Kim, K. (1999). Routing straddle carriers for the loading operation of containers using a beam search algorithm. Elsevier. *Computers & Industrial Engineering, 36*(1), 109–136. doi:10.1016/S0360-8352(99)00005-4

Lazar, I., Motogna, S., & Parv, B. (2010). Behaviour-Driven Development of Foundational UML Components. Department of Computer Science, Babes-Bolyai University. doi:10.1016/j.entcs.2010.07.007

Leitch, R., & Day, Ch. (2000). *Action research and reflective practice: towards a holistic view*. Taylor & Francis. https://www.tandfonline.com/doi/ref/10.1080/09650790000200108?scroll=top

Lockwood, R. (2018). *Introduction The Relational Data Model*. Retrieved January 29, 2019, from http://www.jakobsens.dk/Nekrologer.htm

Loginovskiy, O. V., Dranko, O. I., & Hollay, A. V. (2018). *Mathematical Models for Decision-Making on Strategic Management of Industrial Enterprise in Conditions of Instability*. Conference: Internationalization of Education in Applied Mathematics and Informatics for HighTech Applications (EMIT 2018), Leipzig, Germany.

Merson, P. (2009). *Data Model as an Architectural View. TECHNICAL NOTE. CMU/SEI-2009-TN-024*. Research, Technology, and System Solutions.

Moore, J. (2014). *Java programming with lambda expressions-A mathematical example demonstrates the power of lambdas in Java 8*. Retrieved March 10, 2018, from, https://www.javaworld.com/article/2092260/java-se/java-programming-with-lambda-expressions.html

Nijboer, F., Morin, F., Carmien, S., Koene, R., Leon, E., & Hoffman, U. (2009). Affective brain-computer interfaces: Psychophysiological markers of emotion in healthy persons and in persons with amyotrophic lateral sclerosis. In *3rd International Conference on Affective Computing and Intelligent Interaction and Workshops*. IEEE. 10.1109/ACII.2009.5349479

OASIS. (2014). *ISO/IEC and OASIS Collaborate on E-Business Standards-Standards Groups Increase Cross-Participation to Enhance Interoperability*. https://www.oasis-open.org/news/pr/isoiec-and-oasis-collaborate-on-e-business-standards

OMNI-SCI. (2021). *Data Science - A Complete Introduction*. OMNI-SCI. https://www.omnisci.com/learn/data-science

Sarkar, D. (2018). *Get Smarter with Data Science — Tackling Real Enterprise Challenges. Take your Data Science Projects from Zero to Production*. Towards Data Science. https://towardsdatascience.com/get-smarter-with-data-science-tackling-real-enterprise-challenges-67ee001f6097

SAS. (2021). *History of Big Data*. SAS. https://www.sas.com/en_us/insights/big-data/what-is-big-data.html

Schmelzer, R. (2021). *Data science vs. machine learning vs. AI: How they work together*. TechTarget. https://searchbusinessanalytics.techtarget.com/feature/Data-science-vs-machine-learning-vs-AI-How-they-work-together?track=NL-1816&ad=937375&asrc=EM_NLN_144787081&utm_medium=EM&utm_source=NLN&utm_campaign=20210114_An%20open%20source%20database%20comparison

Sikander, M., & Khiyal, H. (2018). *Computational Models for Upgrading Traditional Agriculture*. Preston University. https://www.researchgate.net/publication/326080886

Simonin, J., Bertin, E., Traon, Y., Jezequel, J.-M., & Crespi, N. (2010). Business and Information System Alignment: A Formal Solution for Telecom Services. In *2010 Fifth International Conference on Software Engineering Advances*. IEEE. 10.1109/ICSEA.2010.49

Tamr. (2014). *The Evolution of ETL*. Tamr. http://www.tamr.com/evolution-etl/

The Open Group. (2011). *Open Group Standard-TOGAF® Guide, Version 9.1*. The Open Group.

The Open Group. (2013b). *ArchiMate®*. Retrieved May 03, 2014, from https://www.opengroup.org/subjectareas/*Entity*/archimate

The Open Group. (2014a). *The Open Group's Architecture Framework-Building blocks-Module 13*. www.open-group.com/TOGAF

TOGAF Modelling. (2015a). *Data dissemination view*. TOGAF Modelling. http://www.TOGAF-modelling.org/models/data-architecture-menu/data-dissemination-diagrams-menu.html

Trad, A., & Kalpić, D. (2020a). *Using Applied Mathematical Models for Business Transformation. Author Book*. IGI-Global. doi:10.4018/978-1-7998-1009-4

Trujillo, J., & Luj'an-Mora, S. (2003). *A UML Based Approach for Modelling ETL Processes in Data Warehouses*. Dept. de Lenguajes y Sistemas Inform'aticos. Universidad de Alicante.

Vella, A., Corne, D., & Murphy, C. (2009). *Hyper-heuristic decision tree induction*. Sch. of MACS, Heriot-Watt Univ.

Veneberg, R. (2014). *Combining enterprise architecture and operational data-to better support decision-making. University of Twente*.

KEY TERMS AND DEFINITIONS

TRADf: Is this research's framework.

Chapter 10
Future Trends and Challenges of UAV:
Conclusion

Tharun V.
Karunya Institute of Technology and Sciences, India

Parthiban S.
Karunya Institute of Technology and Sciences, India

Thusnavis Bella Marry
Karunya Institute of Technology and Sciences, India

Martin Sagayam K.
Karunya Institute of Technology and Sciences, India

Ahmed A. Elngar
https://orcid.org/0000-0001-6124-7152
Beni-Suef University, Egypt

ABSTRACT

In this chapter, the design, modeling, and control of a UAV was presented. The conceptual design stages of the UAV were analyzed in detail. UAVs as observers in the sky will remain important for the indefinite future. Agriculture, water quality monitoring, disease detection, crop monitoring, yield predictions, and drought monitoring are just a few of the applications. Healthcare, microbiological and laboratory samples, drugs, vaccines, emergency medical supplies, and patient transportation can all be delivered using drones.

DOI: 10.4018/978-1-7998-8763-8.ch010

INTRODUCTION

Concluding Thoughts

In this thesis, the design, modelling, and control of a UAV was presented. The conceptual design stages of the UAV were analyzed in detail. UAVs will continue to be useful as sky watchers indefinitely. UAVs are an effective and successful equipment in the field, and they play an important part in the military. Drones have been used to transport IEDs and destroy hostile locations. Agriculture, Drones, tree-based remote sensing applications, water quality monitoring, disease detection, crop monitoring, yield forecasting, and drought monitoring are just a few of the data sources. Drones can carry health care, microbiological and laboratory samples, medications, vaccines, emergency medical supplies, and patient transportation. Road Traffic Congestion Control: Congestion occurs when a high number of cars disrupts the regular flow of traffic, resulting in increased travel time. Drone Shakti, across all sectors, the country will see the usage of massive, unmanned aircraft systems weighing more than 150 kilos. They will grow easier to operate. Finally, obstacles, restrictions, and suggestions were investigated.

Comments on UAV Applications

In this thesis we have briefed clearly about the UAV applications. In the Military, UAVs have been utilised as an effective and successful equipment and they play an important part in the military. It was originally designed as a military instrument for reconnaissance and surveillance activities as mentioned in figure 1.

Figure 1. RQ-1 Predator reconnaissance and surveillance

Combat Operations which risk human life can be changed by the UAV. This also talks about the challenges that we are facing during the research. Different challenges faced by UAV which prevents from reaching the goal of the mission. This section covers the main issues that impact the UAV's performance, including as battery life, collision avoidance, communication, and security. Their limits are also explored, along with some recommendations for getting acceptable outcomes. The portable electronics revolution was made possible by the development of lithium-ion battery technology in the early 1990s. The technology is now being developed for use in electric vehicles and grid storage. In comparison to lead-acid batteries, lithium-ion batteries are more expensive. Surveillance in the risk area which costs human life at certain times. It can be prevented using the drone where there is no need for human intervention for the mission. Parallelly both prevention and destruction are possible by UAV. At the frontline where transportation is a huge challenge because of the climatic conditions where the soldiers suffer, the basic needs like food, medicine, ammunition and other essentials have been transported by drones which is easier, safer and cost efficient. Remote sensing has been by Drone for the upgrade of troops and Combat Vehicles. The fast improvement and spread of unmanned aerial vehicles (UAVs) as a remote sensing platform has resulted in technological breakthroughs in metropolitan areas and remote sensing social networks.

Data from ground sensors is collected and sent to ground base stations. Drones equipped with sensors can be used to monitor disaster management and other potential natural catastrophes. In numerous components, the UAV's security is critical. Intruders can assault them, posing a cybersecurity risk to UAV systems. Several assaults can be launched against the communication links between the various components of the UAV system. Furthermore, UAVs may be subjected to direct assaults, such as signal spoofing and hacking, which can result in serious damage to the UAV system. Ground Control Station (GCS) assaults are dangerous because they allow attackers to transmit orders to equipment and capture all data from the UAV, potentially causing the UAV to fail. By gathering real-time data and distributing it at a cheap cost, drones have accelerated the development of various industrial, commercial, and recreational purposes. Telecommunication drones are used for telemetry and remote diagnosis and treatment. As shown in Figure 2, drones may transfer medical problems, microbiological and laboratory samples, medications, vaccines, emergency medical supplies, and patient transportation (b). EpiPens, poison antidotes, and oxygen masks are just a few examples of life-saving equipment.

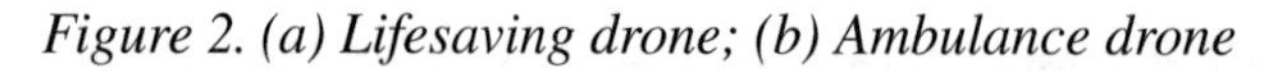

Figure 2. (a) Lifesaving drone; (b) Ambulance drone

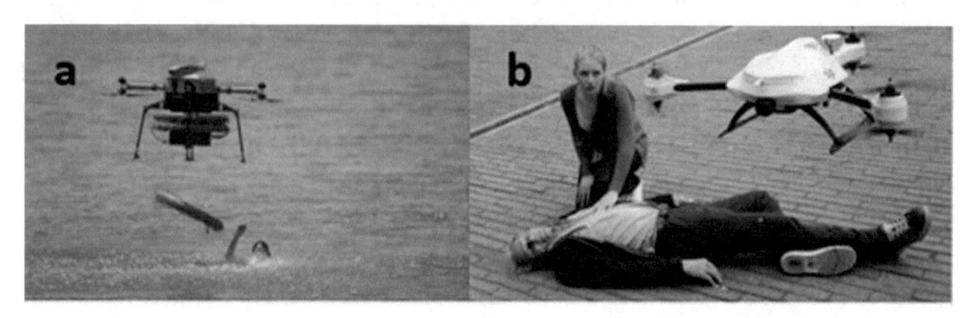

Drones may be used to provide goods and services while also allowing for speedier returns to properly equipped labs, minimising human work and time. People who have gone missing or have been injured at sea can be rescued, as seen in Figure 2. (a),

In the mountains, the desert, or the woods People at home are receiving telemedicine, immunizations, prescription medications, and medical supplies through drones. Cloud and internet of things (IoT) integration: It offers a low-cost approach to link heterogeneous devices and meet expanding data demands in healthcare applications, including seamless application deployment and rendering service (Dempsey and Rasmussen, 2010).

Congestion is a widespread global phenomenon caused by high population density, the expansion of motor vehicles, and accompanying infrastructure. Congestion has been defined in several ways by researchers. Travel demand exceeds available road capacity, according to the most common definition of traffic congestion. Congestion occurs when a significant number of automobiles impede the regular flow of traffic, resulting in increased travel time. Congestion is also defined as an increase in the cost of utilising the road as a result of a disruption in the regular flow of traffic. This is about the incident which happened before. A drone flew beyond visual line of sight (BVLOS) in the Vikarabad area of Telangana in September 2021, carrying a box of life-saving medications and vaccinations. The state's *Medicines from the Sky* project, which aims to increase health care access in distant areas, was launched with this six-kilometre flight completed in around five minutes. Many elements influence the act of creating. After the liberalisation of the Drone Rule in 2021, the business received a significant boost with the announcement of Drone Shakti in the Union Budget. The government stated that 'drone as a service' will be encouraged to help Drone Shakti. Across the board, the country will see the usage of big, unmanned aircraft systems weighing more than 150 kilos. UAV collision avoidance is tough, yet it is necessary to avoid mishaps. UAVs crash into obstacles that are either moving or immovable.

In this thesis we have seen the UAVs in various configurations that are routinely employed in various applications, such as hazardous materials management or operation. Quadcopters appear to have been used in all recent advancements in the field of small autonomous drones. They are well-known for their little drones due to their mechanical simplicity.

A quadrotor is an unmanned aerial vehicle with two pairs of counter-rotating rotors and propellers positioned on the top of a square frame. Vertical take-offs and landings, similar to those of standard helicopters, are possible. The quadcopter is designed to allow either inside or outdoor flights as a fundamental study parameter due to user safety and protection. The criterion for meeting the scheme should be swift and dependable at the proper moment (Arjomandi et al., 2006).

One of the most responsive drone designs is the X-copter. "X" was replaced by quad-, hexa-, and octa- based on the number of propellers utilised. As a result, two types of UAV collision avoidance algorithms have recently emerged: Collision avoidance methods include the geometric method, which uses a geometric strategy to identify a flying route without colliding, the path planning approach, which uses a geometric strategy to identify a flying route without colliding, and the vision-based approach, which gathers images from cameras installed on the UAV to avoid the collision problem. Collision prevention system that works together. The authors designed a route management system for the drone path (RMS). This strategy can increase drone safety at airports by designating specific paths for drones. Environmental conditions are also a difficulty for UAVs, since they can cause deviations from preset courses or even cause the UAV to crash.

FUTURE RESEARCH TRENDS

This thesis talks about the future trends of the drone like how it's going to work on the basis of technology. Some governments are already building the infrastructure required for large-scale drone operations. India is rapidly catching up. Drones were recently deployed by the Rajasthan government's agriculture department to spray insecticides and control locust swarms in the state.

India is also aggressively using unmanned aerial vehicles for national defence. India's development in commercial drone use has been minimal thus far, with only a few trials pizza and medication delivery. The Ministry of Civil Aviation has set aside INR 120 crore for a PLI initiative to help the drone manufacturing industry grow. All of this has the potential to boost drone manufacturing and usage (Sun et al., 2019).

Drone Shakti is an initiative to promote and facilitate drones as a service through startups. A Spike in the use of drones will also open doors of employment for skilled youths. PM flagged off 100 Kisan drones in different cities and towns across the nation to spray pesticides in farms. A large number of unresolved problems necessitate fresh, effective solutions. New UAV system potential is highlighted, including security and privacy, battery charging, machine learning, and other intriguing future research issues and directions.

A swarm is a collection of insects that work together to achieve a significant or desirable outcome. A group of UAVs works together to achieve a certain purpose in swarm UAV systems. Each UAV has a minor task that is linked to a larger objective; this notion was primarily created by the military for surveillance purposes. Swarm UAVs have the ability to distribute duties and coordinate the operation of several drones without the need for human intervention (Li et al., 2018).

Figure 3. Working of Swarm UAVs

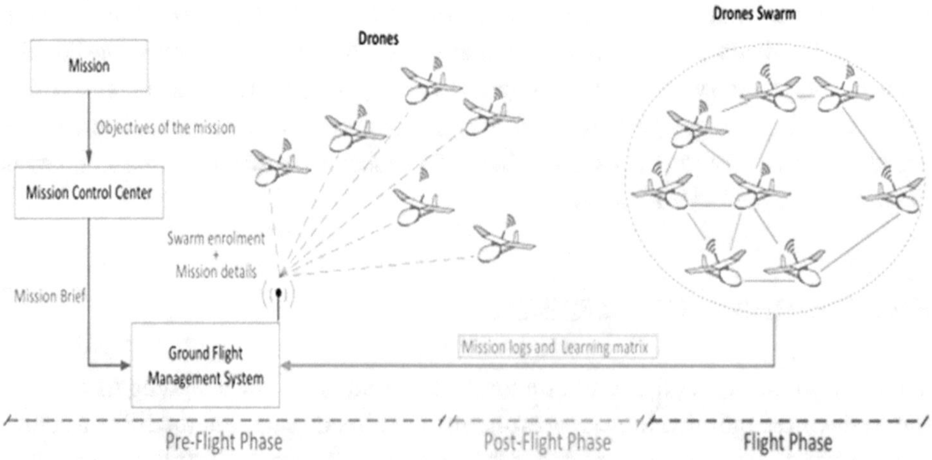

Time savings, man-hour reductions, labour reductions, and other cost reductions are all advantages of swarm UAV systems. Artificial Intelligence (AI) systems can learn from data thanks to machine learning. Deep learning algorithms are a subset of Machine Learning algorithms that include learning representations at different levels of hierarchy. Deep learning approaches based on image sensors have been used to extract features in a variety of applications utilising various technologies (e.g., monocular RGB camera, RGB-D sensors, infrared, etc.). UAV motion control is another deep learning-based application. Machine learning and deep learning algorithms have recently gained a lot of attention in fields such as resource allocation, obstacle avoidance, tracking, path planning, and battery scheduling, to name a few. Because they may be vulnerable to intruder attacks that threaten the privacy and confidentiality of the data they collect, new onboard technologies such as blockchain and physical layer security are required to ensure security and privacy.

This is about an occurrence that occurred previously. An assault using a radio connection: data can be hijacked and decrypted, and a control channel can be disguised - a strategy that was used to hack American drones in Iraq. The attackers, strangely, utilised a Russian application called Sky Grabber. It is imperative to improve security (Tahir et al., 2019).

Path planning is the process of determining the best route from one location to another while avoiding impediments. It is one of the most important technologies for increasing the autonomy of unmanned aerial vehicles (UAV) [Figure 4]

Figure 4. The First Unmanned Aerial Vehicle

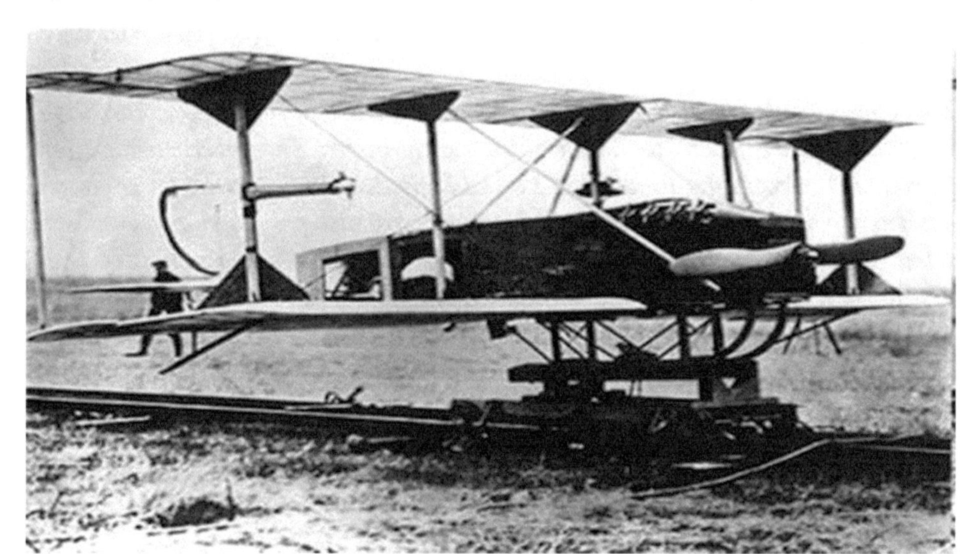

UAV Mission Route Optimization, Tracking, and Path Planning

Algorithms for planning should be improved. A multi-objective optimization approach is used during path planning to avoid obstacles, find the shortest path, and save energy. Recent advancements in battery technology, such as upgraded lithium–ion batteries and hydrogen fuel cells, as well as green energy sources like solar energy, are all worth noting. The photovoltaic (PV) effect is used to convert sunlight to energy. This current is then either utilised immediately or stored in a battery, which then powers the system.

The key advantages of battery-powered drones are their ability to charge practically anywhere, their ability to be moved almost anywhere, and their ease of recharging by just replacing the battery pack. The downsides include a limited number of recharge cycles and poor energy density (Radoglou-Grammatikis et al., 2020).

Experts have shifted away from conventional microwave communication to optical communication technology as the demand for high-definition (HD) remote detection using unmanned aerial vehicles (UAVs) grows. Given the large data amount of an HD photo taken by a UAV, the demand for a communication system with a high data rate keeps growing. The optical communication approach may be used to satisfy the high data rate demand. The Drone Shakti initiative has been aimed to promote and facilitate drones as a service through startups. Drone Shakti will be facilitated by the government through various apps and for 'Drone-As-A-Service (DrAAS)'.

The government has also approved drone use in the agriculture sector, which includes spraying pesticides and fertilisers with drones. The government's effort, in conjunction with the Production-Linked Incentive (PLI) scheme for drones, is designed to boost domestic drone production. It will also provide opportunities for talented teenagers to find work. According to reports, India's drone industry would develop at a CAGR of 20.9 percent from 2020 to 2026. According to statistics, the worldwide commercial drone industry is predicted to reach USD 27.4 billion in 2021, and USD 43 billion in 2025 (Shakhatreh et al., 2019).

India is preparing to enter this fast-growing sector, with the Ministry of Civil Aviation showing interest in making India the world's Drone Hub.

REFERENCES

Arjomandi, M., Agostino, S., Mammone, M., Nelson, M., & Zhou, T. (2006). *Classification of unmanned aerial vehicles. Report for Mechanical Engineering class*. University of Adelaide.

Dempsey, M. E., & Rasmussen, S. (2010). *Eyes of the army–US Army roadmap for unmanned aircraft systems 2010–2035*. US Army UAS Center of Excellence.

Li, B., Fei, Z., & Zhang, Y. (2018). UAV communications for 5G and beyond: Recent advances and future trends. *IEEE Internet of Things Journal*, *6*(2), 2241–2263. doi:10.1109/JIOT.2018.2887086

Radoglou-Grammatikis, P., Sarigiannidis, P., Lagkas, T., & Moscholios, I. (2020). A compilation of UAV applications for precision agriculture. *Computer Networks*, *172*, 107148. doi:10.1016/j.comnet.2020.107148

Shakhatreh, H., Sawalmeh, A. H., Al-Fuqaha, A., Dou, Z., Almaita, E., Khalil, I., Othman, N. S., Khreishah, A., & Guizani, M. (2019). Unmanned aerial vehicles (UAVs): A survey on civil applications and key research challenges. *IEEE Access: Practical Innovations, Open Solutions*, *7*, 48572–48634. doi:10.1109/ACCESS.2019.2909530

Sun, X., Ng, D. W. K., Ding, Z., Xu, Y., & Zhong, Z. (2019). Physical layer security in UAV systems: Challenges and opportunities. *IEEE Wireless Communications*, *26*(5), 40–47. doi:10.1109/MWC.001.1900028

Tahir, A., Böling, J., Haghbayan, M. H., Toivonen, H. T., & Plosila, J. (2019). Swarms of unmanned aerial vehicles—A survey. *Journal of Industrial Information Integration*, *16*, 100106. doi:10.1016/j.jii.2019.100106

Compilation of References

Ackerman, E., & Koziol, M. (2019). The blood is here: Zipline's medical delivery drones are changing the game in Rwanda. *IEEE Spectrum*, *56*(5), 24–31.

Adamu, A. A., Wang, D., Salau, A. O., & Ajayi, O. (2020). An integrated IoT system pathway for smart cities. *International Journal on Emerging Technologies*, *11*(1), 1–9.

ADDITIONAL READING

Afrin, S., Khan, A. T., Mahia, M., Ahsan, R., Mishal, M. R., Ahmed, W., & Rahman, R. M. (2018). Analysis of Soil Properties and Climatic Data to Predict Crop Yields and Cluster Different Agricultural Regions of Bangladesh. *IEEE/ACIS 17th International Conference on Computer and Information Science (ICIS).* 10.1109/ICIS.2018.8466397

Afrin, T., & Yodo, N. (2020). A survey of road traffic congestion measures towards a sustainable and resilient transportation system. *Sustainability*, *12*(11), 4660. doi:10.3390u12114660

Aftabuzzaman, M. (2007, September). Measuring traffic congestion-a critical review. In *30th Australasian transport research forum* (pp. 1-16). London, UK: ETM Group.

Agarwal, A., Singh, A. K., Kumar, S., & Singh, D. (2018, December). Critical analysis of classification techniques for precision agriculture monitoring using satellite and drone. In *2018 IEEE 13th International Conference on Industrial and Information Systems (ICIIS)* (pp. 83-88). IEEE.

Agarwal, P., Singh, V., Saini, G. L., & Panwar, D. (2019). Sustainable Smart-farming framework: smart farming. In *Smart farming technologies for sustainable agricultural development* (pp. 147–173). IGI Global. doi:10.4018/978-1-5225-5909-2.ch007

Agustina, J., & Clavell, G. G. (2011). The impact of CCTV on fundamental rights and crime prevention strategies: The case of the Control commission of video surveillance devices. *Computer Law & Security Review*, *27*(2), 168–174.

Al Murad, M., Khan, A. L., & Muneer, S. (2020). Silicon in horticultural crops: Cross-talk, signaling, and tolerance mechanism under salinity stress. *Plants*, *9*(4), 460. doi:10.3390/plants9040460 PMID:32268477

Alexandrie, G. (2017). Surveillance cameras and crime: A review of randomized and natural experiments. *Journal of Scandinavian Studies in Criminology and Crime Prevention, 18*(2), 210–222.

Al-Fuqaha, A., Guizani, M., Mohammadi, M., Aledhari, M., & Ayyash, M. (2015). Internet of things: A survey on enabling technologies, protocols, and applications. *IEEE Communications Surveys and Tutorials, 17*(4), 2347–2376. doi:10.1109/COMST.2015.2444095

Aliev, K., Jawaid, M. M., Narejo, S., Pasero, E., & Pulatov, A. (2018). Internet of plants application for smart agriculture. *International Journal of Advanced Computer Science and Applications, 9*(4). Advance online publication. doi:10.14569/IJACSA.2018.090458

Allred, B., Martinez, L., Fessehazion, M. K., Rouse, G., Williamson, T. N., Wishart, D., Koganti, T., Freeland, R., Eash, N., Batschelet, A., & Featheringill, R. (2020). Overall results and key findings on the use of UAV visible-color, multispectral, and thermal infrared imagery to map agricultural drainage pipes. *Agricultural Water Management, 232*, 106036. doi:10.1016/j.agwat.2020.106036

Alsalam, B. H. Y., Morton, K., Campbell, D., & Gonzalez, F. (2017, March). *Autonomous UAV with vision based on-board decision making for remote sensing and precision agriculture. In 2017 IEEE Aerospace Conference*. IEEE.

Alvarado, E. (2021). *237 ways drone applications revolutionize business*. Drone Industry Insights.

Alzenad, M., El-Keyi, A., Lagum, F., & Yanikomeroglu, H. (2017). 3-D placement of an unmanned aerial vehicle base station (UAV-BS) for energy-efficient maximal coverage. *IEEE Wireless Communications Letters, 6*(4), 434–437. doi:10.1109/LWC.2017.2700840

Ammad Uddin, M., Mansour, A., Le Jeune, D., Ayaz, M., & Aggoune, E. H. M. (2018). UAV-assisted dynamic clustering of wireless sensor networks for crop health monitoring. *Sensors (Basel), 18*(2), 555. doi:10.339018020555 PMID:29439496

Aneke, M., & Wang, M. (2016). Energy storage technologies and real life applications–A state of the art review. *Applied Energy, 179*, 350–377. doi:10.1016/j.apenergy.2016.06.097

Anwar, N., Izhar, M. A., & Najam, F. A. (2018, July). Construction monitoring and reporting using drones and unmanned aerial vehicles (UAVs). In *The Tenth International Conference on Construction in the 21st Century (CITC-10)* (pp. 2-4). Academic Press.

Arjomandi, M., Agostino, S., Mammone, M., Nelson, M., & Zhou, T. (2006). *Classification of unmanned aerial vehicles. Report for Mechanical Engineering class*. University of Adelaide.

Arjomandi, M., Agostino, S., Mammone, M., Nelson, M., & Zhou, T. (2006). *Classification of unmanned aerial vehicles. Report for Mechanical Engineering Class*. University of Adelaide.

Arjun, K. M. (2013). Indian Agriculture - Status, Importance and Role in Indian Economy. *International Journal of Agriculture and Food Science Technology, 4*(4), 343–346.

Arnett, G. (2015). *The numbers behind the worldwide trade in drones*. Retrieved from https://www.theguardian.com/news/datablog/2015/mar/16/numbers-behind-worldwide-trade-in-drones-uk-israel

Atoev, S., Kwon, K. R., Lee, S. H., & Moon, K. S. (2017, November). Data analysis of the MAVLink communication protocol. In *2017 International Conference on Information Science and Communications Technologies (ICISCT)* (pp. 1-3). IEEE. 10.1109/ICISCT.2017.8188563

Attaran, M., & Deb, P. (2018). Machine learning: The new'big thing'for competitive advantage. *International Journal of Knowledge Engineering and Data Mining*, *5*(4), 277–305.

Atzori, L., Iera, A., & Morabito, G. (2010). The internet of things: A survey. *Computer Networks*, *54*(15), 2787–2805. doi:10.1016/j.comnet.2010.05.010

Avizienis, A., Laprie, J. C., Randell, B., & Landwehr, C. (2004). Basic concepts and taxonomy of dependable and secure computing. *IEEE Transactions on Dependable and Secure Computing*, *1*(1), 11–33. doi:10.1109/TDSC.2004.2

Ayaz, M., Ammad-Uddin, M., Sharif, Z., Mansour, A., & Aggoune, E. H. M. (2019). Internet-of-Things (IoT)-based smart agriculture: Toward making the fields talk. *IEEE Access: Practical Innovations, Open Solutions*, *7*, 129551–129583. doi:10.1109/ACCESS.2019.2932609

Azevedo, C. L., Cardoso, J. L., Ben-Akiva, M., Costeira, J. P., & Marques, M. (2014). Automatic vehicle trajectory extraction by aerial remote sensing. *Procedia: Social and Behavioral Sciences*, *111*, 849–858. doi:10.1016/j.sbspro.2014.01.119

Bacco, M., Berton, A., Ferro, E., Gennaro, C., Gotta, A., Matteoli, S., ... & Zanella, A. (2018). Smart farming: Opportunities, challenges and technology enablers. *2018 IoT Vertical and Topical Summit on Agriculture-Tuscany (IOT Tuscany),* 1-6.

Bacco, M., Berton, A., Gotta, A., & Caviglione, L. (2018). IEEE 802.15. 4 air-ground UAV communications in smart farming scenarios. *IEEE Communications Letters*, *22*(9), 1910–1913. doi:10.1109/LCOMM.2018.2855211

Balaji, B., Chennupati, S. K., Chilakalapudi, S. R. K., Katuri, R., & Mareedu, K. (2018). Design of UAV (drone) for crop, weather monitoring and for spraying fertilizers and pesticides. *Int J Res Trends Innov*, *3*(3), 42–47.

Bateman, J. (2017). China Drone Maker DJI: Alone atop the Unmanned Skies. *CNBC, 1*, 1.

Beecham Research. (2016). *An Introduction to LPWA Public Service Categories: Matching Services to IoT Applications*. Author.

Bellavista, P., Cardone, G., Corradi, A., & Foschini, L. (2013). Convergence of MANET and WSN in IoT urban scenarios. *IEEE Sensors Journal*, *13*(10), 3558–3567. doi:10.1109/JSEN.2013.2272099

Berni, J. A. J., Zarco-Tejada, P. J., Suarez, L., González-Dugo, V., & Fereres, E. (2009). Remote sensing of vegetation from UAV platforms using lightweight multispectral and thermal imaging sensors. *The International Archives of the Photogrammetry, Remote Sensing and Spatial Information Sciences*, *38*(6), 6.

Beycioglu, A., Comak, B., & Akcaabat, D. (2017). Evaluation of pH value by using image processing. *Acta Physica Polonica A*, *132*(3-II), 1142–1144. doi:10.12693/APhysPolA.132.1142

Bhandari, S., Raheja, A., Chaichi, M. R., Green, R. L., Do, D., Pham, F. H., ... Espinas, A. (2018, June). Lessons learned from uav-based remote sensing for precision agriculture. In *2018 International Conference on Unmanned Aircraft Systems (ICUAS)* (pp. 458-467). IEEE. 10.1109/ICUAS.2018.8453445

Bishop, C. M. (2003). *Neural Networks for Pattern Recognition* (Indian edition). Oxford University Press.

Biswas, K. K., & Basu, S. K. (2011, December). Gesture recognition using microsoft kinect®. In *The 5th international conference on automation, robotics and applications* (pp. 100-103). IEEE.

Bobick, A. F., & Davis, J. W. (2001). The recognition of human movement using temporal templates. *IEEE Transactions on Pattern Analysis and Machine Intelligence*, *23*(3), 257–267.

Boken, V. K. (2016). Potential of soil —moisture-estimating technology for monitoring crop yields and assessing drought impacts-case studies in the United States. *IEEE Technological Innovations in ICT for Agriculture and Rural Development (TIAR).*

Bristeau, P. J., Callou, F., Vissière, D., & Petit, N. (2011). The navigation and control technology inside the ar. drone micro uav. *IFAC Proceedings Volumes, 44*(1), 1477-1484.

Bulling, A., Blanke, U., & Schiele, B. (2014). A tutorial on human activity recognition using body-worn inertial sensors. *ACM Computing Surveys*, *46*(3), 1–33.

Burgard, W., Moors, M., Stachniss, C., & Schneider, F. E. (2005). Coordinated multi-robot exploration. *IEEE Transactions on Robotics*, *21*(3), 376–386.

Camarinha-Matos, L. M. (2012). *Scientific research methodologies and techniques- Unit 2: Scientific method. Unit 2: Scientific methodology. PhD program in electrical and computer engineering*. Uninova.

Cano, E., Horton, R., Liljegren, C., & Bulanon, D. M. (2017). Comparison of small unmanned aerial vehicles performance using image processing. *Journal of Imaging*, *3*(1), 4. doi:10.3390/jimaging3010004

Cary, L., & Coyne, J. (2011). ICAO unmanned aircraft systems (UAS), circular 328. *UVS International. Blyenburgh & Co*, *2012*, 112–115.

Chapman, S. C., Merz, T., Chan, A., Jackway, P., Hrabar, S., Dreccer, M. F., Holland, E., Zheng, B., Ling, T., & Jimenez-Berni, J. (2014). Pheno-copter: A low-altitude, autonomous remote-sensing robotic helicopter for high-throughput field-based phenotyping. *Agronomy (Basel)*, *4*(2), 279–301. doi:10.3390/agronomy4020279

Chatys, R., & Koruba, Z. (2005). Gyroscope-based control and stabilization of unmanned aerial mini-vehicle (mini-UAV). *Aviation*, *9*(2), 10–16.

Chlingaryan, A., Sukkarieh, S., & Whelan, B. (2018). Machine learning approaches for crop yield prediction and nitrogen status estimation in precision agriculture: A review. *Computers and Electronics in Agriculture*, *151*, 61–69. doi:10.1016/j.compag.2018.05.012

Cho, A., Kim, J., Lee, S., & Kee, C. (2011). Wind estimation and airspeed calibration using a UAV with a single-antenna GPS receiver and pitot tube. *IEEE Transactions on Aerospace and Electronic Systems*, *47*(1), 109–117. doi:10.1109/TAES.2011.5705663

Chouhan, S. S., Singh, U. P., & Jain, S. (2020). Applications of computer vision in plant pathology: A survey. *Archives of Computational Methods in Engineering*, *27*(2), 611–632. doi:10.100711831-019-09324-0

Chourabi, H., Nam, T., Walker, S., Gil-Garcia, J. R., Mellouli, S., Nahon, K., . . . Scholl, H. J. (2012, January). Understanding smart cities: An integrative framework. In *2012 45th Hawaii international conference on system sciences* (pp. 2289-2297). IEEE.

Ciriza, R., Sola, I., Albizua, L., Álvarez-Mozos, J., & González-Audícana, M. (2017). Automatic Detection of Uprooted Orchards Based on Orthophoto Texture Analysis. *Remote Sensing*, *9*(5), 492. doi:10.3390/rs9050492

Clancy, L. (2019). *Functional programming reaches for stardom in finance*. CME Group. https://www.risk.net/risk-management/6395366/functional-programming-reaches-for-stardom-in-finance

Coombes, M., Chen, W. H., & Liu, C. (2018, July). Fixed wing UAV survey coverage path planning in wind for improving existing ground control station software. In *2018 37th Chinese Control Conference (CCC)* (pp. 9820-9825). IEEE. 10.23919/ChiCC.2018.8482722

Crabit, A., Colin, F., Bailly, J. S., Ayroles, H., & Garnier, F. (2011). Soft water level sensors for characterizing the hydrological behaviour of agricultural catchments. *Sensors (Basel)*, *11*(5), 4656–4673. doi:10.3390110504656 PMID:22163868

Dadshani, S., Kurakin, A., Amanov, S., Hein, B., Rongen, H., Cranstone, S., Blievernicht, U., Menzel, E., Léon, J., Klein, N., & Ballvora, A. (2015). Non-invasive assessment of leaf water status using a dual-mode microwave resonator. *Plant Methods*, *11*(1), 1–10. doi:10.118613007-015-0054-x PMID:25918549

Daellenbach, H., McNickle, D., & Dye, Sh. (2012). *Management Science. Decision-making through systems thinking* (2nd ed.). Plagrave Macmillian.

Dantzig, G. (1949). *Programming of Interdependent Activities: II Mathematical Model*. Academic Press.

De Freitas, E. P., Heimfarth, T., Netto, I. F., Lino, C. E., Pereira, C. E., Ferreira, A. M., . . . Larsson, T. (2010, October). UAV relay network to support WSN connectivity. In International Congress on Ultra Modern Telecommunications and Control Systems (pp. 309-314). IEEE. doi:10.1109/ICUMT.2010.5676621

De Haan, W., & Loader, I. (2002). On the emotions of crime, punishment and social control. *Theoretical Criminology*, *6*(3), 243–253.

Della Croce, F., & T'kindt, V. (2002). A Recovering Beam Search algorithm for the one-machine dynamic total completion time scheduling problem. *The Journal of the Operational Research Society*, *53*(11), 1275–1280. doi:10.1057/palgrave.jors.2601389

Dempsey, M. E., & Rasmussen, S. (2010). *Eyes of the army–US Army roadmap for unmanned aircraft systems 2010–2035*. US Army UAS Center of Excellence.

Doering, D., Benenmann, A., Lerm, R., de Freitas, E. P., Muller, I., Winter, J. M., & Pereira, C. E. (2014). Design and optimization of a heterogeneous platform for multiple uav use in precision agriculture applications. *IFAC Proceedings Volumes, 47*(3), 12272-12277.

Dominguez, L., D'Amato, J. P., Perez, A., Rubiales, A., & Barbuzza, R. (2018, June). Running License Plate Recognition (LPR) algorithms on smart survillance cameras. A feasibility analysis. In *2018 13th Iberian Conference on Information Systems and Technologies (CISTI)* (pp. 1-5). IEEE.

Doree, J. (2015). *Data Science, Data Architecture-Predictive Analytics Methodology: Building a Data Map for a Greenfield Data Science Initiative*. Datasciencearchitect. https://datasciencearchitect.wordpress.com/tag/togaf/

Doward, J. (2019, June 9). *Military drone crashes raise fears for civilians*. https://www.theguardian.com/world/2019/jun/09/two-military-drones-crashing-every-month

Dricus. (2015). *Top 8 solar powered drone (UAV) developing companies*. https://sinovoltaics.com/technology/top8-leading-companiesdeveloping-solar-powered-drone-uav-technology

DroneDeploy. (2018, June 7). https://www.dronedeploy.com/blog/rise-drones-construction/

Dung, N. D., & Rohacs, J. (2018, November). The drone-following models in smart cities. In *2018 IEEE 59th international scientific conference on power and electrical engineering of Riga Technical University (RTUCON)* (pp. 1-6). IEEE.

Duque Domingo, J., Cerrada, C., Valero, E., & Cerrada, J. A. (2017). An improved indoor positioning system using RGB-D cameras and wireless networks for use in complex environments. *Sensors (Basel)*, *17*(10), 2391.

E-business definition. (2014). Accessed and reviewed in April 2019. http://dictionary.reference.com/browse/e-business

Elarab, M., Ticlavilca, A. M., Torres-Rua, A. F., Maslova, I., & McKee, M. (2015). Estimating chlorophyll with thermal and broadband multispectral high resolution imagery from an unmanned aerial system using relevance vector machines for precision agriculture. *International Journal of Applied Earth Observation and Geoinformation*, *43*, 32–42. doi:10.1016/j.jag.2015.03.017

Elsenbeiss, H., & Sauerbier, M. (2011). Investigation of uav systems and flight modes for photogrammetric applications. *The Photogrammetric Record*, *26*(136), 400–421. doi:10.1111/j.1477-9730.2011.00657.x

Estrada-López, J. J., Castillo-Atoche, A. A., Vázquez-Castillo, J., & Sánchez-Sinencio, E. (2018). Smart soil parameters estimation system using an autonomous wireless sensor network with dynamic power management strategy. *IEEE Sensors Journal*, *18*(21), 8913–8923. doi:10.1109/JSEN.2018.2867432

Eze, K. G., Sadiku, M. N., & Musa, S. M. (2018). 5G wireless technology: A primer. *International Journal of Scientific Engineering and Technology*, *7*(7), 62–64.

Fakhrulddin, S. S., Gharghan, S. K., Al-Naji, A., & Chahl, J. (2019). An advanced first aid system based on an unmanned aerial vehicles and a wireless body area sensor network for elderly persons in outdoor environments. *Sensors (Basel)*, *19*(13), 2955.

FAO in India. (2017) Retrieved from https://www.fao.org/india/fao-inindia/india-at-a-glance/en/2017

Feng, X., Yan, F., & Liu, X. (2019). Study of wireless communication technologies on Internet of Things for precision agriculture. *Wireless Personal Communications*, *108*(3), 1785–1802. doi:10.100711277-019-06496-7

Fotouhi, A., Qiang, H., Ding, M., Hassan, M., Giordano, L. G., Garcia-Rodriguez, A., & Yuan, J. (2019). Survey on UAV cellular communications: Practical aspects, standardization advancements, regulation, and security challenges. *IEEE Communications Surveys and Tutorials*, *21*(4), 3417–3442. doi:10.1109/COMST.2019.2906228

Friedman, L., & McCabe, D. (2020). Interior Dept. Grounds Its Drones Over Chinese Spying Fears. *The New York Times*. Retrieved from https://www.nytimes.com/2020/01/29/technology/interior-chinese-drones.html

Fujimori, A., Ukigai, Y. U., Santoki, S., & Oh-hara, S. (2018). Autonomous flight control system of quadrotor and its application to formation control with mobile robot. *IFAC-PapersOnLine*, *51*(22), 343–347. doi:10.1016/j.ifacol.2018.11.565

Fu, S., Saeidi, H., Sand, E., Sadrfaidpour, B., Rodriguez, J., Wang, Y., & Wagner, J. (2016, July). A haptic interface with adjustable feedback for unmanned aerial vehicles (UAVs)-model, control, and test. In *2016 American Control Conference (ACC)* (pp. 467-472). IEEE.

Galkin, B., Kibilda, J., & DaSilva, L. A. (2019). UAVs as mobile infrastructure: Addressing battery lifetime. *IEEE Communications Magazine*, *57*(6), 132–137. doi:10.1109/MCOM.2019.1800545

Galway, D., Etele, J., & Fusina, G. (2011). Modeling of urban wind field effects on unmanned rotorcraft flight. *Journal of Aircraft*, *48*(5), 1613–1620.

Ganeshamurthy, A. N., Kalaivanan, D., & Rajendiran, S. (2020). Carbon sequestration potential of perennial horticultural crops in Indian tropics. In *Carbon Management in Tropical and Sub-Tropical Terrestrial Systems* (pp. 333–348). Springer. doi:10.1007/978-981-13-9628-1_20

Gantala, A., Nehru, K., Telagam, N., Anjaneyulu, P., & Swathi, D. (2017). Human tracking system using beagle board-xm. *International Journal of Applied Engineering Research: IJAER*, *12*(16), 5665–5669.

Gantala, A., Vijaykumar, G., Telagam, N., & Anjaneyulu, P. (2017). Design of Smart Sensor Using Linux-2.6. 29 Kernel. *International Journal of Applied Engineering Research: IJAER*, *12*, 7891–7896.

Gao, J., Nuyttens, D., Lootens, P., He, Y., & Pieters, J. G. (2018). Recognising weeds in a maize crop using a random forest machine-learning algorithm and near-infrared snapshot mosaic hyperspectral imagery. *Biosystems Engineering*, *170*, 39–50. doi:10.1016/j.biosystemseng.2018.03.006

Gartner. (2020). *Three Essentials for Starting and Supporting Master Data Management*. ID G00730039. Gartner Inc. https://www.gartner.com/doc/reprints?id=1-24MIGFGU&ct=201119&st=sb

George, J., PB, S., & Sousa, J. B. (2011). Search strategies for multiple UAV search and destroy missions. *Journal of Intelligent & Robotic Systems*, *61*(1), 355–367. doi:10.100710846-010-9486-8

Ge, Y., Thomasson, J. A., & Sui, R. (2011). Remote sensing of soil properties in precision agriculture: A review. *Frontiers of Earth Science*, *5*(3), 229–238. doi:10.100711707-011-0175-0

Ghosh, A., & Dey, S. (2021). "Sensing the Mind": An Exploratory Study About Sensors Used in E-Health and M-Health Applications for Diagnosis of Mental Health Condition. In *Efficient Data Handling for Massive Internet of Medical Things* (pp. 269–292). Springer. doi:10.1007/978-3-030-66633-0_12

Giorgetti, A., Lucchi, M., Tavelli, E., Barla, M., Gigli, G., Casagli, N., Chiani, M., & Dardari, D. (2016). A robust wireless sensor network for landslide risk analysis: System design, deployment, and field testing. *IEEE Sensors Journal*, *16*(16), 6374–6386. doi:10.1109/JSEN.2016.2579263

Glaroudis, D., Iossifides, A., & Chatzimisios, P. (2020). Survey, comparison and research challenges of IoT application protocols for smart farming. *Computer Networks*, *168*, 107037. doi:10.1016/j.comnet.2019.107037

Glynn, T. J. (1981). Psychological sense of community: Measurement and application. *Human Relations*, *34*(9), 789–818.

GSMA Intelligence. (2017). *Global Mobile Trends 2017*. Retrieved from https://www.gsmaintelligence.com/research/?file=3df1b7d57b1e63a0cbc3d585feb82dc2&utm_source=Triggermail&utm_medium=email&utm_campaign=Post%20Blast%20%28bii-apps-and-platforms%29:%20Apple%20drops%20in-app%20tipping%20tax%20%E2%80%94%20Two-thirds%20of%20the

Gubbi, J., Buyya, R., Marusic, S., & Palaniswami, M. (2013). Internet of Things (IoT): A vision, architectural elements, and future directions. *Future Generation Computer Systems*, *29*(7), 1645–1660. doi:10.1016/j.future.2013.01.010

Gupta, L., Jain, R., & Vaszkun, G. (2015). Survey of important issues in UAV communication networks. *IEEE Communications Surveys and Tutorials*, *18*(2), 1123–1152. doi:10.1109/COMST.2015.2495297

Guru99. (2021). *Data Science Tutorial for Beginners: What is, Basics & Process*. https://www.guru99.com/data-science-tutorial.html

Ha, T. (2010). The UAV Continuous Coverage Problem. *Theses and Dissertations*. 2094. https://scholar.afit.edu/etd/2094

Hachem, S., Mallet, V., Ventura, R., Pathak, A., Issarny, V., Raverdy, P. G., & Bhatia, R. (2015, March). Monitoring noise pollution using the urban civics middleware. In *2015 IEEE First International Conference on Big Data Computing Service and Applications* (pp. 52-61). IEEE. 10.1109/BigDataService.2015.16

Hallinan, A. J. Jr. (1993). A review of the Weibull distribution. *Journal of Quality Technology*, *25*(2), 85–93. doi:10.1080/00224065.1993.11979431

Hallion, R. (2003). *Taking flight: Inventing the aerial age, from antiquity through the First World War*. Oxford University Press.

Han, J., Kamber, M., & Pei, J. (2012). *Data mining: concepts and techniques*. Elsevier.

Hansen, R. L. (2016). *Traffic monitoring using UAV Technology*. The American Surveyor.

Hayat, S., Yanmaz, E., & Muzaffar, R. (2016). Survey on unmanned aerial vehicle networks for civil applications: A communications viewpoint. *IEEE Communications Surveys and Tutorials*, *18*(4), 2624–2661. doi:10.1109/COMST.2016.2560343

Haydon, F. S. (2000). *Military Ballooning during the Early Civil War*. JHU Press.

Hii, M. S. Y., Courtney, P., & Royall, P. G. (2019). An evaluation of the delivery of medicines using drones. *Drones (Basel)*, *3*(3), 52. doi:10.3390/drones3030052

Horowitz, M. C. (2020). Do emerging military technologies matter for international politics? *Annual Review of Political Science*, *23*(1), 385–400. doi:10.1146/annurev-polisci-050718-032725

Hu, J., Bhowmick, P., Jang, I., Arvin, F., & Lanzon, A. (2021). A decentralized cluster formation containment framework for multirobot systems. *IEEE Transactions on Robotics*, *37*(6), 1936–1955. doi:10.1109/TRO.2021.3071615

Hu, J., & Lanzon, A. (2018). An innovative tri-rotor drone and associated distributed aerial drone swarm control. *Robotics and Autonomous Systems*, *103*, 162–174. doi:10.1016/j.robot.2018.02.019

Hunter, M. C., Smith, R. G., Schipanski, M. E., Atwood, L. W., & Mortensen, D. A. (2017). Agriculture in 2050: Recalibrating targets for sustainable intensification. *Bioscience*, *67*(4), 386–391. doi:10.1093/biosci/bix010

IBM. (2015a). *Modelling the Entity data architecture*. IBM.

IBM. (2019). *AI and Data Management-Delivering Data-Driven Business Transformation and Operational Efficiencies*. IBM USA.

Idries, A., Mohamed, N., Jawhar, I., Mohamed, F., & Al-Jaroodi, J. (2015, March). Challenges of developing UAV applications: A project management view. In 2015 International Conference on Industrial Engineering and Operations Management (IEOM) (pp. 1-10). IEEE. doi:10.1109/NTMS.2014.6814041

IMD. (2015). *IMD business school and Cisco join forces on digital business transformation*. IMD.

Isah, E. C., Asuzu, M. C., & Okojie, O. H. (1997). Occupational health hazards in manufacturing industries in Nigeria. *J Community Med Primary Health Care*, *9*, 26–34.

Islam, N., Ray, B., & Pasandideh, F. (2020, August). Iot based smart farming: Are the lpwan technologies suitable for remote communication? In *2020 IEEE International Conference on Smart Internet of Things (SmartIoT)* (pp. 270-276). IEEE. 10.1109/SmartIoT49966.2020.00048

Jain, K. P., & Mueller, M. W. (2020, May). Flying batteries: In-flight battery switching to increase multirotor flight time. In *2020 IEEE International Conference on Robotics and Automation (ICRA)* (pp. 3510-3516). IEEE.

Jain, P. K. (1991). On blackbody radiation. *Physics Education*, *26*(3), 190.

Jain, R. (2016). *Wireless protocols for iot part ii: Ieee 802.15. 4 wireless personal area networks*. IEEE.

Jamali, J., Bahrami, B., Heidari, A., & Allahverdizadeh, P. (2020). IoT Architecture. In J. Jamali, B. Bahrami, A. Heidari, P. Allahverdizadeh, & F. Norouzi (Eds.), *Towards the internet of things* (pp. 9–31). Springer International Publishing.

Järvinen, P. (2007). Action Research is Similar to Design Science. *Quality & Quantity*, *41*(1), 37–54. Retrieved August 10, 2018, from https://link.springer.com/article/10.1007/s11135-005-5427-1

Jaszkiewicz, J., & Sowiñski, R. (1999). The 'Light Beam Search' approach - An overview of methodology and applications. *European Journal of Operational Research*, *113*(2), 300–314. doi:10.1016/S0377-2217(98)00218-5

Jawad, A. M., Jawad, H. M., Nordin, R., Gharghan, S. K., Abdullah, N. F., & Abu-Alshaeer, M. J. (2019). Wireless power transfer with magnetic resonator coupling and sleep/active strategy for a drone charging station in smart agriculture. *IEEE Access: Practical Innovations, Open Solutions*, *7*, 139839–139851. doi:10.1109/ACCESS.2019.2943120

Jiang, F., & Swindlehurst, A. L. (2010, December). Dynamic UAV relay positioning for the ground-to-air uplink. In 2010 IEEE *Globecom Workshops*. IEEE.

Ju, C., & Son, H. I. (2018). Multiple UAV systems for agricultural applications: Control, implementation, and evaluation. *Electronics (Basel)*, *7*(9), 162. doi:10.3390/electronics7090162

Kanezaki, A. (2018). Unsupervised Image Segmentation by Backpropagation. *IEEE International Conference on Acoustics, Speech and Signal Processing (ICASSP).*

Kanistras, K., Martins, G., Rutherford, M. J., & Valavanis, K. P. (2013, May). A survey of unmanned aerial vehicles (UAVs) for traffic monitoring. In *2013 International Conference on Unmanned Aircraft Systems (ICUAS)* (pp. 221-234). IEEE. 10.1109/ICUAS.2013.6564694

Kaplan, P. (2013). *Naval Aviation in the Second World War*. Pen and Sword.

Kapoor, A. (2021). *Artificial Intelligence and Machine Learning: 5 Trends to Watch Out for in 2021.* AI Zone. DZone. https://dzone.com/articles/artificial-intelligence-amp-machine-learning-5-dev

Kardasz, P., Doskocz, J., Hejduk, M., Wiejkut, P., & Zarzycki, H. (2016). Drones and possibilities of their using. *Journal of Civil and Environmental Engineering*, *6*(3), 1–7.

Kassambara, L., Sara, M. F., & Kassambara, V. (2018, March 10). *Multiple Linear Regression in R.* STHDA. http://www.sthda.com/english/articles/40-regression-analysis/168-multiple-linear-regression-in-r

Kawasaki, M., & Hasuike, T. (2017). A recommendation system by collaborative filtering including information and characteristics on users and items. *IEEE Symposium Series on Computational Intelligence (SSCI).* 10.1109/SSCI.2017.8280983

Kelley, K. (2020). *9 types of machine learning algorithms with a cheat sheet.* Techtarget. https://searchenterpriseai.techtarget.com/

Khadka & Lamichhane. (2016). The Relationship between Soil pH and Micronutrients, Western Nepal. *International Journal of Agriculture Innovation and Research*, *4*(5).

Khaki, Saeed, & Wang. (2019). Crop yield prediction using deep neural networks. *Frontiers in Plant Science, 10.*

Khan, M. A., Ectors, W., Bellemans, T., Janssens, D., & Wets, G. (2017). UAV-based traffic analysis: A universal guiding framework based on literature survey. *Transportation Research Procedia*, *22*, 541–550. doi:10.1016/j.trpro.2017.03.043

Khofiyah, N. A., Sutopo, W., & Nugroho, B. D. A. (2019). Technical feasibility battery lithium to support unmanned aerial vehicle (UAV): A technical review. In *Proceedings of the International Conference on Industrial Engineering and Operations Management* (*Vol. 2019*, pp. 3591-3601). Academic Press.

Kim, J., Kim, S., Ju, C., & Son, H. I. (2019). Unmanned aerial vehicles in agriculture: A review of perspective of platform, control, and applications. *IEEE Access: Practical Innovations, Open Solutions*, *7*, 105100–105115. doi:10.1109/ACCESS.2019.2932119

Kim, K., & Kim, K. (1999). Routing straddle carriers for the loading operation of containers using a beam search algorithm. Elsevier. *Computers & Industrial Engineering*, *36*(1), 109–136. doi:10.1016/S0360-8352(99)00005-4

Kong, Y., & Fu, Y. (2022). Human action recognition and prediction: A survey. *International Journal of Computer Vision*, *130*(5), 1366–1401. doi:10.100711263-022-01594-9

Koparan, C., Koc, A. B., Privette, C. V., & Sawyer, C. B. (2018). In situ water quality measurements using an unmanned aerial vehicle (UAV) system. *Water (Basel)*, *10*(3), 264.

Koparan, C., Koc, A. B., Privette, C. V., & Sawyer, C. B. (2020). Adaptive water sampling device for aerial robots. *Drones*, *4*(1), 5.

Kumar, M. A., Telagam, N., Mohankumar, N., Ismail, M., & Rajasekar, T. (2020). Design and implementation of real-time amphibious unmanned aerial vehicle system for sowing seed balls in the agriculture field. *Int J Emerg Technol*, *11*(2), 213–218.

Kumar, S., & Kanniga, E. (2018). Literature Survey on Unmanned aerial Vehicle. *International Journal of Pure and Applied Mathematics*, *119*(12), 4381–4387.

Kumar, V., Vimal, B. K., Kumar, R., Kumar, M., & Kumar, M. (2014, January). Determination of soil pH by using digital image processing technique. *Journal of Applied and Natural Science*, *6*(1), 14–18. doi:10.31018/jans.v6i1.368

La Vigne, N. G., Lowry, S. S., Markman, J. A., & Dwyer, A. M. (2011). *Evaluating the use of public surveillance cameras for crime control and prevention*. US Department of Justice, Office of Community Oriented Policing Services. Urban Institute, Justice Policy Center. doi:10.1037/e718202011-001

Lazar, I., Motogna, S., & Parv, B. (2010). Behaviour-Driven Development of Foundational UML Components. Department of Computer Science, Babes-Bolyai University. doi:10.1016/j.entcs.2010.07.007

Le Mouël, C., & Forslund, A. (2017). How can we feed the world in 2050? A review of the responses from global scenario studies. *European Review of Agriculture Economics*, *44*(4), 541–591. doi:10.1093/erae/jbx006

Le, N. T., Hossain, M. A., Islam, A., Kim, D. Y., Choi, Y. J., & Jang, Y. M. (2016). Survey of promising technologies for 5G networks. *Mobile Information Systems*.

Lee Nelan, A. (2015). *Image Processing system for soil characterization*. Academic Press.

Leitch, R., & Day, Ch. (2000). *Action research and reflective practice: towards a holistic view*. Taylor & Francis. https://www.tandfonline.com/doi/ref/10.1080/09650790000200108?scroll=top

Liang, M., & Delahaye, D. (2019, October). Drone fleet deployment strategy for large scale agriculture and forestry surveying. In 2019 IEEE Intelligent Transportation Systems Conference (ITSC) (pp. 4495-4500). IEEE. doi:10.1109/ITSC.2019.8917235

Li, B., Fei, Z., & Zhang, Y. (2018). UAV communications for 5G and beyond: Recent advances and future trends. *IEEE Internet of Things Journal*, *6*(2), 2241–2263. doi:10.1109/JIOT.2018.2887086

Liew, C. F., DeLatte, D., Takeishi, N., & Yairi, T. (2017). *Recent developments in aerial robotics: A survey and prototypes overview.* arXiv preprint arXiv:1711.10085.

Liu, B. L., Wang, L., Lee, S. J., Liu, J., Qin, F., & Jiao, Z. L. (Eds.). (2015). *Contemporary Logistics in China: Proliferation and Internationalization*. Springer.

Liu, Z., Huang, J., Wang, Q., Wang, Y., & Fu, J. (2013, June). Real-time barrier lakes monitoring and warning system based on wireless sensor network. In *2013 Fourth International Conference on Intelligent Control and Information Processing (ICICIP)* (pp. 551-554). IEEE. 10.1109/ICICIP.2013.6568136

Lockwood, R. (2018). *Introduction The Relational Data Model*. Retrieved January 29, 2019, from http://www.jakobsens.dk/Nekrologer.htm

Loginovskiy, O. V., Dranko, O. I., & Hollay, A. V. (2018). *Mathematical Models for Decision-Making on Strategic Management of Industrial Enterprise in Conditions of Instability*. Conference: Internationalization of Education in Applied Mathematics and Informatics for HighTech Applications (EMIT 2018), Leipzig, Germany.

Lv, M., Xiao, S., Yu, T., & He, Y. (2019). Influence of UAV flight speed on droplet deposition characteristics with the application of infrared thermal imaging. *International Journal of Agricultural and Biological Engineering*, *12*(3), 10–17. doi:10.25165/j.ijabe.20191203.4868

Maleki, H., Zurek, R., Howard, J. N., & Hallmark, J. A. (2016). Lithium ion cell/batteries electromagnetic field reduction in phones for hearing aid compliance. *Batteries*, *2*(2), 19.

Mancini, F., Dubbini, M., Gattelli, M., Stecchi, F., Fabbri, S., & Gabbianelli, G. (2013). Using unmanned aerial vehicles (UAV) for high-resolution reconstruction of topography: The structure from motion approach on coastal environments. *Remote Sensing*, *5*(12), 6880–6898. doi:10.3390/rs5126880

McAlexander, J. H., Schouten, J. W., & Koenig, H. F. (2002). Building brand community. *Journal of Marketing*, *66*(1), 38–54.

Mckinnon, A. C. (2016). The possible impact of 3D printing and drones on last-mile logistics: An exploratory study. *Built Environment*, *42*(4), 617–629.

McNabb, M. (2020). *Drones Get the Lights Back on Faster for Florida Communities*. Retrieved from https://dronelife.com/2020/02/28/drones-get-the-lights-back-on-faster-for-florida-communities/

Medela, A., Cendón, B., Gonzalez, L., Crespo, R., & Nevares, I. (2013, July). IoT multiplatform networking to monitor and control wineries and vineyards. In *2013 Future Network & Mobile Summit* (pp. 1-10). IEEE.

Meola, C., & Carlomagno, G. M. (2004). Recent advances in the use of infrared thermography. *Measurement Science & Technology*, *15*(9), R27.

Merson, P. (2009). *Data Model as an Architectural View. TECHNICAL NOTE. CMU/SEI-2009-TN-024*. Research, Technology, and System Solutions.

Mikesh, R. C. (1973). Japan's World War II balloon bomb attacks on North America. *Smithsonian Annals of Flight*. Advance online publication. doi:10.5479i.AnnalsFlight.9

Miller, M. (2020). *DOJ bans use of grant funds for certain foreign-made drones*. Retrieved from https://thehill.com/policy/cybersecurity/520269-justice-department-issues-policy-banning-use-of-grant-funds-for-certain/

Mogili, U. R., & Deepak, B. B. V. L. (2018). Review on application of drone systems in precision agriculture. *Procedia Computer Science*, *133*, 502–509. doi:10.1016/j.procs.2018.07.063

Mohamed, N., Al-Jaroodi, J., Jawhar, I., Idries, A., & Mohammed, F. (2020). Unmanned aerial vehicles applications in future smart cities. *Technological Forecasting and Social Change*, *153*, 119293. doi:10.1016/j.techfore.2018.05.004

Mohammed, F., Idries, A., Mohamed, N., Al-Jaroodi, J., & Jawhar, I. (2014, March). Opportunities and challenges of using UAVs for dubai smart city. In *2014 6th International Conference on New Technologies, Mobility and Security (NTMS)* (pp. 1-4). IEEE.

Mohammed, F., Idries, A., Mohamed, N., Al-Jaroodi, J., & Jawhar, I. (2014, May). UAVs for smart cities: Opportunities and challenges. In *2014 International Conference on Unmanned Aircraft Systems (ICUAS)* (pp. 267-273). IEEE.

Mohan, A., & Poobal, S. (2018). Crack detection using image processing: A critical review and analysis. *Alexandria Engineering Journal*, *57*(2), 787–798. doi:10.1016/j.aej.2017.01.020

Moore, J. (2014). *Java programming with lambda expressions-A mathematical example demonstrates the power of lambdas in Java 8*. Retrieved March 10, 2018, from, https://www.javaworld.com/article/2092260/java-se/java-programming-with-lambda-expressions.html

Moribe, T., Okada, H., Kobayashl, K., & Katayama, M. (2018, January). Combination of a wireless sensor network and drone using infrared thermometers for smart agriculture. In 2018 15th IEEE Annual Consumer Communications & Networking Conference (CCNC) (pp. 1-2). IEEE. doi:10.1109/CCNC.2018.8319300

Motwani, S. (2020, September). Tactical Drone for Point-to-Point data delivery using Laser-Visible Light Communication (L-VLC). In *2020 3rd International Conference on Advanced Communication Technologies and Networking (CommNet)* (pp. 1-8). IEEE.

Mozaffari, M., Saad, W., Bennis, M., & Debbah, M. (2016). Unmanned aerial vehicle with underlaid device-to-device communications: Performance and tradeoffs. *IEEE Transactions on Wireless Communications*, *15*(6), 3949–3963. doi:10.1109/TWC.2016.2531652

Muchiri, G. N., & Kimathi, S. (2022, April). A review of applications and potential applications of UAV. In *Proceedings of the Sustainable Research and Innovation Conference* (pp. 280-283). Academic Press.

Mukherjee, A., Misra, S., Sukrutha, A., & Raghuwanshi, N. S. (2020). Distributed aerial processing for IoT-based edge UAV swarms in smart farming. *Computer Networks*, *167*, 107038. doi:10.1016/j.comnet.2019.107038

Mulligan, C. E., & Olsson, M. (2013). Architectural implications of smart city business models: An evolutionary perspective. *IEEE Communications Magazine*, *51*(6), 80–85. doi:10.1109/MCOM.2013.6525599

Murphy, J. D. (2005). *Military aircraft, origins to 1918: An illustrated history of their impact*. ABC-CLIO.

Murugan, D., Garg, A., Ahmed, T., & Singh, D. (2016, December). Fusion of drone and satellite data for precision agriculture monitoring. In *2016 11th International Conference on Industrial and Information Systems (ICIIS)* (pp. 910-914). IEEE.

Murugan, D., Garg, A., & Singh, D. (2017). Development of an adaptive approach for precision agriculture monitoring with drone and satellite data. *IEEE Journal of Selected Topics in Applied Earth Observations and Remote Sensing*, *10*(12), 5322–5328. doi:10.1109/JSTARS.2017.2746185

Muttin, F. (2011). Umbilical deployment modeling for tethered UAV detecting oil pollution from ship. *Applied Ocean Research*, *33*(4), 332–343. doi:10.1016/j.apor.2011.06.004

Nanalyze. (2019). *How Autonomous Drone Flights Will Go Beyond Line of Sight*. Retrieved from https://www.nanalyze.com/2019/12/autonomous-drone-flights/

National Cancer Institute. (n.d.a). *About Cancer / EMF radiation and Cancer Risk*. https://www.cancer.gov/about-cancer/causes-prevention/risk/radiation/electromagnetic-fields-fact-sheet#r1

National Cancer Institute. (n.d.b). *About Cancer / Cellphones and Cancer Risk*. https://www.cancer.gov/about-cancer/causes-prevention/risk/radiation/cell-phones-fact-sheet

National Portal of India. (2021). https://www.india.gov.in/topics/agriculture

Navulur, S., & Prasad, M. G. (2017). Agricultural management through wireless sensors and internet of things. *Iranian Journal of Electrical and Computer Engineering*, *7*(6), 3492. doi:10.11591/ijece.v7i6.pp3492-3499

Nijboer, F., Morin, F., Carmien, S., Koene, R., Leon, E., & Hoffman, U. (2009). Affective brain-computer interfaces: Psychophysiological markers of emotion in healthy persons and in persons with amyotrophic lateral sclerosis. In *3rd International Conference on Affective Computing and Intelligent Interaction and Workshops*. IEEE. 10.1109/ACII.2009.5349479

Nukala, R., Panduru, K., Shields, A., Riordan, D., Doody, P., & Walsh, J. (2016, June). Internet of Things: A review from 'Farm to Fork'. In *2016 27th Irish signals and systems conference (ISSC)* (pp. 1-6). IEEE.

Nummenmaa, L., Glerean, E., Hari, R., & Hietanen, J. K. (2014). Bodily maps of emotions. *Proceedings of the National Academy of Sciences of the United States of America*, *111*(2), 646–651. doi:10.1073/pnas.1321664111 PMID:24379370

Nurkin, T., Bedard, K., Clad, J., Scott, C., & Grevatt, J. (2018). *China's Advanced Weapons Systems*. Jane's by IHS Markit.

OASIS. (2014). *ISO/IEC and OASIS Collaborate on E-Business Standards-Standards Groups Increase Cross-Participation to Enhance Interoperability*. https://www.oasis-open.org/news/pr/isoiec-and-oasis-collaborate-on-e-business-standards

Oliveira, H. C., Guizilini, V. C., Nunes, I. P., & Souza, J. R. (2018). Failure Detection in Row Crops From UAV Images Using Morphological Operators. *IEEE Geoscience and Remote Sensing Letters*, *15*(7), 991–995. doi:10.1109/LGRS.2018.2819944

Omid, M., & Torbjorn, N. (2019). Drones in manufacturing: Exploring opportunities for research and practice. *Journal of Manufacturing Technology Management*.

OMNI-SCI. (2021). *Data Science - A Complete Introduction*. OMNI-SCI. https://www.omnisci.com/learn/data-science

Orfanus, D., De Freitas, E. P., & Eliassen, F. (2016). Self-organization as a supporting paradigm for military UAV relay networks. *IEEE Communications Letters*, *20*(4), 804–807. doi:10.1109/LCOMM.2016.2524405

Orlando, F., Movedi, E., Coduto, D., Parisi, S., Brancadoro, L., Pagani, V., Guarneri, T., & Confalonieri, R. (2016). Estimating leaf area index (LAI) in vineyards using the PocketLAI smart-app. *Sensors (Basel)*, *16*(12), 2004. doi:10.339016122004 PMID:27898028

Osuch, A., Przygodziński, P., Rybacki, P., Osuch, E., Kowalik, I., Piechnik, L., Przygodziński, A., & Herkowiak, M. (2020). Analysis of the Effectiveness of Shielded Band Spraying in Weed Control in Field Crops. *Agronomy (Basel)*, *10*(4), 475. doi:10.3390/agronomy10040475

Panchasara, H., Samrat, N. H., & Islam, N. (2021). Greenhouse gas emissions trends and mitigation measures in australian agriculture sector—A review. *Revista de Agricultura (Piracicaba)*, *11*, 85.

Patel, K. D., Jayaraman, C. S., & Maurya, S. K. (2015). Selection of BLDC Motor and Propeller for Autonomous Amphibious Unmanned Aerial Vehicle. *WSEAS Transactions on Systems and Control*, *10*, 179–185.

Pawar & Chillarge. (2018). Soil Toxicity Prediction and Recommendation System Using Data Mining In Precision Agriculture. *3rd International Conference for Convergence in Technology (I2CT).*

Peck, A. (2020). *Coronavirus Spurs Percepto's Drone-in-a-Box Surveillance Solution.* Retrieved from https://insideunmannedsystems.com/unintended-consequences-coronavirus-spurs-perceptos-drone-in-a-box-surveillance-solution/

Peterson, A. (2013). *States are competing to be the Silicon Valley of drones.* Retrieved from https://www.washingtonpost.com/news/the-switch/wp/2013/08/19/states-are-competing-to-be-the-silicon-valley-of-drones/

Pobkrut, T., Eamsa-Ard, T., & Kerdcharoen, T. (2016, June). Sensor drone for aerial odor mapping for agriculture and security services. In *2016 13th International Conference on Electrical Engineering/Electronics, Computer, Telecommunications and Information Technology (ECTI-CON)* (pp. 1-5). IEEE. 10.1109/ECTICon.2016.7561340

Pudumalar, S., Ramanujam, E., Rajashree, R. H., Kavya, C., Kiruthika, T., & Nisha, J. (2017). Crop recommendation system for precision agriculture. *Eighth International Conference on Advanced Computing (ICoAC).* 10.1109/ICoAC.2017.7951740

Puri, V., Nayyar, A., & Raja, L. (2017). Agriculture drones: A modern breakthrough in precision agriculture. *Journal of Statistics and Management Systems, 20*(4), 507–518. doi:10.1080/09720510.2017.1395171

Radha, D., Kumar, M. A., Telagam, N., & Sabarimuthu, M. (2021). *Smart Sensor Network-Based Autonomous Fire Extinguish Robot Using IoT.* Academic Press.

Radoglou-Grammatikis, P., Sarigiannidis, P., Lagkas, T., & Moscholios, I. (2020). A compilation of UAV applications for precision agriculture. *Computer Networks, 172*, 107148. doi:10.1016/j.comnet.2020.107148

Raghuraman, S., Bahirat, K., & Prabhakaran, B. (2015, June). Evaluating the efficacy of RGB-D cameras for surveillance. In *2015 IEEE International Conference on Multimedia and Expo (ICME)* (pp. 1-6). IEEE.

Rao, B., Gopi, A. G., & Maione, R. (2016). The societal impact of commercial drones. *Technology in Society, 45*, 83–90. doi:10.1016/j.techsoc.2016.02.009

Reed, T., & Kidd, J. (2019). *Global traffic scorecard.* INRIX Research.

Renner, S. L. (2016). *Broken Wings: The Hungarian Air Force, 1918–45.* Indiana University Press. doi:10.2307/j.ctt2005t4h

Rodrigues, M., Pigatto, D. F., Fontes, J. V., Pinto, A. S., Diguet, J. P., & Branco, K. R. (2017, May). UAV integration into IoIT: opportunities and challenges. In ICAS 2017 (Vol. 95). Academic Press.

Rodríguez-Moreno, I., Martínez-Otzeta, J. M., Sierra, B., Rodriguez, I., & Jauregi, E. (2019). Video activity recognition: State-of-the-art. *Sensors (Basel), 19*(14), 3160.

Ryoo, M. S. (2011, November). Human activity prediction: Early recognition of ongoing activities from streaming videos. In *2011 International Conference on Computer Vision* (pp. 1036-1043). IEEE.

Saad, E. W., Vian, J. L., Vavrina, M. A., Nisbett, J. A., & Wunsch, D. C. (2014). *Vehicle base station*. Academic Press.

Saari, H., Akujärvi, A., Holmlund, C., Ojanen, H., Kaivosoja, J., Nissinen, A., & Niemeläinen, O. (2017). Visible, very near IR and short wave IR hyperspectral drone imaging system for agriculture and natural water applications. *The International Archives of the Photogrammetry, Remote Sensing and Spatial Information Sciences*, *42*(W3), 165–170. doi:10.5194/isprs-archives-XLII-3-W3-165-2017

Saha, A. K., Saha, J., Ray, R., Sircar, S., Dutta, S., Chattopadhyay, S. P., & Saha, H. N. (2018, January). IOT-based drone for improvement of crop quality in agricultural field. In *2018 IEEE 8th Annual Computing and Communication Workshop and Conference (CCWC)* (pp. 612-615). IEEE. 10.1109/CCWC.2018.8301662

Sahota, H., Kumar, R., & Kamal, A. (2011). A wireless sensor network for precision agriculture and its performance. *Wireless Communications and Mobile Computing*, *11*(12), 1628–1645. doi:10.1002/wcm.1229

Salaan, C. J., Tadakuma, K., Okada, Y., Sakai, Y., Ohno, K., & Tadokoro, S. (2019). Development and experimental validation of aerial vehicle with passive rotating shell on each rotor. *IEEE Robotics and Automation Letters*, *4*(3), 2568–2575. doi:10.1109/LRA.2019.2894903

Samir, K. C., & Lutz, W. (2017). The human core of the shared socioeconomic pathways: Population scenarios by age, sex and level of education for all countries to 2100. *Global Environmental Change*, *42*, 181–192. doi:10.1016/j.gloenvcha.2014.06.004 PMID:28239237

Sarkar, D. (2018). *Get Smarter with Data Science — Tackling Real Enterprise Challenges. Take your Data Science Projects from Zero to Production*. Towards Data Science. https://towardsdatascience.com/get-smarter-with-data-science-tackling-real-enterprise-challenges-67ee001f6097

SAS. (2021). *History of Big Data*. SAS. https://www.sas.com/en_us/insights/big-data/what-is-big-data.html

Sawe, B. E. (2017). *Countries Most Dependent on Agriculture*. Retrieved from https://www.worldatlas.com/articles/countries most-dependent-onagriculture.html

Schenkelberg, F. (2016, January). How reliable does a delivery drone have to be? In *2016 annual reliability and maintainability symposium (RAMS)* (pp. 1-5). IEEE.

Schmelzer, R. (2021). *Data science vs. machine learning vs. AI: How they work together*. TechTarget. https://searchbusinessanalytics.techtarget.com/feature/Data-science-vs-machine-learning-vs-AI-How-they-work-together?track=NL-1816&ad=937375&asrc=EM_NLN_144787081&utm_medium=EM&utm_source=NLN&utm_campaign=20210114_An%20open%20source%20database%20comparison

Semsch, E., Jakob, M., Pavlicek, D., & Pechoucek, M. (2009, September). Autonomous UAV surveillance in complex urban environments. In *2009 IEEE/WIC/ACM International Joint Conference on Web Intelligence and Intelligent Agent Technology* (Vol. 2, pp. 82-85). IEEE. 10.1109/WI-IAT.2009.132

Shafian, S., Rajan, N., Schnell, R., Bagavathiannan, M., Valasek, J., Shi, Y., & Olsenholler, J. (2018). Unmanned aerial systems-based remote sensing for monitoring sorghum growth and development. *PLoS One*, *13*(5), e0196605. doi:10.1371/journal.pone.0196605 PMID:29715311

Shakhatreh, H., Sawalmeh, A. H., Al-Fuqaha, A., Dou, Z., Almaita, E., Khalil, I., Othman, N. S., Khreishah, A., & Guizani, M. (2019). Unmanned aerial vehicles (UAVs): A survey on civil applications and key research challenges. *IEEE Access: Practical Innovations, Open Solutions*, *7*, 48572–48634. doi:10.1109/ACCESS.2019.2909530

Sharma, Jain, Gupta, & Chowdary. (2021). Machine Learning Applications for Precision Agriculture: A Comprehensive Review. Institute of Electrical and Electronical Engineers.

Sharma, A., Vanjani, P., Paliwal, N., Basnayaka, C. M. W., Jayakody, D. N. K., Wang, H. C., & Muthuchidambaranathan, P. (2020). Communication and networking technologies for UAVs: A survey. *Journal of Network and Computer Applications*, *168*, 102739. doi:10.1016/j.jnca.2020.102739

Shekara, Balasubramani, & Shukla. (2016). *Farmer's Handbook on Basic Agriculture*. Agricultural Department and Sciences.

Shi, X., An, X., Zhao, Q., Liu, H., Xia, L., Sun, X., & Guo, Y. (2019). State-of-the-art internet of things in protected agriculture. *Sensors (Basel)*, *19*(8), 1833. doi:10.339019081833 PMID:30999637

Shotton, J., Fitzgibbon, A., Cook, M., Sharp, T., Finocchio, M., Moore, R., & Blake, A. (2011, June). Real-time human pose recognition in parts from single depth images. In *CVPR 2011* (pp. 1297–1304). IEEE.

Sikander, M., & Khiyal, H. (2018). *Computational Models for Upgrading Traditional Agriculture*. Preston University. https://www.researchgate.net/publication/326080886

Simonin, J., Bertin, E., Traon, Y., Jezequel, J.-M., & Crespi, N. (2010). Business and Information System Alignment: A Formal Solution for Telecom Services. In *2010 Fifth International Conference on Software Engineering Advances*. IEEE. 10.1109/ICSEA.2010.49

Singh, A., Patil, D., & Omkar, S. N. (2018). Eye in the sky: Real-time drone surveillance system (dss) for violent individuals identification using scatternet hybrid deep learning network. In *Proceedings of the IEEE conference on computer vision and pattern recognition workshops* (pp. 1629-1637). 10.1109/CVPRW.2018.00214

Siriborvornratanakul, T. (2018). An automatic road distress visual inspection system using an onboard in-car camera. *Advances in Multimedia*, *2018*. doi:10.1155/2018/2561953

Sivakumar, M., & TYJ, N. M. (2021). A literature survey of unmanned aerial vehicle usage for civil applications. *Journal of Aerospace Technology and Management*, •••, 13.

Somanaidu, U., Telagam, N., Nehru, K., & Menakadevi, N. (2018). USRP 2901 based FM transceiver with large file capabilities in virtual and remote laboratory. *iJOE, 14*(10).

Statista Research Department. (2019). *Military drones (UAS/UAV): estimated U.S. and global R&D budget 2014-2023*. Statista Research Department.

Stöcker, C., Bennett, R., Nex, F., Gerke, M., & Zevenbergen, J. (2017). Review of the current state of UAV regulations. *Remote Sensing*, *9*(5), 459. doi:10.3390/rs9050459

Sugiura, R., Noguchi, N., & Ishii, K. (2005). Remote-sensing technology for vegetation monitoring using an unmanned helicopter. *Biosystems Engineering*, *90*(4), 369–379. doi:10.1016/j.biosystemseng.2004.12.011

Suhonen, J., Kohvakka, M., Kaseva, V., Hämäläinen, T. D., & Hännikäinen, M. (2012). *Low-power wireless sensor networks: protocols, services and applications*. Springer Science & Business Media. doi:10.1007/978-1-4614-2173-3

Sundmaeker, H., Verdouw, C., Wolfert, S., Freire, L. P., Vermesan, O., & Friess, P. (2016). Internet of food and farm 2020. In *Digitising the Industry-Internet of Things connecting physical, digital and virtual worlds*. River Publishers.

Sun, X., Ng, D. W. K., Ding, Z., Xu, Y., & Zhong, Z. (2019). Physical layer security in UAV systems: Challenges and opportunities. *IEEE Wireless Communications*, *26*(5), 40–47. doi:10.1109/MWC.001.1900028

Support Vector Machines in R. (n.d.). *DataCamp Community*. https://www.datacamp.com/community/tutorials/support-vector-machines-r

Suriyarachchi, C., Waidyasekara, K. G. A. S., & Madhusanka, N. (2019, June). Integrating Internet of Things (IoT) and facilities manager in smart buildings: A conceptual framework. In *The 7th World Construction Symposium 2018: Built Asset Sustainability: Rethinking Design Construction and Operation* (Vol. 29, pp. 325-334). Academic Press.

Sylvester, G. (Ed.). (2018). E-agriculture in action: drones for agriculture. Food and Agriculture Organization of the United Nations and International Telecommunication Union.

Tahir, A., Böling, J., Haghbayan, M. H., Toivonen, H. T., & Plosila, J. (2019). Swarms of unmanned aerial vehicles—A survey. *Journal of Industrial Information Integration*, *16*, 100106. doi:10.1016/j.jii.2019.100106

Tamr. (2014). *The Evolution of ETL*. Tamr. http://www.tamr.com/evolution-etl/

Telagam, N., Kandasamy, N., & Nanjundan, M. (2017). Smart sensor network based high quality air pollution monitoring system using labview. *International Journal of Online Engineering*, *13*(08), 79–87. doi:10.3991/ijoe.v13i08.7161

Telagam, N., Kandasamy, N., Nanjundan, M., & Arulanandth, T. S. (2017). Smart Sensor Network based Industrial Parameters Monitoring in IOT Environment using Virtual Instrumentation Server. *Int. J. Online Eng.*, *13*(11), 111–119. doi:10.3991/ijoe.v13i11.7630

Telagam, N., Lakshmi, S., & Nehru, K. (2019). Ber analysis of concatenated levels of encoding in GFDM system using labview. *Indonesian Journal of Electrical Engineering and Computer Science*, *14*(1), 80–91. doi:10.11591/ijeecs.v14.i1.pp77-87

Telagam, N., & Manoharan, A. (2020). Multi user based performance analysis in upcoming 5G Techniques. *Journal of Critical Reviews.*, *7*(12). Advance online publication. doi:10.31838/jcr.07.12.152

Telagam, N., Nanjundan, M., Kandasamy, N., & Naidu, S. (2017). Cruise Control of Phase Irrigation Motor Using SparkFun Sensor. *Int. J. Online Eng.*, *13*(8), 192–198. doi:10.3991/ijoe.v13i08.7318

Telegam, N., Lakshmi, S., & Nehru, K. (2019). USRP 2901-based SISO-GFDM transceiver design experiment in virtual and remote laboratory. *International Journal of Electrical Engineering Education*, 0020720919857620.

Tewari, V. K., Kumar, A. A., Kumar, S. P., Pandey, V., & Chandel, N. S. (2013). Estimation of plant nitrogen content using digital image processing. *Agricultural Engineering International: CIGR Journal*, *15*(2), 78–86.

Tezza, D., & Andujar, M. (2019). The state-of-the-art of human–drone interaction: A survey. *IEEE Access: Practical Innovations, Open Solutions*, *7*, 167438–167454. doi:10.1109/ACCESS.2019.2953900

The Open Group. (2011). *Open Group Standard-TOGAF® Guide, Version 9.1*. The Open Group.

The Open Group. (2013b). *ArchiMate®*. Retrieved May 03, 2014, from https://www.opengroup.org/subjectareas/*Entity*/archimate

The Open Group. (2014a). *The Open Group's Architecture Framework-Building blocks-Module 13*. www.open-group.com/TOGAF

Thotakuri, A., Kalyani, T., & Vucha, M., MC, C., & Nagarjuna, T. (2017). Survey on Robot Vision: Techniques, Tools and Methodologies. *International Journal of Applied Engineering Research: IJAER*, *12*(17), 6887–6896.

Tice, B. P. (1991). Unmanned aerial vehicles: The force multiplier of the 1990s. *Airpower Journal*, *5*(1), 41–55.

TOGAF Modelling. (2015a). *Data dissemination view*. TOGAF Modelling. http://www.TOGAF-modelling.org/models/data-architecture-menu/data-dissemination-diagrams-menu.html

Torres, M., Llamas, I., Torres, B., Toral, L., Sampedro, I., & Bejar, V. (2020). Growth promotion on horticultural crops and antifungal activity of Bacillus velezensis XT1. *Applied Soil Ecology*, *150*, 103453. doi:10.1016/j.apsoil.2019.103453

Torres-Ruiz, M., Juárez-Hipólito, J. H., Lytras, M. D., & Moreno-Ibarra, M. (2016, July). Environmental noise sensing approach based on volunteered geographic information and spatio-temporal analysis with machine learning. In *International Conference on Computational Science and Its Applications* (pp. 95-110). Springer. 10.1007/978-3-319-42089-9_7

Torres-Sánchez, J., López-Granados, F., De Castro, A. I., & Peña-Barragán, J. M. (2013). Configuration and specifications of an unmanned aerial vehicle (UAV) for early site specific weed management. *PLoS One*, *8*(3), e58210. doi:10.1371/journal.pone.0058210 PMID:23483997

Trad, A., & Kalpić, D. (2020a). *Using Applied Mathematical Models for Business Transformation. Author Book*. IGI-Global. doi:10.4018/978-1-7998-1009-4

Trujillo, J., & Luj'an-Mora, S. (2003). *A UML Based Approach for Modelling ETL Processes in Data Warehouses*. Dept. de Lenguajes y Sistemas Inform'aticos. Universidad de Alicante.

Tzounis, A., Katsoulas, N., Bartzanas, T., & Kittas, C. (2017). Internet of Things in agriculture, recent advances and future challenges. *Biosystems Engineering*, *164*, 31–48. doi:10.1016/j.biosystemseng.2017.09.007

Udland, M. (2015, October 12). *World Economic Forum: Why labour is becoming more expensive*. https://www.weforum.org/agenda/2015/10/why-labour-is-becoming-more-expensive/

United Nations. (2017). *World Population Prospects: The 2017 Revision*. Department of Economic and Social Affairs. Retrieved from https://www.un.org/development/desa/en/news/population/world-population-prospects-2017.html

United Nations. (2019). *World Urbanization Prospects: The 2018 Revision*. United Nations.

Veenadhari, S., Misra, B., & Singh, C. (2014). Machine learning approach for forecasting crop yield based on climatic parameters. *International Conference on Computer Communication and Informatics*. 10.1109/ICCCI.2014.6921718

Vella, A., Corne, D., & Murphy, C. (2009). *Hyper-heuristic decision tree induction*. Sch. of MACS, Heriot-Watt Univ.

Vellidis, G., Garrick, V., Pocknee, S., Perry, C., Kvien, C., & Tucker, M. (2007, June). How wireless will change agriculture. In *Precision Agriculture '07–Proceedings of the Sixth European Conference on Precision Agriculture (6ECPA), Skiathos, Greece* (pp. 57-67). Academic Press.

Veneberg, R. (2014). *Combining enterprise architecture and operational data-to better support decision-making. University of Twente.*

Vergouw, B., Nagel, H., Bondt, G., & Custers, B. (2016). Drone technology: Types, payloads, applications, frequency spectrum issues and future developments. In *The future of drone use* (pp. 21–45). TMC Asser Press. doi:10.1007/978-94-6265-132-6_2

Villa-Henriksen, A., Edwards, G. T., Pesonen, L. A., Green, O., & Sørensen, C. A. G. (2020). Internet of Things in arable farming: Implementation, applications, challenges and potential. *Biosystems Engineering*, *191*, 60–84. doi:10.1016/j.biosystemseng.2019.12.013

Vrigkas, M., Nikou, C., & Kakadiaris, I. A. (2015). A review of human activity recognition methods. *Frontiers in Robotics and AI*, *2*, 28.

Vroegindeweij, B. A., van Wijk, S. W., & van Henten, E. (2014). *Autonomous unmanned aerial vehicles for agricultural applications*. Academic Press.

Vu, Q., Raković, M., Delic, V., & Ronzhin, A. (2018, September). Trends in development of UAV-UGV cooperation approaches in precision agriculture. In *International Conference on Interactive Collaborative Robotics* (pp. 213-221). Springer. 10.1007/978-3-319-99582-3_22

Walter, A., Finger, R., Huber, R., & Buchmann, N. (2017). Smart farming is key to developing sustainable agriculture. *Proceedings of the National Academy of Sciences of the United States of America*, *114*(24), 6148–6150. doi:10.1073/pnas.1707462114 PMID:28611194

Wang, L., Misra, G., & Bai, X. (2019). AK Nearest neighborhood-based wind estimation for rotary-wing VTOL UAVs. *Drones*, *3*(2), 31.

Wang, Y., Chen, K. S., Mishler, J., Cho, S. C., & Adroher, X. C. (2011). A review of polymer electrolyte membrane fuel cells: Technology, applications, and needs on fundamental research. *Applied Energy*, *88*(4), 981–1007. doi:10.1016/j.apenergy.2010.09.030

Weiger, R. L. (2007). *Military unmanned aircraft systems in support of homeland security*. Army War Coll Carlisle Barracks PA.

Wolfert, S., Ge, L., Verdouw, C., & Bogaardt, M. J. (2017). Big data in smart farming–a review. *Agricultural Systems*, *153*, 69–80. doi:10.1016/j.agsy.2017.01.023

Wong, S. (2019). Decentralised, off-grid solar pump irrigation systems in developing countries—Are they pro-poor, pro-environment and pro-women? In *Climate change-resilient agriculture and agroforestry* (pp. 367–382). Springer. doi:10.1007/978-3-319-75004-0_21

Wuest, T., Weimer, D., Irgens, C., & Thoben, K. D. (2016). Machine learning in manufacturing: Advantages, challenges, and applications. *Production & Manufacturing Research*, *4*(1), 23–45.

Yan, H., Xu, L. D., Bi, Z., Pang, Z., Zhang, J., & Chen, Y. (2015). An emerging technology–wearable wireless sensor networks with applications in human health condition monitoring. *Journal of Management Analytics*, *2*(2), 121–137.

Zanella, A., Bui, N., Castellani, A., Vangelista, L., & Zorzi, M. (2014). Internet of things for smart cities. *IEEE Internet of Things Journal*, *1*(1), 22–32. doi:10.1109/JIOT.2014.2306328

Zavatta, G. (2014). Agriculture Remains Central to the World Economy, 60% of the population Depends on Agriculture for Survival. Expo2015, Milan, Italy.

Zhai, Z., Martínez Ortega, J. F., Lucas Martínez, N., & Rodríguez-Molina, J. (2018). A mission planning approach for precision farming systems based on multi-objective optimization. *Sensors (Basel)*, *18*(6), 1795. doi:10.339018061795 PMID:29865251

Zhang, J., Basso, B., Price, R. F., Putman, G., & Shuai, G. (2018). Estimating plant distance in maize using Unmanned Aerial Vehicle (UAV). *PLoS One*, *13*(4), e0195223. doi:10.1371/journal.pone.0195223 PMID:29677204

Zhang, Y., Wang, L., & Duan, Y. (2016). Agricultural information dissemination using ICTs: A review and analysis of information dissemination models in China. *Information Processing in Agriculture*, *3*(1), 17–29. doi:10.1016/j.inpa.2015.11.002

Zheng, R., Zhang, T., Liu, Z., & Wang, H. (2016). An EIoT system designed for ecological and environmental management of the Xianghe Segment of China's Grand Canal. *International Journal of Sustainable Development and World Ecology*, *23*(4), 372–380. doi:10.1080/13504509.2015.1124470

Related References

To continue our tradition of advancing information science and technology research, we have compiled a list of recommended IGI Global readings. These references will provide additional information and guidance to further enrich your knowledge and assist you with your own research and future publications.

Aasi, P., Rusu, L., & Vieru, D. (2017). The Role of Culture in IT Governance Five Focus Areas: A Literature Review. *International Journal of IT/Business Alignment and Governance, 8*(2), 42-61. https://doi.org/ doi:10.4018/IJITBAG.2017070103

Abdrabo, A. A. (2018). Egypt's Knowledge-Based Development: Opportunities, Challenges, and Future Possibilities. In A. Alraouf (Ed.), *Knowledge-Based Urban Development in the Middle East* (pp. 80–101). Hershey, PA: IGI Global. doi:10.4018/978-1-5225-3734-2.ch005

Abu Doush, I., & Alhami, I. (2018). Evaluating the Accessibility of Computer Laboratories, Libraries, and Websites in Jordanian Universities and Colleges. *International Journal of Information Systems and Social Change*, *9*(2), 44–60. doi:10.4018/IJISSC.2018040104

Adegbore, A. M., Quadri, M. O., & Oyewo, O. R. (2018). A Theoretical Approach to the Adoption of Electronic Resource Management Systems (ERMS) in Nigerian University Libraries. In A. Tella & T. Kwanya (Eds.), *Handbook of Research on Managing Intellectual Property in Digital Libraries* (pp. 292–311). Hershey, PA: IGI Global. doi:10.4018/978-1-5225-3093-0.ch015

Afolabi, O. A. (2018). Myths and Challenges of Building an Effective Digital Library in Developing Nations: An African Perspective. In A. Tella & T. Kwanya (Eds.), *Handbook of Research on Managing Intellectual Property in Digital Libraries* (pp. 51–79). Hershey, PA: IGI Global. doi:10.4018/978-1-5225-3093-0.ch004

Agarwal, P., Kurian, R., & Gupta, R. K. (2022). Additive Manufacturing Feature Taxonomy and Placement of Parts in AM Enclosure. In S. Salunkhe, H. Hussein, & J. Davim (Eds.), *Applications of Artificial Intelligence in Additive Manufacturing* (pp. 138–176). IGI Global. https://doi.org/10.4018/978-1-7998-8516-0.ch007

Al-Alawi, A. I., Al-Hammam, A. H., Al-Alawi, S. S., & AlAlawi, E. I. (2021). The Adoption of E-Wallets: Current Trends and Future Outlook. In Y. Albastaki, A. Razzaque, & A. Sarea (Eds.), *Innovative Strategies for Implementing FinTech in Banking* (pp. 242–262). IGI Global. https://doi.org/10.4018/978-1-7998-3257-7.ch015

Alsharo, M. (2017). Attitudes Towards Cloud Computing Adoption in Emerging Economies. *International Journal of Cloud Applications and Computing*, *7*(3), 44–58. doi:10.4018/IJCAC.2017070102

Amer, T. S., & Johnson, T. L. (2017). Information Technology Progress Indicators: Research Employing Psychological Frameworks. In A. Mesquita (Ed.), *Research Paradigms and Contemporary Perspectives on Human-Technology Interaction* (pp. 168–186). Hershey, PA: IGI Global. doi:10.4018/978-1-5225-1868-6.ch008

Andreeva, A., & Yolova, G. (2021). Liability in Labor Legislation: New Challenges Related to the Use of Artificial Intelligence. In B. Vassileva & M. Zwilling (Eds.), *Responsible AI and Ethical Issues for Businesses and Governments* (pp. 214–232). IGI Global. https://doi.org/10.4018/978-1-7998-4285-9.ch012

Anohah, E. (2017). Paradigm and Architecture of Computing Augmented Learning Management System for Computer Science Education. *International Journal of Online Pedagogy and Course Design*, *7*(2), 60–70. doi:10.4018/IJOPCD.2017040105

Anohah, E., & Suhonen, J. (2017). Trends of Mobile Learning in Computing Education from 2006 to 2014: A Systematic Review of Research Publications. *International Journal of Mobile and Blended Learning*, *9*(1), 16–33. doi:10.4018/IJMBL.2017010102

Arbaiza, C. S., Huerta, H. V., & Rodriguez, C. R. (2021). Contributions to the Technological Adoption Model for the Peruvian Agro-Export Sector. *International Journal of E-Adoption*, *13*(1), 1–17. https://doi.org/10.4018/IJEA.2021010101

Bailey, E. K. (2017). Applying Learning Theories to Computer Technology Supported Instruction. In M. Grassetti & S. Brookby (Eds.), *Advancing Next-Generation Teacher Education through Digital Tools and Applications* (pp. 61–81). Hershey, PA: IGI Global. doi:10.4018/978-1-5225-0965-3.ch004

Baker, J. D. (2021). Introduction to Machine Learning as a New Methodological Framework for Performance Assessment. In M. Bocarnea, B. Winston, & D. Dean (Eds.), *Handbook of Research on Advancements in Organizational Data Collection and Measurements: Strategies for Addressing Attitudes, Beliefs, and Behaviors* (pp. 326–342). IGI Global. https://doi.org/10.4018/978-1-7998-7665-6.ch021

Banerjee, S., Sing, T. Y., Chowdhury, A. R., & Anwar, H. (2018). Let's Go Green: Towards a Taxonomy of Green Computing Enablers for Business Sustainability. In M. Khosrow-Pour (Ed.), *Green Computing Strategies for Competitive Advantage and Business Sustainability* (pp. 89–109). Hershey, PA: IGI Global. doi:10.4018/978-1-5225-5017-4.ch005

Basham, R. (2018). Information Science and Technology in Crisis Response and Management. In M. Khosrow-Pour, D.B.A. (Ed.), Encyclopedia of Information Science and Technology, Fourth Edition (pp. 1407-1418). Hershey, PA: IGI Global. doi:10.4018/978-1-5225-2255-3.ch121

Batyashe, T., & Iyamu, T. (2018). Architectural Framework for the Implementation of Information Technology Governance in Organisations. In M. Khosrow-Pour, D.B.A. (Ed.), Encyclopedia of Information Science and Technology, Fourth Edition (pp. 810-819). Hershey, PA: IGI Global. doi:10.4018/978-1-5225-2255-3.ch070

Bekleyen, N., & Çelik, S. (2017). Attitudes of Adult EFL Learners towards Preparing for a Language Test via CALL. In D. Tafazoli & M. Romero (Eds.), *Multiculturalism and Technology-Enhanced Language Learning* (pp. 214–229). Hershey, PA: IGI Global. doi:10.4018/978-1-5225-1882-2.ch013

Bergeron, F., Croteau, A., Uwizeyemungu, S., & Raymond, L. (2017). A Framework for Research on Information Technology Governance in SMEs. In S. De Haes & W. Van Grembergen (Eds.), *Strategic IT Governance and Alignment in Business Settings* (pp. 53–81). Hershey, PA: IGI Global. doi:10.4018/978-1-5225-0861-8.ch003

Bhardwaj, M., Shukla, N., & Sharma, A. (2021). Improvement and Reduction of Clustering Overhead in Mobile Ad Hoc Network With Optimum Stable Bunching Algorithm. In S. Kumar, M. Trivedi, P. Ranjan, & A. Punhani (Eds.), *Evolution of Software-Defined Networking Foundations for IoT and 5G Mobile Networks* (pp. 139–158). IGI Global. https://doi.org/10.4018/978-1-7998-4685-7.ch008

Bhatt, G. D., Wang, Z., & Rodger, J. A. (2017). Information Systems Capabilities and Their Effects on Competitive Advantages: A Study of Chinese Companies. *Information Resources Management Journal*, *30*(3), 41–57. doi:10.4018/IRMJ.2017070103

Bhattacharya, A. (2021). Blockchain, Cybersecurity, and Industry 4.0. In A. Tyagi, G. Rekha, & N. Sreenath (Eds.), *Opportunities and Challenges for Blockchain Technology in Autonomous Vehicles* (pp. 210–244). IGI Global. https://doi.org/10.4018/978-1-7998-3295-9.ch013

Bhyan, P., Shrivastava, B., & Kumar, N. (2022). Requisite Sustainable Development Contemplating Buildings: Economic and Environmental Sustainability. In A. Hussain, K. Tiwari, & A. Gupta (Eds.), *Addressing Environmental Challenges Through Spatial Planning* (pp. 269–288). IGI Global. https://doi.org/10.4018/978-1-7998-8331-9.ch014

Boido, C., Davico, P., & Spallone, R. (2021). Digital Tools Aimed to Represent Urban Survey. In M. Khosrow-Pour D.B.A. (Ed.), *Encyclopedia of Information Science and Technology, Fifth Edition* (pp. 1181-1195). IGI Global. https://doi.org/10.4018/978-1-7998-3479-3.ch082

Borkar, P. S., Chanana, P. U., Atwal, S. K., Londe, T. G., & Dalal, Y. D. (2021). The Replacement of HMI (Human-Machine Interface) in Industry Using Single Interface Through IoT. In R. Raut & A. Mihovska (Eds.), *Examining the Impact of Deep Learning and IoT on Multi-Industry Applications* (pp. 195–208). IGI Global. https://doi.org/10.4018/978-1-7998-7511-6.ch011

Brahmane, A. V., & Krishna, C. B. (2021). Rider Chaotic Biography Optimization-driven Deep Stacked Auto-encoder for Big Data Classification Using Spark Architecture: Rider Chaotic Biography Optimization. *International Journal of Web Services Research, 18*(3), 42–62. https://doi.org/10.4018/ijwsr.2021070103

Burcoff, A., & Shamir, L. (2017). Computer Analysis of Pablo Picasso's Artistic Style. *International Journal of Art, Culture and Design Technologies, 6*(1), 1–18. doi:10.4018/IJACDT.2017010101

Byker, E. J. (2017). I Play I Learn: Introducing Technological Play Theory. In C. Martin & D. Polly (Eds.), *Handbook of Research on Teacher Education and Professional Development* (pp. 297–306). Hershey, PA: IGI Global. doi:10.4018/978-1-5225-1067-3.ch016

Calongne, C. M., Stricker, A. G., Truman, B., & Arenas, F. J. (2017). Cognitive Apprenticeship and Computer Science Education in Cyberspace: Reimagining the Past. In A. Stricker, C. Calongne, B. Truman, & F. Arenas (Eds.), *Integrating an Awareness of Selfhood and Society into Virtual Learning* (pp. 180–197). Hershey, PA: IGI Global. doi:10.4018/978-1-5225-2182-2.ch013

Carneiro, A. D. (2017). Defending Information Networks in Cyberspace: Some Notes on Security Needs. In M. Dawson, D. Kisku, P. Gupta, J. Sing, & W. Li (Eds.), Developing Next-Generation Countermeasures for Homeland Security Threat Prevention (pp. 354-375). Hershey, PA: IGI Global. https://doi.org/ doi:10.4018/978-1-5225-0703-1.ch016

Carvalho, W. F., & Zarate, L. (2021). Causal Feature Selection. In A. Azevedo & M. Santos (Eds.), *Integration Challenges for Analytics, Business Intelligence, and Data Mining* (pp. 145-160). IGI Global. https://doi.org/10.4018/978-1-7998-5781-5.ch007

Chase, J. P., & Yan, Z. (2017). Affect in Statistics Cognition. In *Assessing and Measuring Statistics Cognition in Higher Education Online Environments: Emerging Research and Opportunities* (pp. 144–187). Hershey, PA: IGI Global. doi:10.4018/978-1-5225-2420-5.ch005

Chatterjee, A., Roy, S., & Shrivastava, R. (2021). A Machine Learning Approach to Prevent Cancer. In G. Rani & P. Tiwari (Eds.), *Handbook of Research on Disease Prediction Through Data Analytics and Machine Learning* (pp. 112–141). IGI Global. https://doi.org/10.4018/978-1-7998-2742-9.ch007

Cifci, M. A. (2021). Optimizing WSNs for CPS Using Machine Learning Techniques. In A. Luhach & A. Elçi (Eds.), *Artificial Intelligence Paradigms for Smart Cyber-Physical Systems* (pp. 204–228). IGI Global. https://doi.org/10.4018/978-1-7998-5101-1.ch010

Cimermanova, I. (2017). Computer-Assisted Learning in Slovakia. In D. Tafazoli & M. Romero (Eds.), *Multiculturalism and Technology-Enhanced Language Learning* (pp. 252–270). Hershey, PA: IGI Global. doi:10.4018/978-1-5225-1882-2.ch015

Cipolla-Ficarra, F. V., & Cipolla-Ficarra, M. (2018). Computer Animation for Ingenious Revival. In F. Cipolla-Ficarra, M. Ficarra, M. Cipolla-Ficarra, A. Quiroga, J. Alma, & J. Carré (Eds.), *Technology-Enhanced Human Interaction in Modern Society* (pp. 159–181). Hershey, PA: IGI Global. doi:10.4018/978-1-5225-3437-2.ch008

Cockrell, S., Damron, T. S., Melton, A. M., & Smith, A. D. (2018). Offshoring IT. In M. Khosrow-Pour, D.B.A. (Ed.), Encyclopedia of Information Science and Technology, Fourth Edition (pp. 5476-5489). Hershey, PA: IGI Global. https://doi.org/ doi:10.4018/978-1-5225-2255-3.ch476

Coffey, J. W. (2018). Logic and Proof in Computer Science: Categories and Limits of Proof Techniques. In J. Horne (Ed.), *Philosophical Perceptions on Logic and Order* (pp. 218–240). Hershey, PA: IGI Global. doi:10.4018/978-1-5225-2443-4.ch007

Dale, M. (2017). Re-Thinking the Challenges of Enterprise Architecture Implementation. In M. Tavana (Ed.), *Enterprise Information Systems and the Digitalization of Business Functions* (pp. 205–221). Hershey, PA: IGI Global. doi:10.4018/978-1-5225-2382-6.ch009

Das, A., & Mohanty, M. N. (2021). An Useful Review on Optical Character Recognition for Smart Era Generation. In A. Tyagi (Ed.), *Multimedia and Sensory Input for Augmented, Mixed, and Virtual Reality* (pp. 1–41). IGI Global. https://doi.org/10.4018/978-1-7998-4703-8.ch001

Dash, A. K., & Mohapatra, P. (2021). A Survey on Prematurity Detection of Diabetic Retinopathy Based on Fundus Images Using Deep Learning Techniques. In S. Saxena & S. Paul (Eds.), *Deep Learning Applications in Medical Imaging* (pp. 140–155). IGI Global. https://doi.org/10.4018/978-1-7998-5071-7.ch006

De Maere, K., De Haes, S., & von Kutzschenbach, M. (2017). CIO Perspectives on Organizational Learning within the Context of IT Governance. *International Journal of IT/Business Alignment and Governance, 8*(1), 32-47. https://doi.org/doi:10.4018/IJITBAG.2017010103

Demir, K., Çaka, C., Yaman, N. D., İslamoğlu, H., & Kuzu, A. (2018). Examining the Current Definitions of Computational Thinking. In H. Ozcinar, G. Wong, & H. Ozturk (Eds.), *Teaching Computational Thinking in Primary Education* (pp. 36–64). Hershey, PA: IGI Global. doi:10.4018/978-1-5225-3200-2.ch003

Deng, X., Hung, Y., & Lin, C. D. (2017). Design and Analysis of Computer Experiments. In S. Saha, A. Mandal, A. Narasimhamurthy, S. V, & S. Sangam (Eds.), Handbook of Research on Applied Cybernetics and Systems Science (pp. 264-279). Hershey, PA: IGI Global. doi:10.4018/978-1-5225-2498-4.ch013

Denner, J., Martinez, J., & Thiry, H. (2017). Strategies for Engaging Hispanic/Latino Youth in the US in Computer Science. In Y. Rankin & J. Thomas (Eds.), *Moving Students of Color from Consumers to Producers of Technology* (pp. 24–48). Hershey, PA: IGI Global. doi:10.4018/978-1-5225-2005-4.ch002

Devi, A. (2017). Cyber Crime and Cyber Security: A Quick Glance. In R. Kumar, P. Pattnaik, & P. Pandey (Eds.), *Detecting and Mitigating Robotic Cyber Security Risks* (pp. 160–171). Hershey, PA: IGI Global. doi:10.4018/978-1-5225-2154-9.ch011

Dhaya, R., & Kanthavel, R. (2022). Futuristic Research Perspectives of IoT Platforms. In D. Jeya Mala (Ed.), *Integrating AI in IoT Analytics on the Cloud for Healthcare Applications* (pp. 258–275). IGI Global. doi:10.4018/978-1-7998-9132-1.ch015

Doyle, D. J., & Fahy, P. J. (2018). Interactivity in Distance Education and Computer-Aided Learning, With Medical Education Examples. In M. Khosrow-Pour, D.B.A. (Ed.), Encyclopedia of Information Science and Technology, Fourth Edition (pp. 5829-5840). Hershey, PA: IGI Global. https://doi.org/ doi:10.4018/978-1-5225-2255-3.ch507

Eklund, P. (2021). Reinforcement Learning in Social Media Marketing. In B. Christiansen & T. Škrinjarić (Eds.), *Handbook of Research on Applied AI for International Business and Marketing Applications* (pp. 30–48). IGI Global. https://doi.org/10.4018/978-1-7998-5077-9.ch003

El Ghandour, N., Benaissa, M., & Lebbah, Y. (2021). An Integer Linear Programming-Based Method for the Extraction of Ontology Alignment. *International Journal of Information Technology and Web Engineering*, *16*(2), 25–44. https://doi.org/10.4018/IJITWE.2021040102

Elias, N. I., & Walker, T. W. (2017). Factors that Contribute to Continued Use of E-Training among Healthcare Professionals. In F. Topor (Ed.), *Handbook of Research on Individualism and Identity in the Globalized Digital Age* (pp. 403–429). Hershey, PA: IGI Global. doi:10.4018/978-1-5225-0522-8.ch018

Fisher, R. L. (2018). Computer-Assisted Indian Matrimonial Services. In M. Khosrow-Pour, D.B.A. (Ed.), Encyclopedia of Information Science and Technology, Fourth Edition (pp. 4136-4145). Hershey, PA: IGI Global. doi:10.4018/978-1-5225-2255-3.ch358

Galiautdinov, R. (2021). Nonlinear Filtering in Artificial Neural Network Applications in Business and Engineering. In Q. Do (Ed.), *Artificial Neural Network Applications in Business and Engineering* (pp. 1–23). IGI Global. https://doi.org/10.4018/978-1-7998-3238-6.ch001

Gardner-McCune, C., & Jimenez, Y. (2017). Historical App Developers: Integrating CS into K-12 through Cross-Disciplinary Projects. In Y. Rankin & J. Thomas (Eds.), *Moving Students of Color from Consumers to Producers of Technology* (pp. 85–112). Hershey, PA: IGI Global. doi:10.4018/978-1-5225-2005-4.ch005

Garg, P. K. (2021). The Internet of Things-Based Technologies. In S. Kumar, M. Trivedi, P. Ranjan, & A. Punhani (Eds.), *Evolution of Software-Defined Networking Foundations for IoT and 5G Mobile Networks* (pp. 37–65). IGI Global. https://doi.org/10.4018/978-1-7998-4685-7.ch003

Garg, T., & Bharti, M. (2021). Congestion Control Protocols for UWSNs. In N. Goyal, L. Sapra, & J. Sandhu (Eds.), *Energy-Efficient Underwater Wireless Communications and Networking* (pp. 85–100). IGI Global. https://doi.org/10.4018/978-1-7998-3640-7.ch006

Gauttier, S. (2021). A Primer on Q-Method and the Study of Technology. In M. Khosrow-Pour D.B.A. (Eds.), *Encyclopedia of Information Science and Technology, Fifth Edition* (pp. 1746-1756). IGI Global. https://doi.org/10.4018/978-1-7998-3479-3.ch120

Ghafele, R., & Gibert, B. (2018). Open Growth: The Economic Impact of Open Source Software in the USA. In M. Khosrow-Pour (Ed.), *Optimizing Contemporary Application and Processes in Open Source Software* (pp. 164–197). Hershey, PA: IGI Global. doi:10.4018/978-1-5225-5314-4.ch007

Ghobakhloo, M., & Azar, A. (2018). Information Technology Resources, the Organizational Capability of Lean-Agile Manufacturing, and Business Performance. *Information Resources Management Journal, 31*(2), 47–74. doi:10.4018/IRMJ.2018040103

Gikandi, J. W. (2017). Computer-Supported Collaborative Learning and Assessment: A Strategy for Developing Online Learning Communities in Continuing Education. In J. Keengwe & G. Onchwari (Eds.), *Handbook of Research on Learner-Centered Pedagogy in Teacher Education and Professional Development* (pp. 309–333). Hershey, PA: IGI Global. doi:10.4018/978-1-5225-0892-2.ch017

Gokhale, A. A., & Machina, K. F. (2017). Development of a Scale to Measure Attitudes toward Information Technology. In L. Tomei (Ed.), *Exploring the New Era of Technology-Infused Education* (pp. 49–64). Hershey, PA: IGI Global. doi:10.4018/978-1-5225-1709-2.ch004

Goswami, J. K., Jalal, S., Negi, C. S., & Jalal, A. S. (2022). A Texture Features-Based Robust Facial Expression Recognition. *International Journal of Computer Vision and Image Processing, 12*(1), 1–15. https://doi.org/10.4018/IJCVIP.2022010103

Hafeez-Baig, A., Gururajan, R., & Wickramasinghe, N. (2017). Readiness as a Novel Construct of Readiness Acceptance Model (RAM) for the Wireless Handheld Technology. In N. Wickramasinghe (Ed.), *Handbook of Research on Healthcare Administration and Management* (pp. 578–595). Hershey, PA: IGI Global. doi:10.4018/978-1-5225-0920-2.ch035

Hanafizadeh, P., Ghandchi, S., & Asgarimehr, M. (2017). Impact of Information Technology on Lifestyle: A Literature Review and Classification. *International Journal of Virtual Communities and Social Networking*, *9*(2), 1–23. doi:10.4018/IJVCSN.2017040101

Haseski, H. İ., Ilic, U., & Tuğtekin, U. (2018). Computational Thinking in Educational Digital Games: An Assessment Tool Proposal. In H. Ozcinar, G. Wong, & H. Ozturk (Eds.), *Teaching Computational Thinking in Primary Education* (pp. 256–287). Hershey, PA: IGI Global. doi:10.4018/978-1-5225-3200-2.ch013

Hee, W. J., Jalleh, G., Lai, H., & Lin, C. (2017). E-Commerce and IT Projects: Evaluation and Management Issues in Australian and Taiwanese Hospitals. *International Journal of Public Health Management and Ethics*, *2*(1), 69–90. doi:10.4018/IJPHME.2017010104

Hernandez, A. A. (2017). Green Information Technology Usage: Awareness and Practices of Philippine IT Professionals. *International Journal of Enterprise Information Systems*, *13*(4), 90–103. doi:10.4018/IJEIS.2017100106

Hernandez, M. A., Marin, E. C., Garcia-Rodriguez, J., Azorin-Lopez, J., & Cazorla, M. (2017). Automatic Learning Improves Human-Robot Interaction in Productive Environments: A Review. *International Journal of Computer Vision and Image Processing*, *7*(3), 65–75. doi:10.4018/IJCVIP.2017070106

Hirota, A. (2021). Design of Narrative Creation in Innovation: "Signature Story" and Two Types of Pivots. In T. Ogata & J. Ono (Eds.), *Bridging the Gap Between AI, Cognitive Science, and Narratology With Narrative Generation* (pp. 363–376). IGI Global. https://doi.org/10.4018/978-1-7998-4864-6.ch012

Hond, D., Asgari, H., Jeffery, D., & Newman, M. (2021). An Integrated Process for Verifying Deep Learning Classifiers Using Dataset Dissimilarity Measures. *International Journal of Artificial Intelligence and Machine Learning*, *11*(2), 1–21. https://doi.org/10.4018/IJAIML.289536

Horne-Popp, L. M., Tessone, E. B., & Welker, J. (2018). If You Build It, They Will Come: Creating a Library Statistics Dashboard for Decision-Making. In L. Costello & M. Powers (Eds.), *Developing In-House Digital Tools in Library Spaces* (pp. 177–203). Hershey, PA: IGI Global. doi:10.4018/978-1-5225-2676-6.ch009

Hu, H., Hu, P. J., & Al-Gahtani, S. S. (2017). User Acceptance of Computer Technology at Work in Arabian Culture: A Model Comparison Approach. In M. Khosrow-Pour (Ed.), *Handbook of Research on Technology Adoption, Social Policy, and Global Integration* (pp. 205–228). Hershey, PA: IGI Global. doi:10.4018/978-1-5225-2668-1.ch011

Huang, C., Sun, Y., & Fuh, C. (2022). Vehicle License Plate Recognition With Deep Learning. In C. Chen, W. Yang, & L. Chen (Eds.), *Technologies to Advance Automation in Forensic Science and Criminal Investigation* (pp. 161-219). IGI Global. https://doi.org/10.4018/978-1-7998-8386-9.ch009

Ifinedo, P. (2017). Using an Extended Theory of Planned Behavior to Study Nurses' Adoption of Healthcare Information Systems in Nova Scotia. *International Journal of Technology Diffusion*, *8*(1), 1–17. doi:10.4018/IJTD.2017010101

Ilie, V., & Sneha, S. (2018). A Three Country Study for Understanding Physicians' Engagement With Electronic Information Resources Pre and Post System Implementation. *Journal of Global Information Management*, *26*(2), 48–73. doi:10.4018/JGIM.2018040103

Ilo, P. I., Nkiko, C., Ugwu, C. I., Ekere, J. N., Izuagbe, R., & Fagbohun, M. O. (2021). Prospects and Challenges of Web 3.0 Technologies Application in the Provision of Library Services. In M. Khosrow-Pour D.B.A. (Ed.), *Encyclopedia of Information Science and Technology, Fifth Edition* (pp. 1767-1781). IGI Global. https://doi.org/10.4018/978-1-7998-3479-3.ch122

Inoue-Smith, Y. (2017). Perceived Ease in Using Technology Predicts Teacher Candidates' Preferences for Online Resources. *International Journal of Online Pedagogy and Course Design*, *7*(3), 17–28. doi:10.4018/IJOPCD.2017070102

Islam, A. Y. (2017). Technology Satisfaction in an Academic Context: Moderating Effect of Gender. In A. Mesquita (Ed.), *Research Paradigms and Contemporary Perspectives on Human-Technology Interaction* (pp. 187–211). Hershey, PA: IGI Global. doi:10.4018/978-1-5225-1868-6.ch009

Jagdale, S. C., Hable, A. A., & Chabukswar, A. R. (2021). Protocol Development in Clinical Trials for Healthcare Management. In M. Khosrow-Pour D.B.A. (Ed.), *Encyclopedia of Information Science and Technology, Fifth Edition* (pp. 1797-1814). IGI Global. https://doi.org/10.4018/978-1-7998-3479-3.ch124

Jamil, G. L., & Jamil, C. C. (2017). Information and Knowledge Management Perspective Contributions for Fashion Studies: Observing Logistics and Supply Chain Management Processes. In G. Jamil, A. Soares, & C. Pessoa (Eds.), *Handbook of Research on Information Management for Effective Logistics and Supply Chains* (pp. 199–221). Hershey, PA: IGI Global. doi:10.4018/978-1-5225-0973-8.ch011

Jamil, M. I., & Almunawar, M. N. (2021). Importance of Digital Literacy and Hindrance Brought About by Digital Divide. In M. Khosrow-Pour D.B.A. (Ed.), *Encyclopedia of Information Science and Technology, Fifth Edition* (pp. 1683-1698). IGI Global. https://doi.org/10.4018/978-1-7998-3479-3.ch116

Janakova, M. (2018). Big Data and Simulations for the Solution of Controversies in Small Businesses. In M. Khosrow-Pour, D.B.A. (Ed.), Encyclopedia of Information Science and Technology, Fourth Edition (pp. 6907-6915). Hershey, PA: IGI Global. doi:10.4018/978-1-5225-2255-3.ch598

Jhawar, A., & Garg, S. K. (2018). Logistics Improvement by Investment in Information Technology Using System Dynamics. In A. Azar & S. Vaidyanathan (Eds.), *Advances in System Dynamics and Control* (pp. 528–567). Hershey, PA: IGI Global. doi:10.4018/978-1-5225-4077-9.ch017

Kalelioğlu, F., Gülbahar, Y., & Doğan, D. (2018). Teaching How to Think Like a Programmer: Emerging Insights. In H. Ozcinar, G. Wong, & H. Ozturk (Eds.), *Teaching Computational Thinking in Primary Education* (pp. 18–35). Hershey, PA: IGI Global. doi:10.4018/978-1-5225-3200-2.ch002

Kamberi, S. (2017). A Girls-Only Online Virtual World Environment and its Implications for Game-Based Learning. In A. Stricker, C. Calongne, B. Truman, & F. Arenas (Eds.), *Integrating an Awareness of Selfhood and Society into Virtual Learning* (pp. 74–95). Hershey, PA: IGI Global. doi:10.4018/978-1-5225-2182-2.ch006

Kamel, S., & Rizk, N. (2017). ICT Strategy Development: From Design to Implementation – Case of Egypt. In C. Howard & K. Hargiss (Eds.), *Strategic Information Systems and Technologies in Modern Organizations* (pp. 239–257). Hershey, PA: IGI Global. doi:10.4018/978-1-5225-1680-4.ch010

Kamel, S. H. (2018). The Potential Role of the Software Industry in Supporting Economic Development. In M. Khosrow-Pour, D.B.A. (Ed.), Encyclopedia of Information Science and Technology, Fourth Edition (pp. 7259-7269). Hershey, PA: IGI Global. doi:10.4018/978-1-5225-2255-3.ch631

Kang, H., Kang, Y., & Kim, J. (2022). Improved Fall Detection Model on GRU Using PoseNet. *International Journal of Software Innovation*, *10*(2), 1–11. https://doi.org/10.4018/IJSI.289600

Kankam, P. K. (2021). Employing Case Study and Survey Designs in Information Research. *Journal of Information Technology Research*, *14*(1), 167–177. https://doi.org/10.4018/JITR.2021010110

Karas, V., & Schuller, B. W. (2021). Deep Learning for Sentiment Analysis: An Overview and Perspectives. In F. Pinarbasi & M. Taskiran (Eds.), *Natural Language Processing for Global and Local Business* (pp. 97–132). IGI Global. https://doi.org/10.4018/978-1-7998-4240-8.ch005

Kaufman, L. M. (2022). Reimagining the Magic of the Workshop Model. In T. Driscoll III, (Ed.), *Designing Effective Distance and Blended Learning Environments in K-12* (pp. 89–109). IGI Global. https://doi.org/10.4018/978-1-7998-6829-3.ch007

Kawata, S. (2018). Computer-Assisted Parallel Program Generation. In M. Khosrow-Pour, D.B.A. (Ed.), Encyclopedia of Information Science and Technology, Fourth Edition (pp. 4583-4593). Hershey, PA: IGI Global. doi:10.4018/978-1-5225-2255-3.ch398

Kharb, L., & Singh, P. (2021). Role of Machine Learning in Modern Education and Teaching. In S. Verma & P. Tomar (Ed.), *Impact of AI Technologies on Teaching, Learning, and Research in Higher Education* (pp. 99-123). IGI Global. https://doi.org/10.4018/978-1-7998-4763-2.ch006

Khari, M., Shrivastava, G., Gupta, S., & Gupta, R. (2017). Role of Cyber Security in Today's Scenario. In R. Kumar, P. Pattnaik, & P. Pandey (Eds.), *Detecting and Mitigating Robotic Cyber Security Risks* (pp. 177–191). Hershey, PA: IGI Global. doi:10.4018/978-1-5225-2154-9.ch013

Khekare, G., & Sheikh, S. (2021). Autonomous Navigation Using Deep Reinforcement Learning in ROS. *International Journal of Artificial Intelligence and Machine Learning*, *11*(2), 63–70. https://doi.org/10.4018/IJAIML.20210701.oa4

Khouja, M., Rodriguez, I. B., Ben Halima, Y., & Moalla, S. (2018). IT Governance in Higher Education Institutions: A Systematic Literature Review. *International Journal of Human Capital and Information Technology Professionals*, *9*(2), 52–67. doi:10.4018/IJHCITP.2018040104

Kiourt, C., Pavlidis, G., Koutsoudis, A., & Kalles, D. (2017). Realistic Simulation of Cultural Heritage. *International Journal of Computational Methods in Heritage Science*, *1*(1), 10–40. doi:10.4018/IJCMHS.2017010102

Köse, U. (2017). An Augmented-Reality-Based Intelligent Mobile Application for Open Computer Education. In G. Kurubacak & H. Altinpulluk (Eds.), *Mobile Technologies and Augmented Reality in Open Education* (pp. 154–174). Hershey, PA: IGI Global. doi:10.4018/978-1-5225-2110-5.ch008

Lahmiri, S. (2018). Information Technology Outsourcing Risk Factors and Provider Selection. In M. Gupta, R. Sharman, J. Walp, & P. Mulgund (Eds.), *Information Technology Risk Management and Compliance in Modern Organizations* (pp. 214–228). Hershey, PA: IGI Global. doi:10.4018/978-1-5225-2604-9.ch008

Lakkad, A. K., Bhadaniya, R. D., Shah, V. N., & Lavanya, K. (2021). Complex Events Processing on Live News Events Using Apache Kafka and Clustering Techniques. *International Journal of Intelligent Information Technologies*, *17*(1), 39–52. https://doi.org/10.4018/IJIIT.2021010103

Landriscina, F. (2017). Computer-Supported Imagination: The Interplay Between Computer and Mental Simulation in Understanding Scientific Concepts. In I. Levin & D. Tsybulsky (Eds.), *Digital Tools and Solutions for Inquiry-Based STEM Learning* (pp. 33–60). Hershey, PA: IGI Global. doi:10.4018/978-1-5225-2525-7.ch002

Lara López, G. (2021). Virtual Reality in Object Location. In A. Negrón & M. Muñoz (Eds.), *Latin American Women and Research Contributions to the IT Field* (pp. 307–324). IGI Global. https://doi.org/10.4018/978-1-7998-7552-9.ch014

Lee, W. W. (2018). Ethical Computing Continues From Problem to Solution. In M. Khosrow-Pour, D.B.A. (Ed.), Encyclopedia of Information Science and Technology, Fourth Edition (pp. 4884-4897). Hershey, PA: IGI Global. doi:10.4018/978-1-5225-2255-3.ch423

Lin, S., Chen, S., & Chuang, S. (2017). Perceived Innovation and Quick Response Codes in an Online-to-Offline E-Commerce Service Model. *International Journal of E-Adoption*, *9*(2), 1–16. doi:10.4018/IJEA.2017070101

Liu, M., Wang, Y., Xu, W., & Liu, L. (2017). Automated Scoring of Chinese Engineering Students' English Essays. *International Journal of Distance Education Technologies*, *15*(1), 52–68. doi:10.4018/IJDET.2017010104

Ma, X., Li, X., Zhong, B., Huang, Y., Gu, Y., Wu, M., Liu, Y., & Zhang, M. (2021). A Detector and Evaluation Framework of Abnormal Bidding Behavior Based on Supplier Portrait. *International Journal of Information Technology and Web Engineering*, *16*(2), 58–74. https://doi.org/10.4018/IJITWE.2021040104

Mabe, L. K., & Oladele, O. I. (2017). Application of Information Communication Technologies for Agricultural Development through Extension Services: A Review. In T. Tossy (Ed.), *Information Technology Integration for Socio-Economic Development* (pp. 52–101). Hershey, PA: IGI Global. doi:10.4018/978-1-5225-0539-6.ch003

Mahboub, S. A., Sayed Ali Ahmed, E., & Saeed, R. A. (2021). Smart IDS and IPS for Cyber-Physical Systems. In A. Luhach & A. Elçi (Eds.), *Artificial Intelligence Paradigms for Smart Cyber-Physical Systems* (pp. 109–136). IGI Global. https://doi.org/10.4018/978-1-7998-5101-1.ch006

Manogaran, G., Thota, C., & Lopez, D. (2018). Human-Computer Interaction With Big Data Analytics. In D. Lopez & M. Durai (Eds.), *HCI Challenges and Privacy Preservation in Big Data Security* (pp. 1–22). Hershey, PA: IGI Global. doi:10.4018/978-1-5225-2863-0.ch001

Margolis, J., Goode, J., & Flapan, J. (2017). A Critical Crossroads for Computer Science for All: "Identifying Talent" or "Building Talent," and What Difference Does It Make? In Y. Rankin & J. Thomas (Eds.), *Moving Students of Color from Consumers to Producers of Technology* (pp. 1–23). Hershey, PA: IGI Global. doi:10.4018/978-1-5225-2005-4.ch001

Mazzù, M. F., Benetton, A., Baccelloni, A., & Lavini, L. (2022). A Milk Blockchain-Enabled Supply Chain: Evidence From Leading Italian Farms. In P. De Giovanni (Ed.), *Blockchain Technology Applications in Businesses and Organizations* (pp. 73–98). IGI Global. https://doi.org/10.4018/978-1-7998-8014-1.ch004

Mbale, J. (2018). Computer Centres Resource Cloud Elasticity-Scalability (CRECES): Copperbelt University Case Study. In S. Aljawarneh & M. Malhotra (Eds.), *Critical Research on Scalability and Security Issues in Virtual Cloud Environments* (pp. 48–70). Hershey, PA: IGI Global. doi:10.4018/978-1-5225-3029-9.ch003

McKee, J. (2018). The Right Information: The Key to Effective Business Planning. In *Business Architectures for Risk Assessment and Strategic Planning: Emerging Research and Opportunities* (pp. 38–52). Hershey, PA: IGI Global. doi:10.4018/978-1-5225-3392-4.ch003

Meddah, I. H., Remil, N. E., & Meddah, H. N. (2021). Novel Approach for Mining Patterns. *International Journal of Applied Evolutionary Computation*, *12*(1), 27–42. https://doi.org/10.4018/IJAEC.2021010103

Mensah, I. K., & Mi, J. (2018). Determinants of Intention to Use Local E-Government Services in Ghana: The Perspective of Local Government Workers. *International Journal of Technology Diffusion*, *9*(2), 41–60. doi:10.4018/IJTD.2018040103

Mohamed, J. H. (2018). Scientograph-Based Visualization of Computer Forensics Research Literature. In J. Jeyasekar & P. Saravanan (Eds.), *Innovations in Measuring and Evaluating Scientific Information* (pp. 148–162). Hershey, PA: IGI Global. doi:10.4018/978-1-5225-3457-0.ch010

Montañés-Del Río, M. Á., Cornejo, V. R., Rodríguez, M. R., & Ortiz, J. S. (2021). Gamification of University Subjects: A Case Study for Operations Management. *Journal of Information Technology Research*, *14*(2), 1–29. https://doi.org/10.4018/JITR.2021040101

Moore, R. L., & Johnson, N. (2017). Earning a Seat at the Table: How IT Departments Can Partner in Organizational Change and Innovation. *International Journal of Knowledge-Based Organizations*, *7*(2), 1–12. doi:10.4018/IJKBO.2017040101

Mukul, M. K., & Bhattaharyya, S. (2017). Brain-Machine Interface: Human-Computer Interaction. In E. Noughabi, B. Raahemi, A. Albadvi, & B. Far (Eds.), *Handbook of Research on Data Science for Effective Healthcare Practice and Administration* (pp. 417–443). Hershey, PA: IGI Global. doi:10.4018/978-1-5225-2515-8.ch018

Na, L. (2017). Library and Information Science Education and Graduate Programs in Academic Libraries. In L. Ruan, Q. Zhu, & Y. Ye (Eds.), *Academic Library Development and Administration in China* (pp. 218–229). Hershey, PA: IGI Global. doi:10.4018/978-1-5225-0550-1.ch013

Nagpal, G., Bishnoi, G. K., Dhami, H. S., & Vijayvargia, A. (2021). Use of Data Analytics to Increase the Efficiency of Last Mile Logistics for Ecommerce Deliveries. In B. Patil & M. Vohra (Eds.), *Handbook of Research on Engineering, Business, and Healthcare Applications of Data Science and Analytics* (pp. 167–180). IGI Global. https://doi.org/10.4018/978-1-7998-3053-5.ch009

Nair, S. M., Ramesh, V., & Tyagi, A. K. (2021). Issues and Challenges (Privacy, Security, and Trust) in Blockchain-Based Applications. In A. Tyagi, G. Rekha, & N. Sreenath (Eds.), *Opportunities and Challenges for Blockchain Technology in Autonomous Vehicles* (pp. 196–209). IGI Global. https://doi.org/10.4018/978-1-7998-3295-9.ch012

Naomi, J. F. M., K., & V., S. (2021). Machine and Deep Learning Techniques in IoT and Cloud. In S. Velayutham (Ed.), *Challenges and Opportunities for the Convergence of IoT, Big Data, and Cloud Computing* (pp. 225-247). IGI Global. https://doi.org/10.4018/978-1-7998-3111-2.ch013

Nath, R., & Murthy, V. N. (2018). What Accounts for the Differences in Internet Diffusion Rates Around the World? In M. Khosrow-Pour, D.B.A. (Ed.), Encyclopedia of Information Science and Technology, Fourth Edition (pp. 8095-8104). Hershey, PA: IGI Global. https://doi.org/ doi:10.4018/978-1-5225-2255-3.ch705

Nedelko, Z., & Potocan, V. (2018). The Role of Emerging Information Technologies for Supporting Supply Chain Management. In M. Khosrow-Pour, D.B.A. (Ed.), Encyclopedia of Information Science and Technology, Fourth Edition (pp. 5559-5569). Hershey, PA: IGI Global. doi:10.4018/978-1-5225-2255-3.ch483

Negrini, L., Giang, C., & Bonnet, E. (2022). Designing Tools and Activities for Educational Robotics in Online Learning. In N. Eteokleous & E. Nisiforou (Eds.), *Designing, Constructing, and Programming Robots for Learning* (pp. 202–222). IGI Global. https://doi.org/10.4018/978-1-7998-7443-0.ch010

Ngafeeson, M. N. (2018). User Resistance to Health Information Technology. In M. Khosrow-Pour, D.B.A. (Ed.), Encyclopedia of Information Science and Technology, Fourth Edition (pp. 3816-3825). Hershey, PA: IGI Global. doi:10.4018/978-1-5225-2255-3.ch331

Nguyen, T. T., Giang, N. L., Tran, D. T., Nguyen, T. T., Nguyen, H. Q., Pham, A. V., & Vu, T. D. (2021). A Novel Filter-Wrapper Algorithm on Intuitionistic Fuzzy Set for Attribute Reduction From Decision Tables. *International Journal of Data Warehousing and Mining*, *17*(4), 67–100. https://doi.org/10.4018/IJDWM.2021100104

Nigam, A., & Dewani, P. P. (2022). Consumer Engagement Through Conditional Promotions: An Exploratory Study. *Journal of Global Information Management*, *30*(5), 1–19. https://doi.org/10.4018/JGIM.290364

Odagiri, K. (2017). Introduction of Individual Technology to Constitute the Current Internet. In *Strategic Policy-Based Network Management in Contemporary Organizations* (pp. 20–96). Hershey, PA: IGI Global. doi:10.4018/978-1-68318-003-6.ch003

Odia, J. O., & Akpata, O. T. (2021). Role of Data Science and Data Analytics in Forensic Accounting and Fraud Detection. In B. Patil & M. Vohra (Eds.), *Handbook of Research on Engineering, Business, and Healthcare Applications of Data Science and Analytics* (pp. 203–227). IGI Global. https://doi.org/10.4018/978-1-7998-3053-5.ch011

Okike, E. U. (2018). Computer Science and Prison Education. In I. Biao (Ed.), *Strategic Learning Ideologies in Prison Education Programs* (pp. 246–264). Hershey, PA: IGI Global. doi:10.4018/978-1-5225-2909-5.ch012

Olelewe, C. J., & Nwafor, I. P. (2017). Level of Computer Appreciation Skills Acquired for Sustainable Development by Secondary School Students in Nsukka LGA of Enugu State, Nigeria. In C. Ayo & V. Mbarika (Eds.), *Sustainable ICT Adoption and Integration for Socio-Economic Development* (pp. 214–233). Hershey, PA: IGI Global. doi:10.4018/978-1-5225-2565-3.ch010

Oliveira, M., Maçada, A. C., Curado, C., & Nodari, F. (2017). Infrastructure Profiles and Knowledge Sharing. *International Journal of Technology and Human Interaction*, *13*(3), 1–12. doi:10.4018/IJTHI.2017070101

Otarkhani, A., Shokouhyar, S., & Pour, S. S. (2017). Analyzing the Impact of Governance of Enterprise IT on Hospital Performance: Tehran's (Iran) Hospitals – A Case Study. *International Journal of Healthcare Information Systems and Informatics*, *12*(3), 1–20. doi:10.4018/IJHISI.2017070101

Otunla, A. O., & Amuda, C. O. (2018). Nigerian Undergraduate Students' Computer Competencies and Use of Information Technology Tools and Resources for Study Skills and Habits' Enhancement. In M. Khosrow-Pour, D.B.A. (Ed.), Encyclopedia of Information Science and Technology, Fourth Edition (pp. 2303-2313). Hershey, PA: IGI Global. https://doi.org/ doi:10.4018/978-1-5225-2255-3.ch200

Özçınar, H. (2018). A Brief Discussion on Incentives and Barriers to Computational Thinking Education. In H. Ozcinar, G. Wong, & H. Ozturk (Eds.), *Teaching Computational Thinking in Primary Education* (pp. 1–17). Hershey, PA: IGI Global. doi:10.4018/978-1-5225-3200-2.ch001

Pandey, J. M., Garg, S., Mishra, P., & Mishra, B. P. (2017). Computer Based Psychological Interventions: Subject to the Efficacy of Psychological Services. *International Journal of Computers in Clinical Practice*, *2*(1), 25–33. doi:10.4018/IJCCP.2017010102

Pandkar, S. D., & Paatil, S. D. (2021). Big Data and Knowledge Resource Centre. In S. Dhamdhere (Ed.), *Big Data Applications for Improving Library Services* (pp. 90–106). IGI Global. https://doi.org/10.4018/978-1-7998-3049-8.ch007

Patro, C. (2017). Impulsion of Information Technology on Human Resource Practices. In P. Ordóñez de Pablos (Ed.), *Managerial Strategies and Solutions for Business Success in Asia* (pp. 231–254). Hershey, PA: IGI Global. doi:10.4018/978-1-5225-1886-0.ch013

Patro, C. S., & Raghunath, K. M. (2017). Information Technology Paraphernalia for Supply Chain Management Decisions. In M. Tavana (Ed.), *Enterprise Information Systems and the Digitalization of Business Functions* (pp. 294–320). Hershey, PA: IGI Global. doi:10.4018/978-1-5225-2382-6.ch014

Paul, P. K. (2018). The Context of IST for Solid Information Retrieval and Infrastructure Building: Study of Developing Country. *International Journal of Information Retrieval Research*, *8*(1), 86–100. doi:10.4018/IJIRR.2018010106

Paul, P. K., & Chatterjee, D. (2018). iSchools Promoting "Information Science and Technology" (IST) Domain Towards Community, Business, and Society With Contemporary Worldwide Trend and Emerging Potentialities in India. In M. Khosrow-Pour, D.B.A. (Ed.), Encyclopedia of Information Science and Technology, Fourth Edition (pp. 4723-4735). Hershey, PA: IGI Global. https://doi.org/ doi:10.4018/978-1-5225-2255-3.ch410

Pessoa, C. R., & Marques, M. E. (2017). Information Technology and Communication Management in Supply Chain Management. In G. Jamil, A. Soares, & C. Pessoa (Eds.), *Handbook of Research on Information Management for Effective Logistics and Supply Chains* (pp. 23–33). Hershey, PA: IGI Global. doi:10.4018/978-1-5225-0973-8.ch002

Pineda, R. G. (2018). Remediating Interaction: Towards a Philosophy of Human-Computer Relationship. In M. Khosrow-Pour (Ed.), *Enhancing Art, Culture, and Design With Technological Integration* (pp. 75–98). Hershey, PA: IGI Global. doi:10.4018/978-1-5225-5023-5.ch004

Prabha, V. D., & R., R. (2021). Clinical Decision Support Systems: Decision-Making System for Clinical Data. In G. Rani & P. Tiwari (Eds.), *Handbook of Research on Disease Prediction Through Data Analytics and Machine Learning* (pp. 268-280). IGI Global. https://doi.org/10.4018/978-1-7998-2742-9.ch014

Pushpa, R., & Siddappa, M. (2021). An Optimal Way of VM Placement Strategy in Cloud Computing Platform Using ABCS Algorithm. *International Journal of Ambient Computing and Intelligence, 12*(3), 16–38. https://doi.org/10.4018/IJACI.2021070102

Qian, Y. (2017). Computer Simulation in Higher Education: Affordances, Opportunities, and Outcomes. In P. Vu, S. Fredrickson, & C. Moore (Eds.), *Handbook of Research on Innovative Pedagogies and Technologies for Online Learning in Higher Education* (pp. 236–262). Hershey, PA: IGI Global. doi:10.4018/978-1-5225-1851-8.ch011

Rahman, N. (2017). Lessons from a Successful Data Warehousing Project Management. *International Journal of Information Technology Project Management, 8*(4), 30–45. doi:10.4018/IJITPM.2017100103

Rahman, N. (2018). Environmental Sustainability in the Computer Industry for Competitive Advantage. In M. Khosrow-Pour (Ed.), *Green Computing Strategies for Competitive Advantage and Business Sustainability* (pp. 110–130). Hershey, PA: IGI Global. doi:10.4018/978-1-5225-5017-4.ch006

Rajh, A., & Pavetic, T. (2017). Computer Generated Description as the Required Digital Competence in Archival Profession. *International Journal of Digital Literacy and Digital Competence*, *8*(1), 36–49. doi:10.4018/IJDLDC.2017010103

Raman, A., & Goyal, D. P. (2017). Extending IMPLEMENT Framework for Enterprise Information Systems Implementation to Information System Innovation. In M. Tavana (Ed.), *Enterprise Information Systems and the Digitalization of Business Functions* (pp. 137–177). Hershey, PA: IGI Global. doi:10.4018/978-1-5225-2382-6.ch007

Rao, A. P., & Reddy, K. S. (2021). Automated Soil Residue Levels Detecting Device With IoT Interface. In V. Sathiyamoorthi & A. Elci (Eds.), *Challenges and Applications of Data Analytics in Social Perspectives* (Vol. S, pp. 123–135). IGI Global. https://doi.org/10.4018/978-1-7998-2566-1.ch007

Rao, Y. S., Rauta, A. K., Saini, H., & Panda, T. C. (2017). Mathematical Model for Cyber Attack in Computer Network. *International Journal of Business Data Communications and Networking*, *13*(1), 58–65. doi:10.4018/IJBDCN.2017010105

Rapaport, W. J. (2018). Syntactic Semantics and the Proper Treatment of Computationalism. In M. Danesi (Ed.), *Empirical Research on Semiotics and Visual Rhetoric* (pp. 128–176). Hershey, PA: IGI Global. doi:10.4018/978-1-5225-5622-0.ch007

Raut, R., Priyadarshinee, P., & Jha, M. (2017). Understanding the Mediation Effect of Cloud Computing Adoption in Indian Organization: Integrating TAM-TOE- Risk Model. *International Journal of Service Science, Management, Engineering, and Technology*, *8*(3), 40–59. doi:10.4018/IJSSMET.2017070103

Rezaie, S., Mirabedini, S. J., & Abtahi, A. (2018). Designing a Model for Implementation of Business Intelligence in the Banking Industry. *International Journal of Enterprise Information Systems*, *14*(1), 77–103. doi:10.4018/IJEIS.2018010105

Rezende, D. A. (2018). Strategic Digital City Projects: Innovative Information and Public Services Offered by Chicago (USA) and Curitiba (Brazil). In M. Lytras, L. Daniela, & A. Visvizi (Eds.), *Enhancing Knowledge Discovery and Innovation in the Digital Era* (pp. 204–223). Hershey, PA: IGI Global. doi:10.4018/978-1-5225-4191-2.ch012

Rodriguez, A., Rico-Diaz, A. J., Rabuñal, J. R., & Gestal, M. (2017). Fish Tracking with Computer Vision Techniques: An Application to Vertical Slot Fishways. In M. S., & V. V. (Eds.), Multi-Core Computer Vision and Image Processing for Intelligent Applications (pp. 74-104). Hershey, PA: IGI Global. https://doi.org/doi:10.4018/978-1-5225-0889-2.ch003

Romero, J. A. (2018). Sustainable Advantages of Business Value of Information Technology. In M. Khosrow-Pour, D.B.A. (Ed.), Encyclopedia of Information Science and Technology, Fourth Edition (pp. 923-929). Hershey, PA: IGI Global. doi:10.4018/978-1-5225-2255-3.ch079

Romero, J. A. (2018). The Always-On Business Model and Competitive Advantage. In N. Bajgoric (Ed.), *Always-On Enterprise Information Systems for Modern Organizations* (pp. 23–40). Hershey, PA: IGI Global. doi:10.4018/978-1-5225-3704-5.ch002

Rosen, Y. (2018). Computer Agent Technologies in Collaborative Learning and Assessment. In M. Khosrow-Pour, D.B.A. (Ed.), Encyclopedia of Information Science and Technology, Fourth Edition (pp. 2402-2410). Hershey, PA: IGI Global. doi:10.4018/978-1-5225-2255-3.ch209

Roy, D. (2018). Success Factors of Adoption of Mobile Applications in Rural India: Effect of Service Characteristics on Conceptual Model. In M. Khosrow-Pour (Ed.), *Green Computing Strategies for Competitive Advantage and Business Sustainability* (pp. 211–238). Hershey, PA: IGI Global. doi:10.4018/978-1-5225-5017-4.ch010

Ruffin, T. R., & Hawkins, D. P. (2018). Trends in Health Care Information Technology and Informatics. In M. Khosrow-Pour, D.B.A. (Ed.), Encyclopedia of Information Science and Technology, Fourth Edition (pp. 3805-3815). Hershey, PA: IGI Global. doi:10.4018/978-1-5225-2255-3.ch330

Sadasivam, U. M., & Ganesan, N. (2021). Detecting Fake News Using Deep Learning and NLP. In S. Misra, C. Arumugam, S. Jaganathan, & S. S. (Eds.), *Confluence of AI, Machine, and Deep Learning in Cyber Forensics* (pp. 117-133). IGI Global. https://doi.org/10.4018/978-1-7998-4900-1.ch007

Safari, M. R., & Jiang, Q. (2018). The Theory and Practice of IT Governance Maturity and Strategies Alignment: Evidence From Banking Industry. *Journal of Global Information Management, 26*(2), 127–146. doi:10.4018/JGIM.2018040106

Sahin, H. B., & Anagun, S. S. (2018). Educational Computer Games in Math Teaching: A Learning Culture. In E. Toprak & E. Kumtepe (Eds.), *Supporting Multiculturalism in Open and Distance Learning Spaces* (pp. 249–280). Hershey, PA: IGI Global. doi:10.4018/978-1-5225-3076-3.ch013

Sakalle, A., Tomar, P., Bhardwaj, H., & Sharma, U. (2021). Impact and Latest Trends of Intelligent Learning With Artificial Intelligence. In S. Verma & P. Tomar (Eds.), *Impact of AI Technologies on Teaching, Learning, and Research in Higher Education* (pp. 172-189). IGI Global. https://doi.org/10.4018/978-1-7998-4763-2.ch011

Sala, N. (2021). Virtual Reality, Augmented Reality, and Mixed Reality in Education: A Brief Overview. In D. Choi, A. Dailey-Hebert, & J. Estes (Eds.), *Current and Prospective Applications of Virtual Reality in Higher Education* (pp. 48–73). IGI Global. https://doi.org/10.4018/978-1-7998-4960-5.ch003

Salunkhe, S., Kanagachidambaresan, G., Rajkumar, C., & Jayanthi, K. (2022). Online Detection and Prediction of Fused Deposition Modelled Parts Using Artificial Intelligence. In S. Salunkhe, H. Hussein, & J. Davim (Eds.), *Applications of Artificial Intelligence in Additive Manufacturing* (pp. 194–209). IGI Global. https://doi.org/10.4018/978-1-7998-8516-0.ch009

Samy, V. S., Pramanick, K., Thenkanidiyoor, V., & Victor, J. (2021). Data Analysis and Visualization in Python for Polar Meteorological Data. *International Journal of Data Analytics*, *2*(1), 32–60. https://doi.org/10.4018/IJDA.2021010102

Sanna, A., & Valpreda, F. (2017). An Assessment of the Impact of a Collaborative Didactic Approach and Students' Background in Teaching Computer Animation. *International Journal of Information and Communication Technology Education*, *13*(4), 1–16. doi:10.4018/IJICTE.2017100101

Sarivougioukas, J., & Vagelatos, A. (2022). Fused Contextual Data With Threading Technology to Accelerate Processing in Home UbiHealth. *International Journal of Software Science and Computational Intelligence*, *14*(1), 1–14. https://doi.org/10.4018/IJSSCI.285590

Scott, A., Martin, A., & McAlear, F. (2017). Enhancing Participation in Computer Science among Girls of Color: An Examination of a Preparatory AP Computer Science Intervention. In Y. Rankin & J. Thomas (Eds.), *Moving Students of Color from Consumers to Producers of Technology* (pp. 62–84). Hershey, PA: IGI Global. doi:10.4018/978-1-5225-2005-4.ch004

Shanmugam, M., Ibrahim, N., Gorment, N. Z., Sugu, R., Dandarawi, T. N., & Ahmad, N. A. (2022). Towards an Integrated Omni-Channel Strategy Framework for Improved Customer Interaction. In P. Lai (Ed.), *Handbook of Research on Social Impacts of E-Payment and Blockchain Technology* (pp. 409–427). IGI Global. https://doi.org/10.4018/978-1-7998-9035-5.ch022

Sharma, A., & Kumar, S. (2021). Network Slicing and the Role of 5G in IoT Applications. In S. Kumar, M. Trivedi, P. Ranjan, & A. Punhani (Eds.), *Evolution of Software-Defined Networking Foundations for IoT and 5G Mobile Networks* (pp. 172–190). IGI Global. https://doi.org/10.4018/978-1-7998-4685-7.ch010

Siddoo, V., & Wongsai, N. (2017). Factors Influencing the Adoption of ISO/IEC 29110 in Thai Government Projects: A Case Study. *International Journal of Information Technologies and Systems Approach, 10*(1), 22–44. doi:10.4018/IJITSA.2017010102

Silveira, C., Hir, M. E., & Chaves, H. K. (2022). An Approach to Information Management as a Subsidy of Global Health Actions: A Case Study of Big Data in Health for Dengue, Zika, and Chikungunya. In J. Lima de Magalhães, Z. Hartz, G. Jamil, H. Silveira, & L. Jamil (Eds.), *Handbook of Research on Essential Information Approaches to Aiding Global Health in the One Health Context* (pp. 219–234). IGI Global. https://doi.org/10.4018/978-1-7998-8011-0.ch012

Simões, A. (2017). Using Game Frameworks to Teach Computer Programming. In R. Alexandre Peixoto de Queirós & M. Pinto (Eds.), *Gamification-Based E-Learning Strategies for Computer Programming Education* (pp. 221–236). Hershey, PA: IGI Global. doi:10.4018/978-1-5225-1034-5.ch010

Simões de Almeida, R., & da Silva, T. (2022). AI Chatbots in Mental Health: Are We There Yet? In A. Marques & R. Queirós (Eds.), *Digital Therapies in Psychosocial Rehabilitation and Mental Health* (pp. 226–243). IGI Global. https://doi.org/10.4018/978-1-7998-8634-1.ch011

Singh, L. K., Khanna, M., Thawkar, S., & Gopal, J. (2021). Robustness for Authentication of the Human Using Face, Ear, and Gait Multimodal Biometric System. *International Journal of Information System Modeling and Design, 12*(1), 39–72. https://doi.org/10.4018/IJISMD.2021010103

Sllame, A. M. (2017). Integrating LAB Work With Classes in Computer Network Courses. In H. Alphin Jr, R. Chan, & J. Lavine (Eds.), *The Future of Accessibility in International Higher Education* (pp. 253–275). Hershey, PA: IGI Global. doi:10.4018/978-1-5225-2560-8.ch015

Smirnov, A., Ponomarev, A., Shilov, N., Kashevnik, A., & Teslya, N. (2018). Ontology-Based Human-Computer Cloud for Decision Support: Architecture and Applications in Tourism. *International Journal of Embedded and Real-Time Communication Systems, 9*(1), 1–19. doi:10.4018/IJERTCS.2018010101

Smith-Ditizio, A. A., & Smith, A. D. (2018). Computer Fraud Challenges and Its Legal Implications. In M. Khosrow-Pour, D.B.A. (Ed.), Encyclopedia of Information Science and Technology, Fourth Edition (pp. 4837-4848). Hershey, PA: IGI Global. doi:10.4018/978-1-5225-2255-3.ch419

Sosnin, P. (2018). Figuratively Semantic Support of Human-Computer Interactions. In *Experience-Based Human-Computer Interactions: Emerging Research and Opportunities* (pp. 244–272). Hershey, PA: IGI Global. doi:10.4018/978-1-5225-2987-3.ch008

Srilakshmi, R., & Jaya Bhaskar, M. (2021). An Adaptable Secure Scheme in Mobile Ad hoc Network to Protect the Communication Channel From Malicious Behaviours. *International Journal of Information Technology and Web Engineering, 16*(3), 54–73. https://doi.org/10.4018/IJITWE.2021070104

Sukhwani, N., Kagita, V. R., Kumar, V., & Panda, S. K. (2021). Efficient Computation of Top-K Skyline Objects in Data Set With Uncertain Preferences. *International Journal of Data Warehousing and Mining, 17*(3), 68–80. https://doi.org/10.4018/IJDWM.2021070104

Susanto, H., Yie, L. F., Setiana, D., Asih, Y., Yoganingrum, A., Riyanto, S., & Saputra, F. A. (2021). Digital Ecosystem Security Issues for Organizations and Governments: Digital Ethics and Privacy. In Z. Mahmood (Ed.), *Web 2.0 and Cloud Technologies for Implementing Connected Government* (pp. 204–228). IGI Global. https://doi.org/10.4018/978-1-7998-4570-6.ch010

Syväjärvi, A., Leinonen, J., Kivivirta, V., & Kesti, M. (2017). The Latitude of Information Management in Local Government: Views of Local Government Managers. *International Journal of Electronic Government Research, 13*(1), 69–85. doi:10.4018/IJEGR.2017010105

Tanque, M., & Foxwell, H. J. (2018). Big Data and Cloud Computing: A Review of Supply Chain Capabilities and Challenges. In A. Prasad (Ed.), *Exploring the Convergence of Big Data and the Internet of Things* (pp. 1–28). Hershey, PA: IGI Global. doi:10.4018/978-1-5225-2947-7.ch001

Teixeira, A., Gomes, A., & Orvalho, J. G. (2017). Auditory Feedback in a Computer Game for Blind People. In T. Issa, P. Kommers, T. Issa, P. Isaías, & T. Issa (Eds.), *Smart Technology Applications in Business Environments* (pp. 134–158). Hershey, PA: IGI Global. doi:10.4018/978-1-5225-2492-2.ch007

Tewari, P., Tiwari, P., & Goel, R. (2022). Information Technology in Supply Chain Management. In V. Garg & R. Goel (Eds.), *Handbook of Research on Innovative Management Using AI in Industry 5.0* (pp. 165–178). IGI Global. https://doi.org/10.4018/978-1-7998-8497-2.ch011

Thompson, N., McGill, T., & Murray, D. (2018). Affect-Sensitive Computer Systems. In M. Khosrow-Pour, D.B.A. (Ed.), Encyclopedia of Information Science and Technology, Fourth Edition (pp. 4124-4135). Hershey, PA: IGI Global. doi:10.4018/978-1-5225-2255-3.ch357

Triberti, S., Brivio, E., & Galimberti, C. (2018). On Social Presence: Theories, Methodologies, and Guidelines for the Innovative Contexts of Computer-Mediated Learning. In M. Marmon (Ed.), *Enhancing Social Presence in Online Learning Environments* (pp. 20–41). Hershey, PA: IGI Global. doi:10.4018/978-1-5225-3229-3.ch002

Tripathy, B. K. T. R., S., & Mohanty, R. K. (2018). Memetic Algorithms and Their Applications in Computer Science. In S. Dash, B. Tripathy, & A. Rahman (Eds.), Handbook of Research on Modeling, Analysis, and Application of Nature-Inspired Metaheuristic Algorithms (pp. 73-93). Hershey, PA: IGI Global. https://doi.org/doi:10.4018/978-1-5225-2857-9.ch004

Turulja, L., & Bajgoric, N. (2017). Human Resource Management IT and Global Economy Perspective: Global Human Resource Information Systems. In M. Khosrow-Pour (Ed.), *Handbook of Research on Technology Adoption, Social Policy, and Global Integration* (pp. 377–394). Hershey, PA: IGI Global. doi:10.4018/978-1-5225-2668-1.ch018

Unwin, D. W., Sanzogni, L., & Sandhu, K. (2017). Developing and Measuring the Business Case for Health Information Technology. In K. Moahi, K. Bwalya, & P. Sebina (Eds.), *Health Information Systems and the Advancement of Medical Practice in Developing Countries* (pp. 262–290). Hershey, PA: IGI Global. doi:10.4018/978-1-5225-2262-1.ch015

Usharani, B. (2022). House Plant Leaf Disease Detection and Classification Using Machine Learning. In M. Mundada, S. Seema, S. K.G., & M. Shilpa (Eds.), *Deep Learning Applications for Cyber-Physical Systems* (pp. 17-26). IGI Global. https://doi.org/10.4018/978-1-7998-8161-2.ch002

Vadhanam, B. R. S., M., Sugumaran, V., V., V., & Ramalingam, V. V. (2017). Computer Vision Based Classification on Commercial Videos. In M. S., & V. V. (Eds.), Multi-Core Computer Vision and Image Processing for Intelligent Applications (pp. 105-135). Hershey, PA: IGI Global. https://doi.org/ doi:10.4018/978-1-5225-0889-2.ch004

Vairinho, S. (2022). Innovation Dynamics Through the Encouragement of Knowledge Spin-Off From Touristic Destinations. In C. Ramos, S. Quinteiro, & A. Gonçalves (Eds.), *ICT as Innovator Between Tourism and Culture* (pp. 170–190). IGI Global. https://doi.org/10.4018/978-1-7998-8165-0.ch011

Valverde, R., Torres, B., & Motaghi, H. (2018). A Quantum NeuroIS Data Analytics Architecture for the Usability Evaluation of Learning Management Systems. In S. Bhattacharyya (Ed.), *Quantum-Inspired Intelligent Systems for Multimedia Data Analysis* (pp. 277–299). Hershey, PA: IGI Global. doi:10.4018/978-1-5225-5219-2.ch009

Vassilis, E. (2018). Learning and Teaching Methodology: "1:1 Educational Computing. In K. Koutsopoulos, K. Doukas, & Y. Kotsanis (Eds.), *Handbook of Research on Educational Design and Cloud Computing in Modern Classroom Settings* (pp. 122–155). Hershey, PA: IGI Global. doi:10.4018/978-1-5225-3053-4.ch007

Verma, S., & Jain, A. K. (2022). A Survey on Sentiment Analysis Techniques for Twitter. In B. Gupta, D. Peraković, A. Abd El-Latif, & D. Gupta (Eds.), *Data Mining Approaches for Big Data and Sentiment Analysis in Social Media* (pp. 57–90). IGI Global. https://doi.org/10.4018/978-1-7998-8413-2.ch003

Wang, H., Huang, P., & Chen, X. (2021). Research and Application of a Multidimensional Association Rules Mining Method Based on OLAP. *International Journal of Information Technology and Web Engineering, 16*(1), 75–94. https://doi.org/10.4018/IJITWE.2021010104

Wexler, B. E. (2017). Computer-Presented and Physical Brain-Training Exercises for School Children: Improving Executive Functions and Learning. In B. Dubbels (Ed.), *Transforming Gaming and Computer Simulation Technologies across Industries* (pp. 206–224). Hershey, PA: IGI Global. doi:10.4018/978-1-5225-1817-4.ch012

Wimble, M., Singh, H., & Phillips, B. (2018). Understanding Cross-Level Interactions of Firm-Level Information Technology and Industry Environment: A Multilevel Model of Business Value. *Information Resources Management Journal, 31*(1), 1–20. doi:10.4018/IRMJ.2018010101

Wimmer, H., Powell, L., Kilgus, L., & Force, C. (2017). Improving Course Assessment via Web-based Homework. *International Journal of Online Pedagogy and Course Design, 7*(2), 1–19. doi:10.4018/IJOPCD.2017040101

Wong, S. (2021). Gendering Information and Communication Technologies in Climate Change. In M. Khosrow-Pour D.B.A. (Eds.), *Encyclopedia of Information Science and Technology, Fifth Edition* (pp. 1408-1422). IGI Global. https://doi.org/10.4018/978-1-7998-3479-3.ch096

Wong, Y. L., & Siu, K. W. (2018). Assessing Computer-Aided Design Skills. In M. Khosrow-Pour, D.B.A. (Ed.), Encyclopedia of Information Science and Technology, Fourth Edition (pp. 7382-7391). Hershey, PA: IGI Global. doi:10.4018/978-1-5225-2255-3.ch642

Wongsurawat, W., & Shrestha, V. (2018). Information Technology, Globalization, and Local Conditions: Implications for Entrepreneurs in Southeast Asia. In P. Ordóñez de Pablos (Ed.), *Management Strategies and Technology Fluidity in the Asian Business Sector* (pp. 163–176). Hershey, PA: IGI Global. doi:10.4018/978-1-5225-4056-4.ch010

Yamada, H. (2021). Homogenization of Japanese Industrial Technology From the Perspective of R&D Expenses. *International Journal of Systems and Service-Oriented Engineering*, *11*(2), 24–51. doi:10.4018/IJSSOE.2021070102

Yang, Y., Zhu, X., Jin, C., & Li, J. J. (2018). Reforming Classroom Education Through a QQ Group: A Pilot Experiment at a Primary School in Shanghai. In H. Spires (Ed.), *Digital Transformation and Innovation in Chinese Education* (pp. 211–231). Hershey, PA: IGI Global. doi:10.4018/978-1-5225-2924-8.ch012

Yilmaz, R., Sezgin, A., Kurnaz, S., & Arslan, Y. Z. (2018). Object-Oriented Programming in Computer Science. In M. Khosrow-Pour, D.B.A. (Ed.), Encyclopedia of Information Science and Technology, Fourth Edition (pp. 7470-7480). Hershey, PA: IGI Global. doi:10.4018/978-1-5225-2255-3.ch650

Yu, L. (2018). From Teaching Software Engineering Locally and Globally to Devising an Internationalized Computer Science Curriculum. In S. Dikli, B. Etheridge, & R. Rawls (Eds.), *Curriculum Internationalization and the Future of Education* (pp. 293–320). Hershey, PA: IGI Global. doi:10.4018/978-1-5225-2791-6.ch016

Yuhua, F. (2018). Computer Information Library Clusters. In M. Khosrow-Pour, D.B.A. (Ed.), Encyclopedia of Information Science and Technology, Fourth Edition (pp. 4399-4403). Hershey, PA: IGI Global. doi:10.4018/978-1-5225-2255-3.ch382

Zakaria, R. B., Zainuddin, M. N., & Mohamad, A. H. (2022). Distilling Blockchain: Complexity, Barriers, and Opportunities. In P. Lai (Ed.), *Handbook of Research on Social Impacts of E-Payment and Blockchain Technology* (pp. 89–114). IGI Global. https://doi.org/10.4018/978-1-7998-9035-5.ch007

Zhang, Z., Ma, J., & Cui, X. (2021). Genetic Algorithm With Three-Dimensional Population Dominance Strategy for University Course Timetabling Problem. *International Journal of Grid and High Performance Computing*, *13*(2), 56–69. https://doi.org/10.4018/IJGHPC.2021040104

About the Contributors

Bella Mary I. Thusnavis received her B.E in Electronics and Communication Engineering with distinction from Karunya Institute of Technology, affiliated to Anna University, Chennai, India in 2007 and her M.E in Applied Electronics from Karunya University, India in 2009. She completed her Ph.D degree in Information and Communication Engineering under Anna University, Chennai, India, in 2020. Her research includes Multimedia Image Retrieval for Medical Diagnosis, Statistical Feature Extraction, Optimization algorithm and Embedded Systems. She is currently working as Assistant Professor in Electronics and Communication Engineering, Karunya University, Coimbatore, India. She has 11 international Journal Publications and 19 International and National Conference Publications.

K. Martin Sagayam received his PhD in Electronics and Communication Engineering (Signal image processing using machine learning algorithms) from Karunya University, Coimbatore, India. He received his both ME in Communication Systems and BE in Electronics and Communication Engineering from Anna University, Chennai. Currently, he is working as Assistant Professor in the Department of ECE, Karunya Institute Technology and Sciences, Coimbatore, India. He has authored/ co-authored more number of referred International Journals. He has also presented more number of papers in reputed international and national conferences. He has authored 2 edited book, 2 authored book, book series and more than 15 book chapters with reputed international publishers. He has three Indian patents and two Australian patents for his innovations and intellectual property right. He is an active IEEE member. His area of interest includes Communication systems, signal and image processing, machine learning and virtual reality.

Ahmed A. Elngar currently is an Assistant Professor at Faculty of Computers and Artificial Intelligence, Department of Computer Science, Beni Suef University. Their current project is 'Biometrics, AI-based Security, Image processing, AI-based Control, AI-based Smart Grids'.

* * *

Visali A. L. completed her B.Tech IT at Anna University - MIT Campus. Presently she is working at PayPal as Senior Development Engineer.

Bikalpa Bagui is currently pursuing his B.Tech in the department of Computer science and Engineering at the University of Engineering and Management, Kolkata. His research interests include IoT, Embedded Systems, AdHoc Network, machine learning, Artificial intelligence.

Arighna Basak is presently working as Assistant Professor in the Department of Electronics & Communication Engineering in Brainware University. He has 4 years of professional experience in academics. He received M.Tech Degree in Nanotechnology from Jadavpur University. His current research interest span around study of modeling and RF/Analog performance analysis of advanced nano- scale MOSFETs device structures. He is a member of IEEE. He has authored 4 journal papers in international refereed journals and 3 research papers in international conferences. He is reviewer of a few international journals of repute and some prestigious conferences in India.

Mousumi Biswas currently is pursuing her B.Tech. in Computer Science and Engineering at Brainware University, Kolkata. She completed her Diploma in Information Technology (IT) from Women's Polytechnic, Tripura in 2016. Her research interests include Image Processing, Machine Learning, and Internet of Things.

Jyothi Prasanth D. R. pursued his B.Tech IT at Anna University - MIT Campus. He will be doing his higher studies abroad in the forthcoming academic year.

Manoj G. received his Bachelor of Engineering degree in Electronics and Electrical Engineering with First class from the Bharathiar University, India in 2004 and his Master of Engineering degree in VLSI Design from Karunya University, India in 2006. He completed his PhD degree in the area of VLSI under Anna University, Chennai focusing on ASIC design performance for space applications. He is working in the area of VLSI, IOT, Machine Learning. He is currently working as Assistant Professor in Electronics and Communication Engineering, Karunya University, Coimbatore, India.

Ahona Ghosh is a B.Tech., M.Tech. in Computer Science and Engineering and presently is an AICTE Doctoral Fellow in the Department of Computer Science and Engineering, Maulana Abul Kalam Azad University of Technology, West Bengal (In house). Before joining Ph.D., she was an Assistant Professor in the Department of Computational Science, Brainware University, Barasat, Kolkata. She has published more than 20 papers in international conferences, peer reviewed international journals and book chapters with IEEE, Springer, Elsevier, IGI Global, CRC Press etc. Her research interests include Machine Learning and Internet of Things.

Mohamed K. is an Assistant Professor at Agni College of Technology.

Alfred Kirubaraj is an Assistant Professor in the Department of Electronics and Communication Engineering at Karunya Institute of Technology and Sciences.

Martin Sagayam Kulandairaj is an Assistant Professor in the Department of Electronics and Communication Engineering at Karunya Institute of Technology and Sciences.

Praveen Kumar is a PhD Research Scholar in Management at Karunya Institute of Management, Coimbatore.

G. Nisha Malini is an Accredited Management Teacher (AMT) by the All India Management Teacher (AMT) by the All India Management Association (AIMA) in the field of Human Resource and General Management. She obtained her Graduate Degree in Law from Coimbatore Law College, affiliated to Bharathiar University, Post Graduate Degree in Management with Human Resource specialization from IGNOU, Master of Philosophy in Management with Human Resource specialization from Bharathiar University, and is currently pursuing her Ph.D. in Bharathiar University. Her area of Research is the Scope of Consumer Protection Act in the Medical Service Sector. She is also currently pursuing her Master of Law at the University of Madras in Private study mode. She is known for her exceptional teaching skills and meticulous working style in all the assignments that she undertakes. She is good at multi-tasking and has been commended for her versatility in performing her tasks. She possesses good communication skills and has a reputation for playing mentoring role with perfection.

Arun Kumar Manoharan is an Assistant Professor in Electronics and Communication Engineering at GITAM University, School of technology, Bengaluru Campus.

Caprio Mistry is currently pursuing his B.Tech in the department of Electronics and Communication Engineering at Brainware University, Kolkata. He had done his high school diploma in Pure science from Swarajya senior secondary school, Rajasthan. He had published many research and review articles in many journals. His research interests include IoT, Embedded Systems, AdHoc Network, Nanoscale electronics, mixed-signal circuits, and VLSI.

Navneet Munoth has been serving in his current position as an Assistant Professor in the Department of Architecture and Planning, MANIT Bhopal for the past 11 years. He is also heading as an Honorary Director of the Council of Architecture's Training and Research Centre (COA-TRC) established at MANIT Bhopal for fostering the synergy for boosting the entrepreneurship culture as well as for facilitating skill development among the students, academicians, and professionals. http://www.manit.ac.in/dr-navneet-munoth.

Divya P. S. is an Account Executive at Karunya Institute of Technology and Sciences.

John Paul works at Karunya University in the Department of Electronics & Communication Engineering.

Prince R. was born in Sivakasi, Tamil Nadu on 23 June 1997. He completed his UG degree in Bachelor of Architecture from NIT Bhopal in 2020. He is currently practicing as an Architect in India, registered under the Council of Architecture, Govt of India. Mr. Prince R is a recipient of the Innovative Design Thesis Award, A3 Foundation India. His area of interest includes Smart Building, IoT Home Automation, Mechatronics, Smart Utilitarian Product Design, Sustainability, etc.

Hariharan S. is a student at Karunya Institute of Technology and Sciences.

Jebasingh S. is an Assistant Professor, Department of Mathematics, Karunya Institute of Technology and Sciences, India.

Parthiban S. is a student at Karunya Institute of Technology and Sciences.

Senith S. is an Assistant Professor at Karunya Institute of Technology and Sciences.

Neha Sharma became a Student Member (SM) of IEEE in 2019, a Professional Member (PM) in 2021. She was born in Gwalior, MP on 14 November 1994. She completed her UG degree in Bachelor of Architecture from NIT Bhopal in 2018. In 2020, she completed her M. Tech in Geomatics from CEPT University, Ahmedabad. She is presently pursuing Post Graduate Diploma in Data Science from IIIT Bangalore in association with Upgrad which will be completed by 2022. She is presently working at Mindtree, Chennai Headquarters in the Geospatial Team as Software Engineer since July 2021. She is also the STEP Coordinator and MOVE Coordinator in IEEE India Council 2021. Ms. Sharma is also a life member of the Indian Society of Geomatics (ISG). She is a recipient of the Academic Excellence

Award, Best Capstone Project Award, and Distinguished Student Award for her exemplary performance in her M. Tech program (2018-20). Her interest areas are remote sensing, GIS, geo-programming, machine learning, webGIS, 3D modeling, etc.

Jyoti Singh is a HOD and Assistant Professor, Department of Physics, at Sant Hirdaram Girls College in Bhopal, India.

Sruthi Sreeram pursued her B.Tech IT at Anna University, MIT Campus. Presently working at Fidelity Investments as Software Engineer.

J. Dhalia Sweetlin is working as an Assistant Professor in IT Department, Anna University - MIT Campus. Her research interests include Artificial Intelligence, Medical Image Processing, Soft Computing and Data Mining.

Nagarjuna Telagam is with the Electronics and Communication Engineering Department, GITAM University, Bangalore, India and His research interests are 5G, IoT, VLSI technologies, Virtual Instrumentation. He is Anna University rank holder (20) for Master's Degree in State Level. Certified LabVIEW Associate Developer (CLAD) from National Instruments.

Tharun V. is a student at Karunya Institute of Technology and Sciences.

Poornima Vijaykumar is working as an Assistant Professor at St. Joseph's College of Commerce (Autonomous), Bengaluru with over a decade of teaching experience. She has a Ph.D. from Bharathiar University, Coimbatore. She also holds a Certification in IFRS issued by ACCA, UK. She has participated and presented in several National and International seminars, conferences, FDP workshops and training programmes. She has to her credit several research papers published in reputed journals. She has also delivered many guest lectures for other colleges and universities in her capacity as a resource person. Dr. Poornima was awarded the "Adarsh Vidhya Saraswati Rashtriya Puraskar (National Award of Excellence 2018)" by the Global Management Council, Ahmedabad, and also a distinction of being one of the Best Teacher in the country for her distinguished work in the interest of students and education field. Her motto is "Passion for Excellence in Teaching" and "Sharing Knowledge".

Index

M

P

R

S

T

U

W

Printed in the United States
by Baker & Taylor Publisher Services